*World Survey of Climatology Volume 5*

CLIMATES OF NORTHERN AND WESTERN EUROPE

*World Survey of Climatology*

*World Survey of Climatology Volume 5*

# Climates of Northern and Western Europe

edited by C. C. WALLÉN

*World Meteorological Organisation*
*Geneva (Switzerland)*

ELSEVIER PUBLISHING COMPANY Amsterdam-London-New York 1970

ELSEVIER PUBLISHING COMPANY
335 Jan van Galenstraat
P.O. Box 211, Amsterdam, The Netherlands

ELSEVIER PUBLISHING COMPANY LTD.
Barking, Essex, England

AMERICAN ELSEVIER PUBLISHING COMPANY, INC.
52 Vanderbilt Avenue
New York, New York 10017

Library of Congress Card Number: 68–12479

ISBN 0–444–40705–7

With 60 illustrations and 101 tables

Printed in The Netherlands

# World Survey of Climatology

## List of Contributors to this Volume

R. ARLÉRY
Meteorological Service, Division of Climatology
Paris (France)

G. MANLEY
University of Lancaster, Department of Environmental Sciences
Lancaster (Great Britain)

A. LINÉS ESCARDÓ
National Meteorological Institute, Division of Climatology
Madrid (Spain)

C. C. WALLÉN
World Meteorological Organisation
Geneva (Switzerland)

T. WERNER JOHANNESSEN
Norwegian Meteorological Institute, Division of Climatology
Blindern (Norway)

# Contents

Contents

*Chapter 3.* THE CLIMATE OF THE BRITISH ISLES
by G. Manley

*Chapter 4.* THE CLIMATE OF FRANCE, BELGIUM, THE NETHERLANDS
AND LUXEMBOURG
by R. Arléry

*Chapter 5.* THE CLIMATE OF THE IBERIAN PENINSULA
by A. Linés Escardó

# Chapter 1

# Introduction

C. C. WALLÉN

The mildness of the climate of western Europe represents, indeed, one of the more spectacular anomalies in the world climatic pattern and it is hardly surprising that the peoples in this area, living under so many favourable climatic characteristics, have had such a great influence on the development of world history.

In this climate the first concepts of meteorology as a science originated and the fundamentals of this science were laid down. It is likely that particularly the typical day-to-day variations in the European climate—so different from the more stable climates of polar and tropical regions or more continental ones— gave impetus to the development of an interest in meteorological phenomena and their interpretation.

It is the intention of this volume to describe the historical development of meteorological concepts and the use of them in interpreting the climate of various regions in western Europe as well as to give an up to date description of these climates, as far as possible with available data. This introduction aims at presenting the most spectacular climatic features which are dominant in western Europe as well as to interpret the fundamental reasons for them in terms of the general circulation of the atmosphere.

## Radiation conditions in western Europe

The most basic of all climatic elements governing the distribution and influence of all the others is the radiation from the sun and its balance with the outgoing long-wave radiation from the earth. Let us, therefore, consider first the location of western Europe in relation to radiation conditions.

At 35°N, which is the latitude corresponding to the southernmost rim of western Europe, at the surface of the earth and over the year the total incoming radiation from sun and sky is around 150,000 cal./cm² taking into consideration the average cloudiness and turbidity conditions. Of this amount about 65% is due to direct solar radiation and 35% to diffuse radiation from the sky. The outgoing radiation at this latitude over the year is slightly smaller (2%); thus in the southernmost part of Europe we are still in the area of the earth where the annual incoming radiation surpasses the outgoing.

Coming to the northern rim of western Europe, at latitude 70°N, the total incoming radiation at the earth's surface over the year is only about 70,000 cal./cm². It is interesting to note that at these high latitudes the part of the incoming radiation that is due to diffuse radiation from the sky is much greater than in low latitudes, i.e., amounts to 60% of the total. As the outgoing radiation at this latitude is of the same order of magnitude as at the southern rim of Europe, i.e., 136,000 cal./cm², it is clear that a deficit of about 70,000 cal./cm² year will occur. The average amount of direct solar and diffuse sky radiation at various latitudes of interest to western Europe are shown in the Table I and II, according to BAUR and PHILLIPS (1934, 1935).

According to the same authors the annual mean of the incoming and outgoing radiation at the same latitudes are as shown in Table III.

Already at latitude 40°N the annual amount of outgoing radiation is greater than the global incoming radiation and the deficit increases rapidly with increasing latitude. It is obvious that western Europe with its green vegetation and its land areas mostly suitable for agriculture must enjoy the benefit of certain particular atmospheric processes which balance the great deficit in radiation. These processes are mainly connected with the general circulation of the atmosphere over Europe and of the sea to the west of the continent. As we shall see these processes tend to carry sufficient heat from lower to high latitudes over western Europe in order to balance the mentioned deficit. It should be emphasized that the long summer nights in all parts of the region situated north of latitude 60°N are of fundamental importance for the development of vegetation and the possibilities of cultivation.

TABLE I

AVERAGE AMOUNT OF DIRECT SOLAR RADIATION
(cal./cm² min)

| Date | 30–40° | 40–50° | 50–60° | 60–90° |
|---|---|---|---|---|
| Dec. 21 | 0.082 | 0.036 | 0.013 | 0.001 |
| March 21 | 0.161 | 0.116 | 0.096 | 0.055 |
| June 21 | 0.233 | 0.183 | 0.159 | 0.133 |
| Sept. 21 | 0.183 | 0.131 | 0.079 | 0.028 |

TABLE II

AVERAGE AMOUNT OF DIFFUSE SKY RADIATION
(cal./cm² min)

| Date | 30–40° | 40–50° | 50–60° | 60–90° |
|---|---|---|---|---|
| Dec. 21 | 0.052 | 0.034 | 0.016 | 0.001 |
| March 21 | 0.093 | 0.083 | 0.066 | 0.047 |
| June 21 | 0.125 | 0.126 | 0.122 | 0.153 |
| Sept. 21 | 0.091 | 0.081 | 0.065 | 0.048 |

TABLE III

ANNUAL MEAN OF INCOMING AND OUTGOING RADIATION
(cal./cm² min)

| | 30–40° | 40–50° | 50–60° | 60–90° |
|---|---|---|---|---|
| Incoming | 0.297 | 0.236 | 0.185 | 0.145 |
| Outgoing | 0.291 | 0.269 | 0.253 | 0.240 |

**Air and sea circulation**

The most spectacular and basic phenomenon in the general circulation of the atmosphere in temperate regions, situated between low, heated latitudes and high, cooled latitudes, is the meandering westerly "jet" stream of air with a maximum at a level of around 300 mbar. Although the westerly "jet" stream is persistent the meandering pattern that it takes changes almost from day to day and the weather and climate of western Europe depends largely on how this pattern of upper air circulation centered around the "jet" stream shows up. Upper air data hitherto available show this pattern best at 500 mbar level and we present here two typical 500 mbar charts for the winter season over western Europe together with the two corresponding surface charts.

**Winter**

The first winter chart shows the most typical and most common of all upper air flow patterns over Europe (Fig.1). According to WILLETT and SANDERS (1959) it may be called a typical "*high-index*" *circulation* pattern where the "jet" stream runs between an extended low pressure system in the north and an extended high pressure system in the south practically parallel with the latitude circle and at a comparatively high latitude. The contours at 500 mbar all run with only a slight angle towards the southwest to the latitude circles and a fairly regular stream of air is conveyed at upper levels from southwest to northwest over western Europe. These regular westerlies of the upper atmosphere correspond to a typical pressure distribution at the earth's surface (Fig.2) where the Icelandic low pressure system is strongly developed, has an elongated shape from the southwest to the northeast and is shifted fairly far to the north. At the same time the Azorian high pressure system is well developed, extends into the European continent and often has a connection with the Asiatic high pressure system caused by winter cooling. Due to this distribution of pressure the winds over large parts of Europe at these occasions are at the surface mainly between south and west and so carry comparatively mild air from the Atlantic Sea over the European continent. The mildness of this air is to a considerable degree due to the warm water carried by the Gulf Stream from the southwest to the northeast through the North Atlantic Ocean. The mildness of the climate of western Europe, in fact, has very often been said to be due to the existence of the Gulf Stream. It should be kept in mind, however, first that the course of the Gulf Stream in itself is partly directed by the general circulation of the atmosphere which is basically from southwest to west, and secondly, that it is due to the circulation of the air that the effect of the Gulf Stream as a heat source becomes important to western Europe. Thus, the importance of the Gulf Stream to the European climate, although undoubtedly great, must be considered secondary.

In the typical winter "high-index" circulation the westerly "jet" stream at 500 mbar shifts its location back and forth from a northernmost position fairly close to latitude 65°N to a southernmost position fairly close to latitude 50°N. The "jet" stream at upper levels corresponds at the surface to the well-known "polar front" in the Bergen-school concept, on which the cyclonic activity of the temperate latitudes takes place. The cyclones, created on the polar front in the Atlantic Ocean and moving with the upper westerlies towards and across western Europe, are the basic mechanisms by which

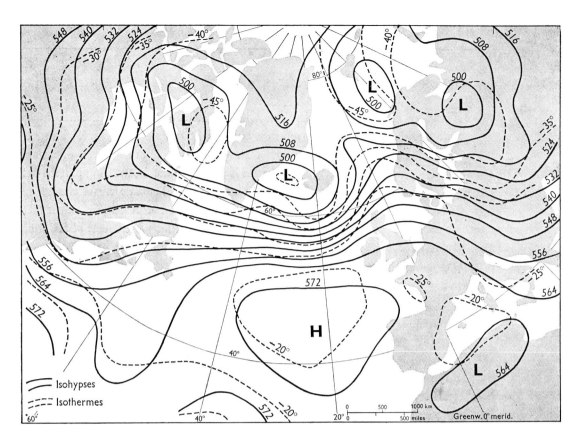

Fig.1. Typical upper air weather situation during the winter, showing westerly to northwesterly zonal circulation at 500 mbar over the Atlantic and western Europe. Date February 26, 1949. (After WALLÉN, 1960.)

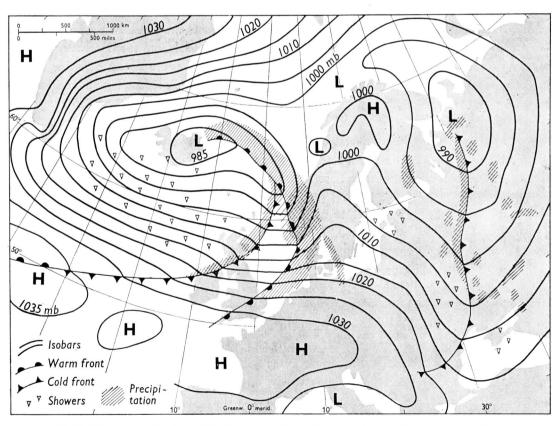

Fig.2. The weather situation of Fig.1 at the earth's surface, showing circulation, fronts and air masses over the Atlantic and western Europe. (After WALLÉN, 1960.)

the import of mild air from the Atlantic Ocean towards the continent takes place. Although as we shall see the above-described zonal type of winter circulation is the most common of all west European weather types which shows up in the mean circulation, it is by no means the only type of circulation that may occur. In fact, air-circulation over Europe, as in all temperate latitudes, shifts more or less regularly from this "high-index" type via several intermediary stages towards the other opposite called the "low-index" type (see Fig.3).

At upper levels the typical "*low-index*" *circulation* type is characterized by a meridional exchange of air from north to south and from south to north over Europe. As seen from Fig.3 and 4, the normal westerly stream of air over Europe is stopped or "blocked" by a warm high pressure system extending, for example, as in the case of the figure, from south-southwest to north-northeast from Biscay over the British Isles towards Scandinavia. With this "blocking" of the westerlies, the upper air "jet" stream still exists but meanders around the high pressure ridges and the low pressure troughs indicating that at the surface the polar front runs in a similar fashion. Consequently the cyclonic activity takes place along south–north tracks or north–south tracks rather than on the more common west–east tracks.

It is reasonable that with the normal westerly type of high-index circulation the winter weather of western Europe becomes fairly mild and cloudy all over the area. Temperature only gradually and slowly decreases from south to north and from west to east. On the contrary weather conditions change very rapidly from west to east in case of the meridional circulation type. The weather of a particular band stretching north–south will, then, depend completely upon where the band is situated in relation to the upper air ridge and trough dominating the circulation. Under proper ridge conditions the weather will be quite clear and sometimes in early or late winter fairly warm during the day due to radiation from the sun but always cold at night because of outgoing radiation. In the high winter season sometimes a coupling occurs between the upper air warm ridge and the anticyclone stretching from Siberia over Russia towards northern and western Europe and created by strong radiational cooling. Western Europe then experiences its hardest winter conditions with low temperatures both day and night accompanied on the eastern coasts of the British Isles and Scandinavia by northeastern winds and widespread snowfall.

Below the upper air trough winter weather conditions vary from cold and unstable weather in connection with the northerly circulation on the westerly side of the surface troughs to mild and rainy or (in northern Europe) even snowy weather in the southerly current on the eastern side of the trough. On the whole the weather is unstable and mostly cloudy below the upper air trough, although it is mainly in connection with the "jet" stream aloft and the polar front at the surface that cyclonic activity creating really abundant precipitation takes place.

**Summer**

In summer the *high-index zonal type of circulation* of the upper air carries fairly cold winds over northwestern Europe. This gives rise to comparatively maritime and cool summer weather and owing to rather strong cyclonic activity rainfall may be abundant in the western parts of northern Europe decreasing, however, eastwards. As in winter

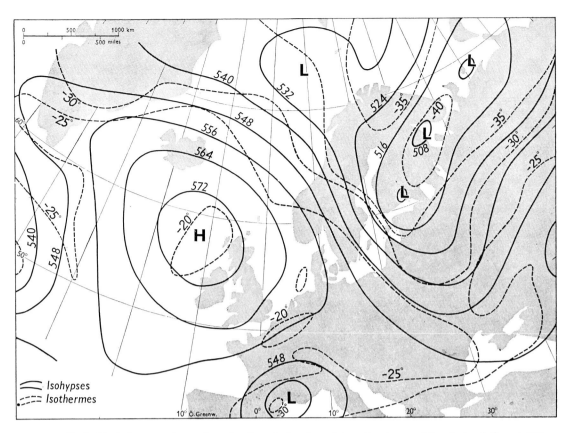

Fig.3. Typical upper air weather situation during the winter, showing meridional circulation at 500 mbar over the eastern Atlantic and western Europe. Date April 9, 1957. (After WALLÉN, 1960.)

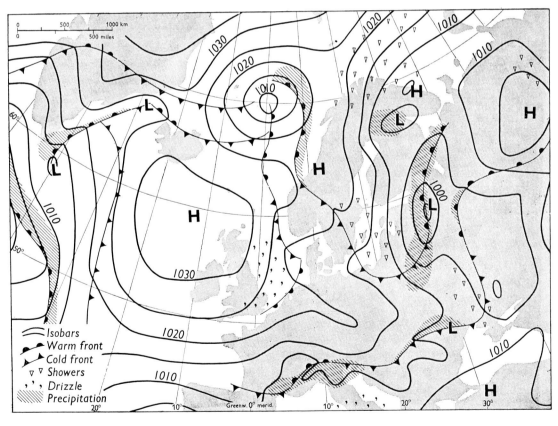

Fig.4. The weather situation of Fig.3 at the earth's surface, showing circulation, fronts and air masses over the eastern Atlantic and western Europe. (After WALLÉN, 1960.)

precipitation is always orographically increased in mountainous areas facing west. In the rear of the cyclones, cool polar air sometimes moves south from the Arctic Sea reaching northern and middle Europe. Owing to the contrast between the warm air, heated over land by insolation and the cool polar air at upper levels convection takes place and may result in frequent showers of rain. Typical high-index circulation summers are thus usually cool and maritime in northern and northwestern Europe. In the central and southern parts of western Europe, however, the Azorian high-pressure system dominates the circulation as the upper air "jet" stream is transferred to the north. In this area the "high-index" circulation, therefore, gives rise to warm and stable summer weather created by the subsiding air in the high-pressure area. Sometimes, parts of northern Europe may experience warm summers due to a high-pressure system which extends over the continent to quite far north.

In case of the *"low-index" circulation* summer weather is again dependent upon the location of the troughs and ridges. Thus when a high-pressure ridge extends from southeastern Europe towards northern Scandinavia the weather is warm and stable in this part of Europe. Connected with this ridge situation is, however, a trough over western Europe where weather in this case is cool and partly very wet. In the southerly current east of the trough and west of the ridge cyclones move northwards giving frequent and abundant precipitation (so-called Vb cyclones). In connection with the cold polar air extending in the trough southwards over heated land areas the weather is very unstable and hence rainshowers are frequent.

**Spring and autumn**

The extreme weather situations of high-index and low-index circulation may occur also in spring and autumn. As late as in the latter part of March the influence of the snow cover is still so strong that in northern Europe the presence of a high pressure system with low index circulation usually produces winter conditions while in southern and western Europe warm spring weather may occur in the same situation. In autumn (October and November), on the other hand, a low index with a high-pressure system usually will bring cloudy and foggy weather over most of Europe with comparatively low temperatures and little precipitation. The high-index, westerly circulation instead creates maritime and wet weather conditions in the north and fairly stable weather in the south at all seasons.

**Circulation of the Mediterranean area**

As far as circulation conditions and climate are concerned the Mediterranean area has a special situation and will be discussed separately in the following.

**Winter**

Winter circulation in the Mediterranean region can be described as an interaction between cyclones moving with upper air westerlies from the Atlantic through the Mediterranean Sea on one side and a more or less pronounced "blocking" of this westerly circulation by the Asiatic winter high pressure system extending towards southwest on the other.

The upper air-circulation connected with the surface cyclones in the Mediterranean is a very complex one and still not too carefully studied. It appears, however, that most of the mid-winter cyclones are connected with a polar-front "jet" branch in the upper air over the Mediterranean Sea, sometimes, when no "blocking" action occurs from the Asiatic high, meandering far into the Near East. Thus in the Mediterranean as in other European latitudes it is possible to distinguish between two principal types of upper air circulation: a zonal and a more or less meridional one. As distinguished from central and northern Europe the zonal circulation is more rare than the meridional one, a fact mainly due to differential heating from the sea and surrounding land in the Mediterranean. This heating causes in favourable regions cyclogenesis and strong deepening of the cyclones changing the circulation pattern aloft from a zonal into a more cellular or meridional one. Cyclogenesis and meridional circulation might also be due to outbreaks in certain regions of cold air from northern and central Europe in the rear of the cyclones. Sometimes, deep upper air troughs are created in connection with such outbreaks of cold air and low level cyclones formed with such troughs have a clear tendency to become stationary or to move slowly. There are two particularly important Mediterranean regions where frequent cyclogenesis from the surface occurs and more or less stable upper air troughs are formed, namely in the Gulf of Genoa and around Cyprus. The former region has a basic influence upon the weather in part of the Mediterranean dealt with in this volume while the second region is of small importance to this region. Very characteristic and famous features of the weather in the Mediterranean region are created in connection with stable upper air troughs and cyclones in the Gulf of Genoa. For long periods in winter strong northerly winds penetrating through the Rhône Valley and spreading along the French Riviera are a famous and quite unpleasant weather phenomenon called "the mistral". On the other hand particularly in autumn and spring the stable cyclones of the Gulf of Genoa create intense southerly winds over large parts of Italy. These winds called "sciroccos" are often unpleasantly hot and humid coming from the Sahara and having passed over the Mediterranean Sea.

As far as the precipitation conditions are concerned there is a fundamental difference between the shallow cyclones rapidly moving with the Mediterranean branch of the polar "jet" stream and the deep cyclones connected with the stagnant upper air troughs. The former type gives rise to comparatively small amounts of frontal precipitation in the Iberian, Appeninian and Balkan peninsulas, while the latter, due to the cold air aloft and its slow-moving nature, may cause considerable amounts of precipitation often of a convective type. It is, therefore, obvious that the Appeninian and northern Balkan peninsulas are more favourably located areas as far as rainfall is concerned, for long periods in winter being under the influence of stagnant cyclones in the Gulf of Genoa. Precipitation amounts in the rainy season are thus much higher in Italy and western Yugoslavia than, for instance, in Spain and Greece.

**Summer**

Summer circulation conditions over the Mediterranean region which prevail from mid-May to mid-October show much greater regularity than do the winter conditions. With the gradual warming up of the Eurasian continent the subtropical high pressure belt and the temperate westerlies are displaced northwards so that in summer the subtropical high pressure system always extends over the Mediterranean region and in the majority of cases also over the southern parts of western Europe. Winds then become weak and mainly from a northwesterly direction over the Mediterranean. Weather is in general sunny and steady.

When in summer, with a typical low-index circulation pattern, western Europe is dominated by a cold trough, the subtropical high pressure system may be confined only to the Mediterranean region. As mentioned earlier weather all the way down to the Pyrenees and the Alps is then characterized by cold northerly winds, considerable amounts of clouds, particularly in the afternoons, and abundant shower precipitation. Generally weather in the Mediterranean is still characterized by sunshine and little cloudiness. On certain occasions, however, the cold air aloft may reach south of the Alps and the Pyrenees and create clouds and some rainshowers usually connected with thunderstorms.

In order to summarize the circulation conditions over western Europe Fig.5 and 6 are shown. The areal distribution of frequency of winter cyclones of less than 1,000 mbar shows typical areas south of Iceland, along latitude 60°N and in the Gulf of Genoa in the Mediterranean. Latitude 50°N is comparatively free of deep cyclones.

In summer deep cyclones are completely lacking south of latitude 50°N and maximum

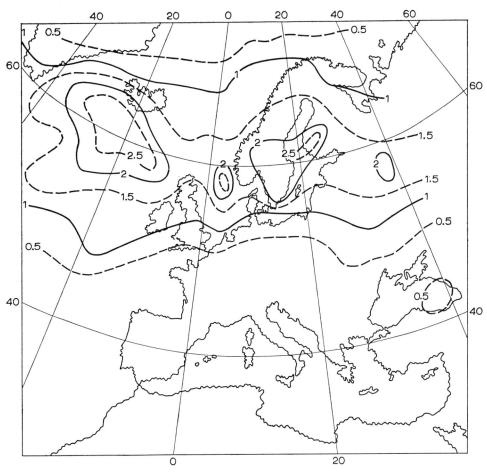

Fig.5. Annual frequency of cyclones with central pressure less than 1,000 mbar. (After SCHEDLER, 1924.)

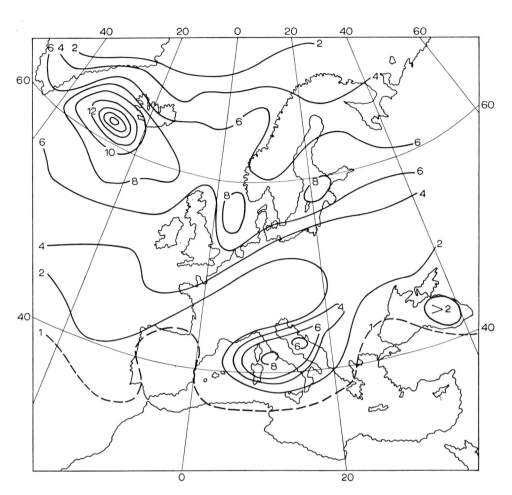

Fig.6. Annual frequency of cyclones with central pressure less than 1,000 mbar in summer. (After SCHEDLER, 1924.)

frequency exists south of Iceland and over southern Scandinavia, i.e., more or less in a zone around latitude 60°N.

## Temperature

The general circulation of the atmosphere at the surface over western Europe is climatologically reflected in the mean temperature distribution (Fig.7, 8) in such a way that the isotherms have a zonal trend in summer but a more meridional one in winter.
This points to the fact that although there is a continuous alternation between periods of zonal and meridional circulation patterns throughout the year, the zonal type is more dominant at the surface in summer and the meridional type in winter. This is a result of a monsoonal effect operating between the continent and the sea which in summer tends to intensify the planetary westerlies but which in winter operates to oppose them in connection with the "blocking" effect of the cold anticyclone over the interior of Eurasia. The contrast in winter between cold continental air over the interior and eastern Europe, and mild, maritime air over western Europe is reflected in a strong positive anomaly

(compared with the latitudes' normal) in temperatures of about 14° in January off the coast of Norway at about 65°–70°N and a negative anomaly over Siberia. In summer, due to the dominating zonal circulation from the relatively cold sea, there is instead a small negative anomaly in the west diminishing gradually eastwards as the influence from the relatively cool maritime air from the Atlantic decreases in favour of continental, warm air in the east.

The isotherms of mean temperatures at station level are shown for January and July in Fig.7 and 8. The extreme contrast between the maritime climate of western Norway and the more continental one over the interior of Scandinavia and Finland causes the steepest temperature gradient in western Europe in winter to occur eastwards from the Atlantic at latitude 65°–70°N. Here mean temperatures in January drop from ±0° to −12° over a distance of about 200 km. The 0° isotherm more or less follows the coast of

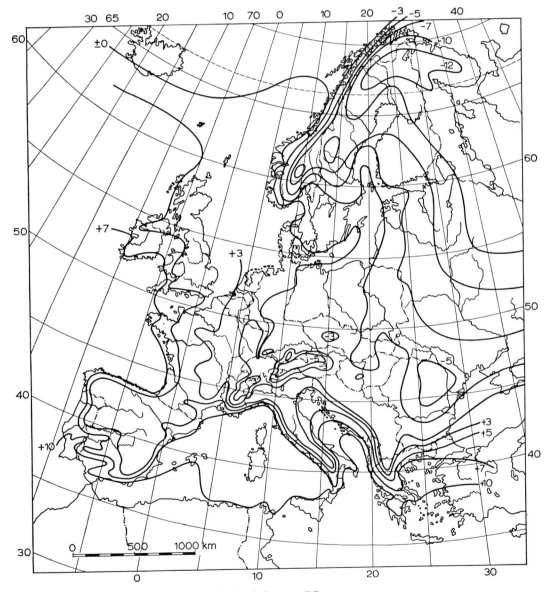

Fig.7. Mean temperature at station level, January (°C).

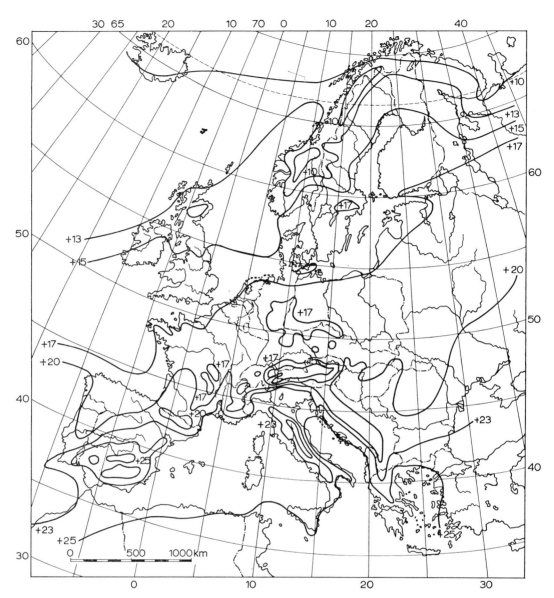

Fig.8. Mean temperature at station level, July (°C).

Norway from 65°N southwards, crosses Denmark and the southern Baltic running later through West Germany more or less along the Rhine to Basel. It follows the western slopes of the Alps and then turns eastwards following the southern slopes of the Alps, continues through the Balkan Mountains and crosses the Black Sea so that a part of Crimea remains on the warmer side. Following the later definition of Köppen for D and C climates this line represents the dividing line in Europe between these two types of climate. A mean temperature in January above +10° in western Europe is found only in the southern parts of the Iberian Peninsula, Sicily, southernmost Greece and Crete.

The Atlantic off the Norwegian coast in July is rather cooler than the continent but no steep gradient occurs in summer between the sea and the Scandinavian inland. The mean temperature field is quite equalized not only in that area but over the whole of western Europe. As far north as northern Finland the mean temperature of July reaches +15°

which in fact is the same as in Ireland and northern England. The $+10°$ isotherm crosses northernmost Scandinavia, indicating that no real summer exists on the Arctic Sea coast of that peninsula. Below $+10°$ mean temperatures are also found above 1,000 m of altitude in the mountains of south Norway and above 750 m in the northern parts of the Scandinavian mountains.

Mean temperatures above $+20°$ are normal in western Europe south of a limit running through the interior of the Iberian Peninsula, southern France and northern Italy on the slopes of the Alps and further on the slopes of the Balkan Mountains returning to cross Hungary and Roumania eastwards. Only in southeastern Spain and Italy as well as in southern Greece does the mean temperature for July exceed $+25°C$.

Also in the southern parts of western Europe the decrease of temperature with altitude is a typical phenomenon. In the higher parts of the Alps, mean temperature drops below $+15°$ above 1,200 m. In the more elevated parts of the Pyrenees and the Balkan Mountains the mean temperature for July is well below $+20°$.

The rapid drop of mean temperature with altitude which is quite conspicuous in spring and autumn also causes a typical decrease of the growing season with increasing altitude. Especially in the maritime parts of northwestern Europe this leads to a remarkably low tree line, for instance, in the mountainous parts of northern England and Scotland. The basic reason for this situation in northwest Europe is the relatively cool climate of the warm season. As the mean temperature in spring and autumn in the northwest, because of the dominating cool maritime air masses, is not too much above the temperature that determines the growing season, i.e., between $+5°$ and $+7°$, it requires only a rather limited increase in altitude to drop the mean temperature below these values.

Due to the fact that all year around the westerly zonal flow aloft dominates western Europe the mean temperature distribution more or less reflects the conditions in connection with that type of general circulation. As earlier stated, the meridional type of upper air circulation, however, is far from uncommon and therefore a discussion of how the temperature conditions in such cases deviate from those in the zonal situations may be appropriate. The meridional and cellular flow situation over western Europe which is most thoroughly studied is the one when a quasi-stationary high in the eastern North Atlantic between $5°W$ and $15°E$ splits the westerly upper air "jet" stream into two branches, one flowing northwards around the "blocking" high and the other southward and into the Mediterranean. According to REX (1950) there is a typical seasonal periodicity in the building up of such "blocking" highs showing a fairly low frequency of "blockings" from June to November, a much higher frequency in December to June and a maximum in April. During the maximum period as much as $40\%$ of the days may be affected by "blocking" while during the other part of the year less than half as many days are affected. A typical westerly "blocking" of this type in winter gives rise to a tongue of positive temperature anomalies on the western and northern sides of the high pressure system where strengthened southerly and southwesterly winds prevail. Most of Scandinavia and the northern Atlantic with Iceland may be warmer then normal in such occasions while negative anomalies are concentrated to the area east of the "blocking" high, i.e., from southern England over the continent to eastern Europe where cold continental polar air moves southward into a well developed pressure trough. There also are periods of negative anomalies in the Mediterranean also effected by the invasion of cold polar air.

Summer "blocking" situations more or less give the same distribution of temperature anomalies as the winter types, only the positive anomalies in the northwest and north are less pronounced while the negative ones over the continent are even stronger than in winter. The maritime air introduced over northern Europe, of course, is fairly cool and may not increase temperature considerably above the normal. The cold polar air transported by northerly winds from the Arctic Sea down into the continental trough, however, is quite cold and as earlier stated gives rise to fairly low temperatures, cloudiness and shower precipitation over central Europe.

The distribution of the anomalies shown for winter and summer is, of course, highly dependent upon the location of the "blocking" high. A small shift towards the west of the high may bring the rear-trough to cover most of western Europe and then leads to below normal temperatures over western Europe at all seasons. A shift eastwards of the

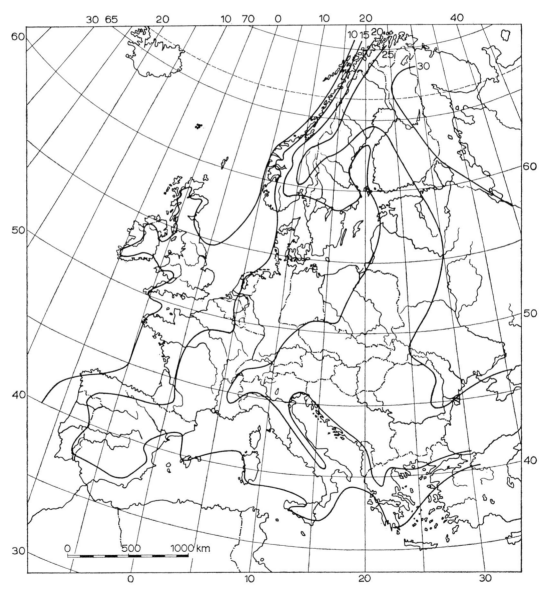

Fig.9. Annual amplitude of mean monthly air temperature (°C).

"blocking" will, on the other hand, lead in winter to negative anomalies over most of Europe due to reflexion from a snow cover and intense longwave radiation during the clear winter nights of the high pressure situations. In summer, the easterly shift will give rise to positive temperature anomalies in connection with an extension of the Azorian high pressure system at the surface over most of the area.

A conspicuous feature in the climate of Europe is the increase in continentality from west to east. This is well demonstrated in the annual amplitude of mean monthly air temperature shown in Fig.9. The increase in this amplitude is from 10°C in the westernmost areas to 25°C in the easternmost parts of the region.

## Precipitation

The precipitation climate of western Europe is characterized by the fairly even distribution over all lowlands and by only a very modest decline as one proceeds from west to east. Most of the lowlands receive an annual amount of precipitation between 500–750 mm as seen from Fig.10. The basic reason for this comparatively homogeneous picture is the fact that most precipitation comes with the cyclones crossing Europe from the Atlantic eastwards. The most conspicuous discrepancies from the homogeneous pattern are, therefore, to be found where the frequency of moving cyclones is largest and where the cyclonic precipitation mechanism is intensified by orographical influences. Generally speaking precipitation due to more intense cyclonic activity is larger in Ireland, Scotland and western Scandinavia than in France, The Netherlands and Germany In these maximum areas, it is further intensified by mountain barriers in Scotland and in Norway. In Scotland annual precipitation locally exceeds 2,000 mm and in western Norway there are regions of more than 3,000 mm. In mountainous parts of western England and Wales annual precipitation exceeds 1,000 mm. In relatively flat areas of western and central continental Europe the precipitation per year is rarely above 750 mm but with orographical intensification it may surpass this figure. Other areas of more precipitation are found in parts of the Mediterranean area and for orographical reasons, of course, in the Pyrenees, the Alps and the Balkan Mountains. In these last areas, particularly those facing westwards and at high altitudes, annual figures are frequently above 3,000 mm.

Areas of low precipitation are generally found on the leeward sides of mountain barriers as, for instance, in south Norway, in Sweden and Finland east of the Scandinavian mountain chain, in the interior of the Iberian Peninsula, east or southeast of the Appenines, in eastern Greece and eastern Roumania etc. Here annual precipitation is even less than 500 mm and semi-arid conditions may prevail.

Seasonal variations of rainfall vary somewhat throughout the region. In the western part of the region the cyclonic type of precipitation is more dominant than further eastwards where the convective type of precipitation gradually becomes of greater significance. As cyclonic activity over western Europe is much more intense in autumn and winter than in summer the precipitation maximum occurs in these seasons in the Atlantic areas and the western central parts. In the eastern, more continental parts of the region, the dominant convective precipitation gives rise to a maximum in late summer; in the southeast maximum occurs in late spring or early summer. The lee-effect of the

Scandinavian mountains is so efficient that convective rain in summer dominates cyclonic rain in interior Scandinavia thus causing a typical summer maximum in that area where one might have expected a winter maximum due to the situation comparatively close to the Atlantic Sea.

As earlier indicated in connection with the circulation conditions the cyclonic activity in the Mediterranean part of the region is completely fixed to the winter season. As the summer period is dominated by high pressure circulation little or no convective activity exists and consequently an extreme precipitation minimum is characteristic of the warm season.

The above-mentioned normal distribution of precipitation is considerably changed in "blocking" situations both in summer and winter conditions.

In winters with typical westerly "blocking" situations, the amount of rainfall greatly exceeds normal in western Iceland and in southwestern Spain but otherwise this type of "blocking" leads to well below normal precipitation amounts in all western Europe. Areas in the center of the "blocking" high may get as little as 10% of normal precipitation. Small areas of rather high rainfall may occur on the northern slopes of the Alps due to the orographic lifting of the northerly winds on the eastern side of the "block". In summer the distribution of precipitation is similar to that in winter, only the area of high precipitation then also covers the northwestern part of the Scandinavian Peninsula. Rainfall in large areas of western Europe is only around 15% of normal precipitation. The frequency maximum of "blocking" situations occurring in spring is reflected in a minimum of precipitation in that same season existing practically all over the region. Only in the southeastern parts of our region where "blocking" situations do not create real droughts do considerable amounts of rainfall occur in late spring.

It should be emphasized once again that the "blocking" situation described here is one of many. The distribution of precipitation in other "blocking" situations where the "block" is located further to the east—as often in winter and sometimes in summer— unfortunately has not been investigated in the same thorough manner as the westerly block and consequently can not be described here.

## Snow cover

Although snow for the main part of western Europe is not a climatic parameter of particular concern a few words are worth mentioning in connection with precipitation. Theoretically speaking, the 0° isotherm for the mean temperature of the coldest month should coincide more or less with the limits for a continuous snow cover during that month. Due to generally unstable weather conditions, however, the snow falling with low temperatures quite often does not stay very long because frequently warm air from the west is brought into the region by a disturbance in the westerlies. In fact statistically it has been

Fig.10. Annual precipitation in Europe. (After BLÜTHGEN, 1958.)
*Total precipitation:* *1* = 0–100 mm; *2* = 100–300 mm; *3* = 300–500 mm; *4* = 500–750 mm; *5* = 750– 1,000 mm; *6* = 1,000–2,000 mm; *7* = 2,000–3,000 mm; *8* = >3,000 mm.
*Periodicity:* *Sp* = spring; *Spl* = late spring; *Se* = early summer; *S* = summer; *A* = autumn; *Al* = late autumn; *We* = early winter; *W* = winter; *Wl* = late winter.
———— = limit of region of predominant rainfall in summer and early summer; – – – – = limit of region of predominant drought in summer; . . . . . . = limit of region with more than 40 days snowfall.

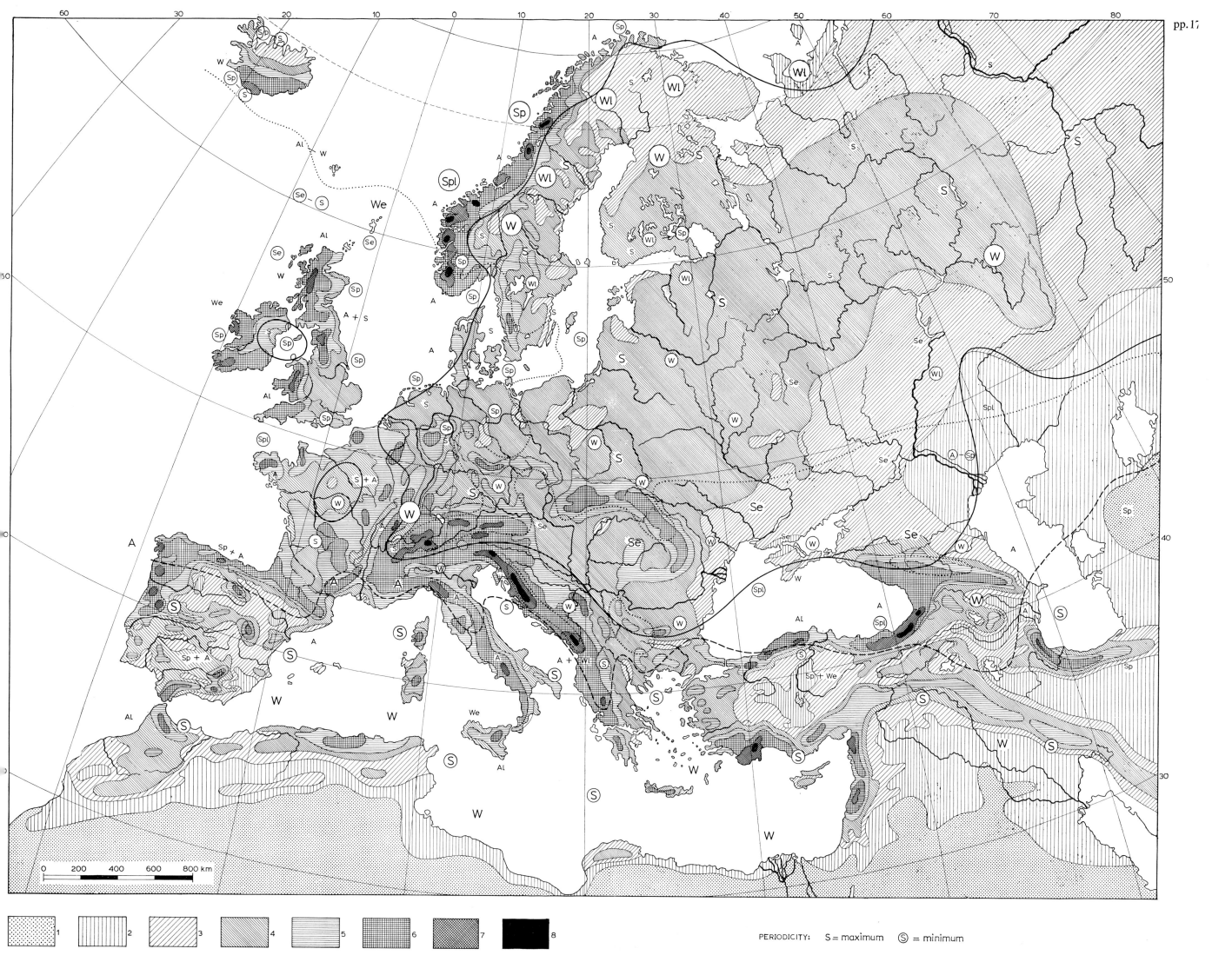

PERIODICITY:   S = maximum   Ⓢ = minimum

found that the monthly mean temperature must stay a couple of degrees below 0° in order for us to count on a permanent snow cover during that month. Looking at the map of isotherms of mean temperatures for the coldest month we, hence, conclude that a permanent snow cover in January or February may exist east of a line running along the west coast of Norway through southern Sweden, more or less along the Oder-Neisse-line, then along the northern slopes of the Alps and again eastwards on the southern side but considerably higher up, continuing through the Balkan Mountains at about 1,000 m and further through Bulgaria towards the Black Sea.

## Conclusion

In this general introduction we have concentrated on giving an overall outline of the circulation conditions over western Europe without entering into details. The normal temperature and precipitation conditions coupled with the various types of circulation have been described but for a thorough understanding of the details and of the distribution of these parameters the reader is referred to the following chapters, as well as for information on other climatic elements and parameters.

## References

BAUR, F. und PHILIPPS, H., 1934. Wärmehaushalt der Lufthülle der Nordhalbkugel im Januar und Juli und zur Zeit der Äquinoxien und Solstizien. I. *Gerlands Beitr. Geophys.*, 42.

BAUR, F. und PHILIPPS, H., 1935. Wärmehaushalt der Lufthülle der Nordhalbkugel im Januar und Juli und zur Zeit der Äquinoxien und Solstizien. II. *Gerlands Beitr. Geophys.*, 45.

BLÜTHGEN, J., 1958. Annual precipitation map of Europe. In: *Grosser Herder Atlas*. Troll, Freiburg.

REX, D. F., 1950. Blocking action in the middle troposphere and its effect upon regional climate. *Tellus*, 2: 275–301.

SCHEDLER, A., 1924. Die Zirkulation im nordatlantischen Ozean und der anliegenden Teilen der Kontinente, dargestellt durch Häufigkeitswerte der Zyklonen. *Ann. Hydrograph. Maritimen Meteorol.*, 52: 1–14.

TREWARTHA, G. T., 1961. *The Earth's Problem Climates*. Univ. Wisconsin Press, Madison, Wisc., 334 pp.

WALLÉN, C. C., 1960. Climate. In: A. SÖMME (Editor), *A Geography of Norden*. Cappelen, Oslo, pp.41–53.

WILLETT, H. C. and SANDERS, F., 1959. *Descriptive Meteorology*, 2nd ed. Academic Press, New York, N.Y., 355 pp.

# Chapter 2

# The Climate of Scandinavia

THOR WERNER JOHANNESSEN

## Introduction

Scandinavia, including Denmark, Finland, Norway and Sweden, constitutes the north-western corner of the Eurasian continent. It is situated between latitudes 54° and 71°N, and to the east borders on the Baltic Sea and the vast land masses of Russia. The extreme eastern border of Scandinavia follows a line close to 30°E meridian and its westernmost point is situated at about 5°E. To the west and north Scandinavia is bounded by the North Sea, the North Atlantic and the Arctic Ocean, respectively, and the warm Gulf Stream passes northeastwards close by the west coast of Norway. Finally, the circulation of the atmosphere over Scandinavia is mostly zonal and characterized by the migratory cyclones within the North Atlantic zone of westerlies.

Because of Scandinavia's high northern latitude and the wide range in latitude the day-length at the extreme southern and northern points differs very much (see Table I).

Similarly, the length of the day varies considerably from summer to winter, and in arctic Scandinavia there is even a season of darkness in winter and a season of midnight sun in summer.

The most outstanding topographical feature of Scandinavia is, by far, the Scandinavian mountain chain running northeast–southwest along the entire Scandinavian Peninsula (including Norway and Sweden). This chain with a few peaks of about 2,500 m in altitude is situated much nearer to the west coast of Norway than to the Gulf of Bothnia. Mountains of more than 2,000 m in elevation are therefore found close behind the coastline in western and northern Norway. Great fjords cut far into these mountains which are also deeply dissected by short valleys leading down to the fjords.

TABLE I

THE LENGTH OF DAY AT THE SOLSTICES AT DIFFERENT LATITUDES IN SCANDINAVIA

| Lat. | Day-length at winter solstice | | Day-length at summer solstice | |
|------|------------------------------|------------------------------|------------------------------|------------------------------|
| | incl. twilight and dawn | excl. twilight and dawn | incl. twilight and dawn | excl. twilight and dawn |
| 55° | 8h 7min | 7h 10min | 18h 30min | 17h 24min |
| 60° | 7h 43min | 5h 42min | 22h 1min | 18h 49min |
| 65° | 6h 22min | 3h 20min | 24h 0min | 21h 56min |
| 70° | 4h 20min | 0h 0min | 24h 0min | 24h 0min |

Behind the coastal mountains and further inland there are large mountain plateaus ranging from about 1,000 m to 1,300 m in elevation in southern Norway and from about 300 m to 500 m in elevation in the interior of Finnmark in northern Norway. The eastern part of southern Norway, on the other hand, slopes slowly towards the southeast and is dissected by several rivers leading down to the Oslofjord or the coast of Skagerrak. Within the northern and western sections of eastern Norway, the river valleys are separated by forested hills of moderate height. But further to the southeast there are large plains with some large lakes. Real lowlands, however, exist only in the extreme southeastern districts and around the Trondheimfjord.

From about 63°N and northwards the Scandinavian mountain chain forms the border of Norway and Sweden and also the watershed between east and west. Hence the western sections of central and northern Sweden are extremely mountainous too, and east of the border there is a broad region of about 400–500 m in elevation, occupied by large lakes. From this lake region, northern Sweden with its large areas of moorlands slopes gently towards the east and is dissected by many large rivers running through large forested areas down to the Gulf of Bothnia.

The uplands of Småland in southern Sweden rise to about 300 m in altitude. But the remainder of southern Sweden and most of central Sweden are regions of flat, low-lying country. The same also applies to Denmark which, in addition, is of very small east–west extension.

Highlands over about 350 m in elevation and with a few peaks over 700 m exist only in northernmost Finland. The remainder of the country, is low and flat, more than half below 200 m in elevation. Large sections in the north are moorlands and in addition more than a thousand large and small lakes are spread throughout the country.

Because of the migratory cyclones and anticyclones within the North Atlantic zone of westerlies and the high northern latitude the weather in Scandinavia is rather variable and subjected to a pronounced annual march. The western coast of Scandinavia is one of the most stormy ones in the world and though precipitation is sufficient, the temperature is a limiting factor for agriculture and frost is frequent in late spring and early autumn and may even occur in summer.

The struggle for life has therefore forced the Scandinavians who in earlier days mainly were engaged in fishing, husbandry and hunting to pay due attention to the weather and to be attentive of its changes. Hence, the interest in weather is of an old date in Scandinavia and already in the last half of the 18th century some individuals and scientific societies and institutions began to observe air temperature and air pressure regularly. Daily records of air temperature thus exist for Copenhagen, Stockholm, Trondheim, Uppsala and Ullensvang from 1751, 1756, 1761, 1773 and 1798, respectively.

During the first half of the 19th century small networks of meteorological stations making regular daily observations were established by different committees in all the Scandinavian countries. Simultaneously with these more private efforts to establish a weather service, the governments in the individual countries were urged to establish official services that could take care of the meteorological observations, forecast the weather and teach meteorology. The University of Oslo complied with this challenge and appointed Dr. Henrik Mohn professor of meteorology in 1866 and that was the first professorship of meteorology in the world.

The Norwegian, Danish, Swedish and Finnish weather services were established in 1866,

1872, 1873 and 1882, respectively, and these institutions have now for many years had divisions for weather forecasting, climatology, aeronautical meteorology, aerology and instruments. In Denmark and Sweden, however, there are independent meteorological services for military purposes and in Denmark also for civil aviation.

Research work has always been carried out at the Scandinavian weather services and a good cooperation has always existed between these services and agriculture, navigation, fishing and forestry. After World War II an extended cooperation has also been established with architects, building engineers, heating and ventilation engineers, town planners and with research workers within various brances of industry.

Besides the domestic meteorological station network Denmark and Norway also operate both surface weather stations and aerological stations in the Arctic. Since 1948 Norway has also operated the ocean weather station "M" at 66°N 02°E, in operation with Sweden, and the present (1966) networks of meteorological stations operated by the individual Scandinavian weather services are as follows:

| Stations | Finland | Sweden | Norway | | Denmark | |
|---|---|---|---|---|---|---|
| | | | domest. | arct. | domest. | Greenl. |
| Synoptic st. | 50 | 180 | 177 | 6 | 48 | 31 |
| Aerological st. | 3 | 3 | 5 | 2 | 1 | 6 |
| Observatories | 2 | 1 | 2 | | | |
| Climatological st. | 100 | 200 | 69 | | 175 | 37 |
| Precipitation st. | 350 | 520 | 462 | | 350 | |

**The climatic controls**

The climatic controls for Scandinavia are very complex and of meteorological as well as of terrestrial nature. First of all the radiation balance and the general circulation of the atmosphere have a decisive influence and provide large, contrasting maritime and continental influences on the weather and climate. Secondly the general relief of the Scandinavian area, the Scandinavian mountain chain, the large differences in topography, elevation, exposure and ground cover as well, make for an extreme diversity of climate.

Considering the radiation balance, there is a surplus of radiant energy from the sun into the lower atmosphere and at the earth's surface, south of about 70°N in summer. In winter, however, there is a deficit in radiant energy, and the annual deficit in radiation balance averages about 40,000 cal./cm² and 67,000 cal./cm² at 55° and 70°N, respectively. On the other hand, the mean annual amounts of precipitation in southern and northern Scandinavia exceed the mean annual evapotranspiration by some 200 mm and 250 mm, respectively. The annual loss of radiant heat to space at these latitudes in Scandinavia, consequently, is reduced by about 1,200 cal./cm² and 1,500 cal./cm² because of condensation. But nevertheless there is a great loss of heat to space in Scandinavia that must be compensated for through advection of heat to the area by air and by sea. The mildness of the climate of Scandinavia, consequently, depends to a great extent on the circulation of the atmosphere within the North Atlantic sector.

**The circulation of the atmosphere and the weather**

As pointed out previously, Scandinavia is exposed to the prevailing westerly and south-westerly winds of the mean Icelandic low-pressure circulation. The westerly or zonal type of circulation, therefore, is the most frequent one and brings in warm, moist Atlantic air over Scandinavia all year. These westerly and southwesterly winds are also the prima-ry causes of the warm Gulf Stream that moves northeastwards close to the west coast of Scandinavia. Because of its high temperature this ocean current has an extremely warming effect on the air blowing over it. Though a secondary agency, the Gulf Stream too becomes an important one by the advection of heat from southern latitudes to Scandinavia.

But over shorter or longer periods of time, the atmospheric circulation over the north-eastern Atlantic and Scandinavia may vary from the prevailing zonal type of circulation. Thus it happens—and sometimes for considerable periods of time—that the middle latitude jetstream is blocked by an anticyclone in the pressure system of the upper air and at the earth's surface. In these cases the zonal flow of air over the North Atlantic and Scandinavia is replaced by a quasi-meridional circulation. The weather in Scandi-navia then mainly depends on the position of the blocking anticyclone and may differ widely from that generally experienced.

Because of its position within the zone of westerlies, Scandinavia comes under the direct influence of the migratory cyclones that form along the Atlantic polar and arctic fronts and traverse the eastern Atlantic. Because of the *zonal type of circulation* Scandinavia is under the influence of air masses of vastly differing properties. They alternate when the fronts pass, and the most frequent conditions over Scandinavia under this type of circulation is a rapid succession of dry and wet spells.

The actual weather in Scandinavia, under the westerly type of circulation, differs widely from one region to another, depending on the intensity and the tracks of the cyclones as well as on the topography and exposure of the region. Cyclones crossing Denmark, for examples pass, one might say, undisturbed because of the flat terrain. The Scandinavian mountain chain, on the other hand, provides an effective barrier against cyclones and front free air currents arriving from directions between north and southwest. Cyclones crossing the Scandinavian mountain chain from one of these directions are, therefore, subjected to an orographic upglide. It is superimposed upon the general frontal upglide and results in increased condensation and precipitation on the western slopes of the chain. A similar but lesser effect, is experienced when cyclones cross the uplands of southern Sweden. Secondly, the Scandinavian mountain chain largely retards the progress of the fronts, with the result that the duration of precipitation on their western slopes is pro-longed. These slopes and the western slopes of the uplands of southern Sweden, there-fore, are generally regions of heavy precipitation and great cloudiness when the circu-lation is westerly.

East of the Scandinavian mountain chain, the westerly winds are subjected to a de-scending motion. Hence, relatively warm föhn winds generally occur because of the westerly type of circulation in Finnish Lapland, northern and central Sweden and in eastern Norway. For this reason and because of the cyclones' tendency to fill when moving over land distant from the source of moisture, the precipitation generally decreases eastwards. But it also happens that the cyclones regenerate over the lakes of central Sweden and the Baltic Sea and cause large amounts of precipitation both in Sweden and southwestern Finland.

In autumn and winter it often happens that the Atlantic polar fronts are displaced far to the south, coincident with a large trough extending from southern Greenland to central Europe. In such situations Scandinavia may also be invaded by warm, moisture-bearing Atlantic air masses from directions between south and east. On crossing the Baltic Sea these air masses receive additional moisture, and when moving in over Scandinavia are exposed to orographical lifting. Heavy frontal rainfalls, sometimes causing large flash floods, often then occur in Denmark, southern Sweden and eastern Norway, while the western side of the Scandinavian mountain chain is favoured by föhn winds and fine weather.

In winter the north–south temperature contrast is much greater than in summer. The *zonal circulation and the cyclonic activity as well*, therefore, *are most intense in winter*. Besides, the sea surface west of Scandinavia is warmer than the air masses blowing over it. Similarly, the evaporation from the sea is comparatively great. Hence, the westerly type of circulation generally brings in mild, moisture-bearing air masses and large temperature anomalies over Scandinavia in winter, especially in the western coastal districts of Norway and in Denmark. Because of the intensified atmospheric circulation, the orographical effects are increased too. Consistent with this, the precipitation on the western slopes of the Scandinavian mountain chain and the uplands of southern Sweden is much heavier and more frequent in winter than in summer. To the east of the Scandinavian mountain chain, however, the warm Atlantic air is cooled by contact with the snow surface or the bar ground with a consequent decrease in precipitation eastwards. However, the frequent invasions of warm, moist air often result in advection fog over the interior regions during winter half year.

Because of the high temperature of the Atlantic air masses in winter, the precipitation caused by the zonal circulation is mainly in the form of rain in Denmark, western Norway, southern and central Sweden and even in southwestern Finland. In the remainder of Scandinavia, however, the air temperatures are low enough for snow to be the normal form of precipitation even when the circulation is westerly or southwesterly-In summer the north–south temperature contrast is weak, and so are the zonal circulation, the Atlantic polar fronts and the cyclonic activity over the East Atlantic. Furthermore, the sea surface of the North Atlantic is colder than that of the continent to the east and that of the warm tropical and modified polar air masses that prevail over the North Atlantic in summer. Besides, the evaporation from the sea surface is small. Hence, the Atlantic air masses invading Scandinavia in summer are relatively cool and dry and are of great thermal stability. For this reason the air temperature and the precipitation are relatively low both in Denmark and in the coastal districts of western and northern Norway by westerly circulation in summer. On the leeward side of the Scandinavian mountain chain, however, these slow-moving air masses are exposed to rapid heating by contact with the earth's surface. High air temperatures and convective precipitation are therefore rather usual phenomena in summer in eastern Norway, Sweden and Finland, even for the westerly type of circulation.

Considering the *meridional types of circulation* there are three types of blocking situations with a perceptible influence on the climatic conditions in Scandinavia in certain seasons of the year. The first one is experienced when a blocking anticyclone builds up over the northeastern Atlantic in spring. These anticyclones are rather persistent and cause large outbreaks of polar or arctic air in Scandinavia from northwest to north for many days in succession and even for weeks. Because of their polar origin, these air masses are very dry, clean and are of great thermal stability. But due to orographical lifting, showers of snow or sleet generally occur on the north and northwestern slopes of the Scandinavian mountain chain and even further inland. The remainder of Scandinavia, to the southeast of this chain, is normally favoured by strong, gusty föhn winds and clear, and blue skies. Hence, the solar and nocturnal radiations are usually large, and in late spring and early summer these outbreaks of cold air may cause severe damage to cereal crops and fruit trees.

As long as there is a snow cover the shower activity inland is generally small. But later in the spring, when the snow cover has disappeared and the insolation increases from day to day, the shower activity becomes more frequent inland too. The weather of the transition period between winter and summer, therefore, is characterized by so-called "April weather". This means slight showers of sleet or rain alternating with short periods of bright sunshine.

In the winter half year and early spring a high-pressure ridge from the Siberian high-pressure cell often extends southwest or west over Scandinavia and becomes persistent for a long time. Scandinavia is then invaded by cold continental air masses that may cause very low air temperatures in all the Scandinavian countries. On crossing the Baltic Sea and its gulfs, however, the air masses receive a good deal of moisture. Therefore, on coasts exposed to northeasterly and easterly winds heavy snowfalls usually occur. But further inland the precipitation generally is slight and falls from low stratus clouds, sometimes for many days. The western and northwestern coastal districts of Norway, on the other hand, are favoured by föhn winds, clear sky and somewhat higher air temperatures than in the inland districts. In arctic Scandinavia, cyclonic activity often occurs along the northern flank of the high-pressure ridge.

The surface winds occurring under this type of circulation are generally very small, so that the air masses over the inland districts are well-exposed to additional cooling by radiation. If the high-pressure ridge becomes stationary for a longer period of time— and that is often the case in late winter and early spring—the conditions are favourable for a very cold spell to develop. Unusual cold and heavy air then accumulates over the mountain plateaus of Norway and western Sweden. Normally the drainage of this air to lower regions takes place very slowly. But if a cyclone passes at some distance from the west or the north coast of Norway, the plateau air is drawn rapidly down the valleys and fjords in large volume. These bora-like winds are very cold and very often cause much arctic sea smoke, especially in the fjords of northern Norway.

The unusually low air temperatures that frequently occur under this type of circulation also favor the formation of sea ice in the Baltic Sea and its gulfs. On the other hand, the persistent northeasterly or easterly winds cause the sea ice to drift out the Danish Sound into the Kattegat and up along the west coast of Sweden. If the cold spell and the easterly winds continue for a longer period of time, the sea ice may drift into the Oslofjord and even down along the coast of Skagerrak.

Sometimes in summer a high-pressure ridge like that discussed above, extends from a warm, deep anticyclone in central Russia, westwards over Scandinavia. Then a warm and dry spell—sometimes of relatively long duration—occurs over most of Scandinavia. This sometimes happens in late September or even in the beginning of October and then may cause a real "Indian summer".

Finally, a really "good weather situation" in Scandinavia first occurs when the core of a deep and warm anticyclone covers the whole area for a longer period of time. In reality this seldom happens, but when it happens, calm weather and almost cloudless skies are experienced in almost the whole of Scandinavia. In this weather situation the temperature conditions are, therefore, extremely influenced by the radiation balance of the earth's surface and the topography. This is especially the case during midsummer and midwinter when the surplus and deficit, respectively, of sun radiation is at its maximum. Hence, extremely high temperatures occur in summer, especially in valley bottoms and depressions in the plains. In winter, on the other hand, the same areas are exposed to some bitterly low temperatures and temperature inversions often occur over low-lying areas. This is especially the case in the interior sections of eastern Norway and of central and northern Sweden.

## Climatic types

There is a continuous change, from the zonal to the meridional types of atmospheric circulation throughout the year. From time to time the latter even dominate and influence the weather and the climatic conditions in Scandinavia for years. But on the average, over many years, the westerly and southwesterly types of circulation predominate and bring in large amounts of heat to Scandinavia.

Thus Scandinavia owes its variable weather and its mild maritime climate to the prevailing warm, moisture-bearing westerlies of the North Atlantic. But this zonal flow of air is considerably influenced by the topography, the Scandinavian mountain chain and the radiation balance of the earth's surface. All these factors make for a diversity of climate, and even though the macro-climate of Scandinavia is mainly maritime, it is far from uniform. Thus no less than five of Köppens' climatic types occur within Scandinavia.

In Denmark, southern Sweden and extreme southern Norway there is a temperate rainy climate without dry seasons but with warm summers (Cfb). North of this zone and south of 60°N there is a narrow zone extending from the east coast of Sweden to the west coast of Norway having a cool forest climate with warm summers (Dfc). The regions west of the Scandinavian mountain chain, between 60° and 70°N, also have a temperate, rainy climate, but with cool summers (Cfc). On the other hand, a tundra climate (ET) occurs in the high mountains of Norway and Sweden, and in arctic Norway. The remainder of Scandinavia, and by far the largest part, however, belongs to the boreal-forest climatic zone with cool summers (Dfc).

Because of the large and pronounced differences in topography, elevation, exposure and ground cover, the local climates of a region may differ widely from its type of macro-climate. This applies particularly to Norway with its low strandflats, great fjords, magnificent mountains, hill lands, valleys and lowlands. No other country of the same area has such extreme climatic contrasts within its borders as Norway.

**Atmospheric pressure and wind**

As pointed out previously, Scandinavia lies within the regime of the semi-permanent troughs of low pressure extending from Iceland towards Novaja Zemlya. In winter the mean pressure gradient between the low pressure centered over Iceland and the high-pressure cell over the Eurasian continent is very strong and associated with strong westerly and southwesterly winds both at the surface and in the upper air. In late spring, however, the Icelandic low becomes shallow and the continental high disappears. By midsummer the mean pressure gradient over Scandinavia is, therefore, weak and results in a feeble and variable circulation.

As a result of the unequal heating of the land masses and the adjacent seas a distinct difference exists between the mean pressure distribution over Scandinavia in winter and summer. Thus, a winter-time ridge of high-pressure occurs over the central and southern parts of the Scandinavian Peninsula and a trough over Finland. In summer, however, the winter-time high-pressure ridge is replaced by local lows over southeastern Norway and northeastern Sweden. But the trough over Finland is more or less persistent throughout the year.

Southwesterly and westerly winds prevail over Denmark, southern and central Sweden and southern Finland all the year. In northern Finland north and northeasterly winds are most frequent. Within the mountainous sections of Norway and Sweden, on the other hand, the surface winds are considerably influenced by the topography and are very irregular.

Because of the annual change in the pressure distribution discussed above, a monsoon-like circulation is experienced in large regions of Scandinavia. Thus northerly winds with an up-fjord component prevail along the western and northwestern coasts of Norway for most of the summer. An on-shore wind component, though often very weak, is also experienced in the coastal districts around the Gulf of Bothnia. In winter, however, the air mainly blows from inland points down-valley and down-fjords following the details of the topography—towards the sea. There it is deviated and blows parallel to the Norwegian coast on the entire tract from the Oslofjord in the south to the Varangerfjord in the north. A similar drainage of cold air also takes place down the many river valleys in northeastern Sweden. North of the Arctic Circle where the differences in temperature between inland and coastal points in Norway are generally extreme during winter, these down-fjord winds are very cold. They resemble "bora", frequently exceed gale force, are felt as an icy blast and generally produce boisterous squalls, sometimes with snow flurries.

The monsoonal circulation discussed above is restricted to the lowest layers of the atmosphere. In the upper air westerly and southwesterly winds dominate all year in Scandinavia. But they are much stronger in winter than in summer, because of the large north–south temperature contrast in winter.

Besides the monsoonal circulation, a land and sea breeze circulation is also experienced in many coastal districts on calm, warm summer days. In Norway these winds are predominant in the Oslofjord and along the coast of Skagerrak. Similarly the land and sea breezes are a well-known feature of the wind regime in summer around the Gulf of Bothnia and along the west coast of Sweden.

The strongest wind velocities occur along the western and northern coasts of Norway and in the high mountain regions; winter is by far the stormiest season in the coastal districts. On the Norwegian coast between Lindesnes in extreme southern Norway and Vardö the mean wind velocities range from 7 m/sec to 10 m/sec in winter, and from 3 m/sec to 5 m/sec in summer. The corresponding velocities in the coastal districts along the Baltic Sea and the Gulf of Bothnia are about 6–6.5 m/sec and 3–3.5 m/sec, respectively. In Denmark, on the other hand, the annual march of the mean velocities are less pronounced, and the annual wind velocities average about 6 m/sec.

There is a marked decrease in the wind velocities up-fjord and inland. But inland the greatest wind velocities usually occur in summer. The mean annual wind velocities in the inland regions, except in the high mountains, seldom exceed 2–3 m/sec, while the annual number of calms in many areas even exceed 50%. At the high mountain station, Finse (1,300 m above M.S.L.), on the Bergen-railway, the annual wind velocity, averages about 6 m/sec. The highest observed wind velocity at inland points is about 35 m/sec, and about 50–55 m/sec at coastal points in North Norway.

The greatest number of days per year with strong gales occurs along the south, west and north coasts of Norway, and averages 14 days at Lista, 22 days at Nordøyan and 12 days at Vardö. The corresponding figures for Vaasa in West Finland and Härnösand in eastern Sweden are 8 and 5, respectively, while Finse has 13 such days.


**Sunshine and radiation**


North of the Arctic Circle the sun does not rise above the horizon for some time in winter (see Table II). But the long time with midnight sun in summer more than compensates for the sunshine hours lost in winter. Therefore, the astronomically possible duration of sunshine increases northwards and is about: 4,440 h/year at 54°N, 4,500 h/year at 62°N, and 4,600 h/year at 70°N.

At many points in Norway and Sweden, however, the possible duration of sunshine is considerably reduced because of the screening effect of mountains and hills. Hence, the possible duration of sunshine in many Norwegian valleys and fjords for instance is less than 55% of what it could have been by free horizon. Besides, the great cloudiness in most parts of Scandinavia throughout the year serves to reduce the actual hours of sunshine and so do the high frequency of rain, snow and fog experienced in many sections of Scandinavia. The effect of these factors upon the duration of sunshine stands out very clearly in Table III, showing the mean monthly and annual hours of sunshine at different points in


TABLE II

THE LENGTH OF THE MIDNIGHT SUN AND THE DARK SEASON AT DIFFERENT LATITUDES IN SCANDINAVIA

| Latitude | Midnight sun | Dark season |
|---|---|---|
| 67°N | June  2–July 10 | |
| 68°N | May 27–July 18 | December 10–January  2 |
| 71°N | May 14–July 31 | November 22–January 21 |

TABLE III

MEAN MONTHLY AND ANNUAL HOURS OF BRIGHT SUNSHINE AT DIFFERENT POINTS IN SCANDINAVIA

| | Jan. | Feb. | Mar. | Apr. | May | June | July | Aug. | Sept. | Oct. | Nov. | Dec. | Year |
|---|---|---|---|---|---|---|---|---|---|---|---|---|---|
| Tylstrup | 40 | 71 | 134 | 190 | 266 | 273 | 264 | 229 | 165 | 102 | 46 | 28 | 1,808 |
| Viborg | 41 | 67 | 134 | 180 | 248 | 247 | 238 | 211 | 159 | 96 | 41 | 30 | 1,692 |
| Copenhagen | 36 | 56 | 118 | 161 | 245 | 245 | 239 | 204 | 157 | 86 | 34 | 19 | 1,600 |
| Göteborg | 48 | 76 | 151 | 201 | 274 | 286 | 285 | 245 | 178 | 108 | 47 | 29 | 1,928 |
| Kjevik | 57 | 103 | 133 | 172 | 208 | 239 | 252 | 203 | 135 | 96 | 56 | 44 | 1,698 |
| Sola | 53 | 86 | 134 | 155 | 209 | 204 | 187 | 160 | 118 | 74 | 46 | 37 | 1,463 |
| Bergen (Florida) | 25 | 62 | 113 | 135 | 186 | 154 | 170 | 127 | 90 | 67 | 28 | 14 | 1,171 |
| Haugastöl | 47 | 100 | 141 | 151 | 204 | 181 | 182 | 149 | 116 | 91 | 40 | 27 | 1,429 |
| Oslo (Blindern) | 47 | 92 | 159 | 170 | 216 | 234 | 221 | 172 | 132 | 87 | 46 | 24 | 1,600 |
| Kise p. Hedmark | 38 | 72 | 162 | 169 | 211 | 232 | 228 | 164 | 129 | 78 | 42 | 11 | 1,536 |
| Stockholm | 41 | 76 | 151 | 208 | 292 | 318 | 295 | 248 | 174 | 103 | 41 | 26 | 1,973 |
| Helsinki | 31 | 63 | 136 | 184 | 270 | 294 | 295 | 251 | 152 | 76 | 30 | 18 | 1,800 |
| Trondheim (Voll) | 16 | 52 | 124 | 162 | 204 | 180 | 181 | 174 | 119 | 72 | 29 | 8 | 1,321 |
| Gisselås | 26 | 61 | 121 | 180 | 252 | 240 | 249 | 204 | 126 | 74 | 29 | 10 | 1,572 |
| Härnösand | 54 | 70 | 144 | 195 | 259 | 251 | 267 | 217 | 145 | 90 | 52 | 27 | 1,771 |
| Bodö | 7 | 32 | 118 | 149 | 175 | 225 | 193 | 195 | 84 | 52 | 22 | — | 1,252 |
| Abisko | — | 36 | 120 | 170 | 195 | 225 | 240 | 160 | 115 | 75 | 5 | — | 1,341 |
| Sodankylä | 8 | 49 | 140 | 191 | 222 | 277 | 278 | 184 | 102 | 60 | 19 | 0.3 | 1,530 |
| Tromsö | 1 | 25 | 115 | 161 | 186 | 213 | 199 | 201 | 97 | 46 | 4 | — | 1,248 |
| Karasjok | — | 34 | 146 | 172 | 172 | 168 | 153 | 105 | 65 | 42 | 3 | — | 1,060 |

Scandinavia. As will be seen from the table, the annual duration of sunshine ranges from about 1,000 h in the least sunny districts to about 2,000 h in the sunniest ones, corresponding to a mean relative duration of 20–40% per year only.

The least annual duration of sunshine is experienced in the cloudy and foggy districts west of the Scandinavian mountain chain and in the uplands of southern Sweden. Riksgränsen and Abisko in northwestern Sweden, for instance, have a mean relative sunshine duration of 20% and 25%, respectively, per year. But there is a marked increase in the annual relative duration of sunshine to the south and to the east. Thus a comparatively great relative duration of sunshine is experienced along the west coast of Sweden, in the southern and southeastern districts of Norway; in Denmark it ranges from about 38% to 42% per year. The greatest relative duration of sunshine per year is experienced in the coastal districts of Sweden, to the north and south of the Stockholm area and interior Finland. As will be shown later these districts are also favoured by the highest annual frequency of clear days occuring in Scandinavia.

The global radiation, like the duration of sunshine, is also considerably reduced because of the great cloudiness and the many screening factors. Thus the mean annual global radiation (see Table IV) ranges from about 88.5 kcal./cm² at Copenhagen to about 50 kcal./cm² at Tromsø, and from 71 kcal./cm² at Bergen to 76.7 kcal./cm² at Helsinki. Though the annual global radiation is subjected to a marked decrease with latitude, the effect of the midnight sun in June and July stands out very clearly in Table IV. Thus Karasjok during these months receives almost the same or even more global radiation than do, for example, Oslo and Stockholm.

The radiation balance of the earth's surface at some selected points have been computed

TABLE IV

THE MEAN DAILY AND ANNUAL GLOBAL RADIATION (cal./cm²) AT DIFFERENT STATIONS IN SCANDINAVIA

|  | Jan. | Feb. | Mar. | Apr. | May. | June | July | Aug. | Sept. | Oct. | Nov. | Dec. | Year |
|---|---|---|---|---|---|---|---|---|---|---|---|---|---|
| Copenhagen | 48.5 | 107.6 | 204.9 | 324.8 | 438.0 | 506.4 | 448.0 | 347.0 | 266.9 | 129.5 | 48.4 | 32.0 | 88,488 |
| Bergen (Florida) | 28.8 | 67.3 | 168.2 | 251.2 | 425.7 | 400.1 | 371.2 | 297.4 | 172.1 | 96.5 | 37.7 | 16.0 | 71,235 |
| Oslo (Blindern) | 26.4 | 73.8 | 189.1 | 291.9 | 433.6 | 422.1 | 396.2 | 300.0 | 182.0 | 113.4 | 32.3 | 16.0 | 75,631 |
| Stockholm | 27.1 | 80.9 | 179.2 | 208.4 | 374.2 | 441.8 | 407.6 | 328.4 | 210.0 | 101.8 | 41.2 | 18.7 | 76,130 |
| Helsinki | 18.6 | 70.3 | 196.7 | 291.0 | 400.5 | 478.2 | 449.5 | 304.4 | 186.6 | 83.5 | 23.1 | 11.6 | 76,747 |
| Trondheim (Voll) | 16.6 | 53.8 | 157.5 | 263.4 | 413.1 | 411.5 | 436.4 | 313.3 | 184.2 | 90.3 | 25.8 | 8.7 | 72,566 |
| Sodankylä | 3.2 | 42.6 | 152.9 | 312.0 | 372.2 | 444.9 | 452.2 | 283.9 | 136.1 | 51.2 | 9.7 | 0.2 | 69,067 |
| Tromsö | — | 20.5 | 102.7 | 215.6 | 294.3 | 400.8 | 396.8 | 237.0 | 120.3 | 41.3 | 4.6 | — | 56,048 |
| Karasjok | — | 22.1 | 120.6 | 248.2 | 363.5 | 436.6 | 415.4 | 271.5 | 123.2 | 45.6 | 4.2 | — | 62,699 |

and are shown in Table V. The effect of the dark season upon the radiation balance in arctic Scandinavia stands out very clearly. At Karasjok, for instance, the daily net loss of heat by radiation in December averages about 102 cal./cm², while only 28 cal./cm² at Copenhagen. Further, the net loss of heat from the earth's surface lasts from about 15 September to 20 April at Karasjok and from only 26 October to about 20 February at Copenhagen. But even though the earth's surface in arctic Scandinavia is subjected to a large deficit in radiation balance in winter, this loss of heat is compensated for by the great amounts of global radiation received when there is midnight sun.

As will be seen from Table V, the mean annual radiation balance of the earth's surface is positive and ranges from about 35 kcal./cm² at Copenhagen to about 0.5 kcal./cm² at Tromsö. Even at points in the high mountain plateaus there is an annual surplus of radiant energy at the earth's surface. The annual radiation balance at Slirå (1,300 m above M.S.L.) and at Røros (628 m above M.S.L.), for example, average 0.5 kcal./cm², and 0.8 kcal./cm², respectively.

## Air temperature

Because of the prevailing westerly and southwesterly winds most of the air masses have been heated or cooled by contact with the warm Gulf Stream before reaching Scandinavia. The regions of Scandinavia that are directly exposed to the mild Atlantic winds, therefore, have remarkably high air temperature in winter considering the latitude. This applies especially to Denmark, extreme southern Sweden and the western and northwestern coasts of Norway. In summer, however, when the Gulf Stream is relatively cold, the air temperatures within these regions do not rise very high.

The warm westerlies cross the Scandinavian mountain chain at higher levels with their physical properties nearly undisturbed. For this reason a maritime influence is experienced even on the temperature regime of the high mountains and mountain plateaus of the Scandinavian Peninsula. Because of the general decrease of temperature with elevation the air temperatures within these regions are, of course, much lower than in the nearby coastal regions. But they are much higher, especially in winter, than would be expected considering the height and latitude.

TABLE V

MEAN DAILY AND ANNUAL RADIATION BALANCE (cal./cm²) OF A GRASS SURFACE THAT IS COVERED BY SNOW IN WINTER

| | Jan. | Feb. | Mar. | Apr. | May | June | July | Aug. | Sept. | Oct. | Nov. | Dec. | Year |
|---|---|---|---|---|---|---|---|---|---|---|---|---|---|
| Copenhagen | − 31.3 | − 14.0 | 55.1 | 150.6 | 223.3 | 275.0 | 246.8 | 167.1 | 96.9 | 19.2 | −17.7 | − 27.9 | 34,975 |
| Bergen (Florida) | − 89.0 | − 60.3 | 8.7 | 102.0 | 222.6 | 220.9 | 200.6 | 140.8 | 64.4 | −11.1 | −67.5 | − 81.0 | 20,045 |
| Oslo (Blindern) | − 86.6 | − 77.2 | −27.2 | 107.1 | 227.2 | 231.2 | 213.2 | 140.5 | 55.6 | − 7.4 | −59.8 | − 77.6 | 19,604 |
| Stockholm | − 75.4 | − 45.2 | 17.1 | 41.0 | 179.7 | 255.7 | 204.1 | 144.3 | 58.7 | −13.3 | −34.8 | − 54.8 | 22,775 |
| Southern Finland | − 93.6 | − 82.1 | − 6.5 | 86.7 | 196.8 | 316.7 | 264.5 | 151.6 | 53.3 | −38.7 | −76.7 | − 88.4 | 21,100 |
| Trondheim (Voll) | −117.3 | − 92.5 | −56.8 | 48.2 | 167.7 | 225.2 | 231.0 | 141.8 | 45.7 | −38.9 | −89.9 | −105.3 | 11,174 |
| Tromsö | −109.5 | − 98.1 | −72.3 | −23.7 | 63.3 | 219.1 | 211.5 | 131.9 | 2.0 | −64.1 | −96.9 | −101.9 | 1,061 |
| Karasjok | −104.7 | −110.8 | −83.9 | −36.3 | 105.3 | 222.0 | 200.9 | 100.1 | −4.2 | −76.1 | −92.8 | −102.3 | 777 |

Even though the Scandinavian mountain chain provides an effective barrier to the mild Atlantic winds, there are nevertheless points in it low enough to allow access of the mild westerly air currents to the interior of Sweden and even to Finland. This is the case, for instance, in the Meraaker Valley east of Trondheim, in the depression extending eastwards from Namsos into Sweden and in the Ofoten by Narvik. Also mild Atlantic air masses from the southwest frequently invade Scandinavia. This is especially true in winter and these air masses have a relatively free access both to eastern Norway, southern and central Sweden and to Finland. On the other hand, the Baltic Sea and its gulfs have considerable maritime influence in the coastal districts of eastern Sweden and Finland as well as in Denmark. The rising of the air temperature in these districts is retarded in spring and early summer when the sea is colder than the land. Likewise the fall of the temperature is delayed in autumn because of the sea being warmer than the land. Besides, the Baltic Sea and its gulfs largely influence the temperature of the air currents moving over them. A weak maritime influence is also experienced on the annual march of air temperature in the surroundings of the many great lakes of Finland, Sweden and Norway. On the whole, a considerable maritime influence is, therefore, experienced even east of the Scandinavian mountain chain, especially in winter when the circulation of the atmosphere is strongest (see Fig.1, 2).

On the leeward side of the Scandinavian mountain chain a downward component is superimposed upon the westerly winds with a consequent dissolution of the clouds. Similarly, the wind velocities within these regions are generally small because of the rough topography and the friction along the earth's surface. The renewal of the air is, therefore, small, while the effect of radiation is large. The climate of these leeward regions is accordingly determined more by radiation than by advection. The temperature regime, therefore, assumes a more continental character with cold winters and warm summers the farther we withdraw eastwards. This longitudinal effect also stands out very clearly in Table VI where the mean temperatures (1931–1960) of the two coldest and warmest months of the year are tabulated for points near to the 61° and 68°N latitude circles, respectively.

As will be seen from the table, the mean temperature of January decreases by 11.8° and 14.9°C from west to east along the 61° and 68°N latitude circles, respectively. The mean temperature in July, on the other hand, increases by only 3.0° and 1.3°C respectively, along the same latitudinal circles. The increasing continental effect on the temperature eastwards is also confirmed by an increase of the amplitude between the mean temperature of the warmest and coldest month of the year. For example, at Bulandet (see Table VI) the mean amplitude is 12.2°C only, while at Punkaharju in eastern Finland it is 26.3°C. Likewise, the amplitude increases by 15.5°C from Glåpen Fyr in the Lofoten Islands to Sodankylä in northern Finland. Compared to the corresponding amplitudes at Irkutsk and Verkojansk which are 38.7° and 62.5°, respectively, the degree of "continentality" experienced in the interior sections of Scandinavia is small.

The temperature regime of Scandinavia is also exposed to a marked latitudinal effect due to radiation. Because of the high frequency of mild westerly and southwesterly winds, however, this effect is small along the west coast in winter. For example, between Borris on the west coast of Jutland (see Table VI) and Gamvik on the coast of Finnmark, the mean temperature of February decreases by only 4.6°C. In summer, on the other hand, there is a high frequency of relatively cold easterly winds along the north coast of

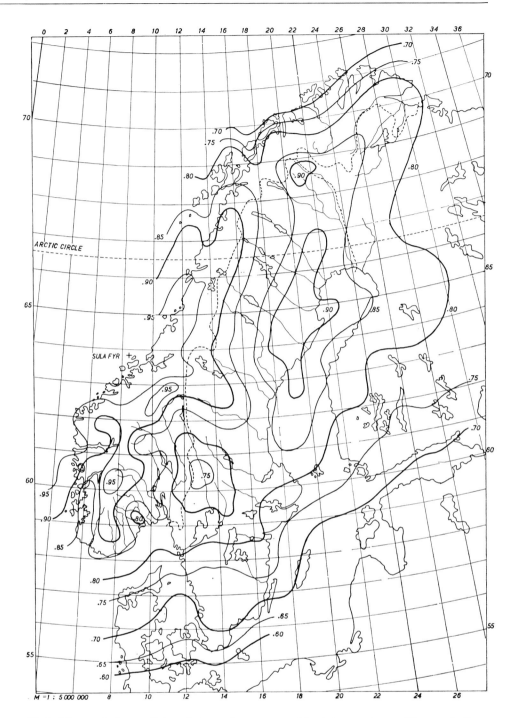

Fig.1. The correlation coefficients between the mean monthly air temperatures of January in the years 1931–1960 at Sula Fyr (63°51′N 8°28′E) and the corresponding temperatures at 63 Norwegian, 45 Swedish, 24 Finnish and 8 Danish weather stations.

Scandinavia and the mean temperature of July decreases by 7.5°C between Borris and Gamvik.

East of the Scandinavian mountain chain, however, the north–south temperature differences are much more extreme. The mean temperature of February, for instance, decreases by no less than 11.6°C from Växjö in southern Sweden to Karasjok in arctic

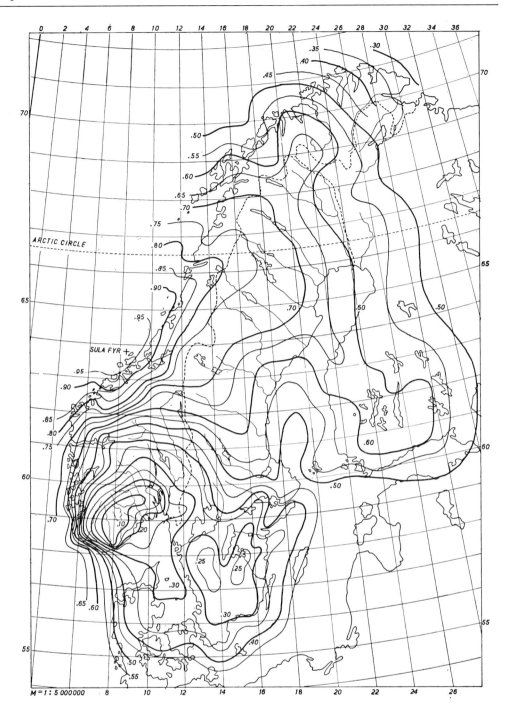

Fig.2. The correlation coefficients between the mean monthly air temperatures of July in the years 1931–1960 at Sula Fyr (63°51′N 8°28′E) and the corresponding temperatures at 63 Norwegian, 45 Swedish, 24 Finnish and 8 Danish weather stations.

Norway. In July, on the other hand, the mean temperature decreases by only 2.6°C, between these two points. The effect on the temperature due to the dark season and the midnight sun is consequently very large. Likewise, the continuous ice cover that generally occurs in the Gulf of Bothnia in winter, considerably influences the temperature of the coastal districts around the gulf. The mean temperature of February at Haparanda,

TABLE VI

LONGITUDINAL AND LATITUDINAL VARIATION OF THE AIR TEMPERATURE OF SCANDINAVIA IN WINTER AND SUMMER

| | Lat. (N) | Long. (E) | H (m) | Mean temperature (°C) | | | | Ampl. (°C) |
|---|---|---|---|---|---|---|---|---|
| | | | | Jan. | Feb. | July | Aug. | |
| *Variation of temperature from west to east along the 61°N latitude circle* | | | | | | | | |
| Bulandet | 61°18′ | 4°35′ | 4 | 2.4 | 2.0 | 13.8 | 14.2 | 12.2 |
| Takle | 61°02′ | 5°23′ | 39 | 1.1 | 0.9 | 14.8 | 14.2 | 13.9 |
| Leikanger | 61°11′ | 6°52′ | 22 | − 0.6 | − 0.6 | 16.0 | 14.9 | 16.6 |
| Fortun | 61°30′ | 7°42′ | 27 | − 5.1 | − 5.1 | 14.8 | 13.7 | 19.9 |
| Luster Sanatorium | 61°26′ | 7°25′ | 484 | − 4.1 | − 4.2 | 13.5 | 12.4 | 17.7 |
| Fanaråken | 61°31′ | 7°54′ | 2,062 | −12.3 | −12.4 | 2.6 | 2.1 | 15.0 |
| Sikkilsdal | 61°29′ | 9°02′ | 1,015 | −10.0 | − 9.6 | 11.3 | 9.8 | 21.3 |
| Rena | 61°08′ | 11°23′ | 225 | −10.5 | − 9.2 | 15.7 | 13.9 | 26.2 |
| Trysil | 61°20′ | 12°15′ | 356 | −10.1 | − 8.6 | 14.4 | 12.9 | 24.5 |
| Engerdal 2 | 61°41′ | 12°01′ | 479 | − 9.5 | − 8.3 | 13.5 | 12.0 | 23.0 |
| Särna | 61°41′ | 13°07′ | 461 | −11.3 | − 9.6 | 14.1 | 12.3 | 25.4 |
| Mora | 61°01′ | 14°35′ | 180 | − 8.5 | − 7.7 | 15.7 | 13.8 | 24.2 |
| Söderhamn | 61°16′ | 17°06′ | 30 | − 5.4 | − 5.2 | 16.2 | 15.0 | 21.6 |
| Mariehamn | 60°07′ | 19°54′ | 15 | − 6.0 | − 6.8 | 16.9 | 15.2 | 22.9 |
| Pori | 61°28′ | 21°48′ | 11 | − 6.4 | − 6.6 | 16.6 | 15.0 | 23.0 |
| Punkaharju | 61°48′ | 29°19′ | 88 | − 9.4 | − 9.5 | 16.8 | 15.2 | 26.3 |
| *Variation of temperature from west to east along the 68°N latitude circle* | | | | | | | | |
| Glåpen Fyr | 67°53′ | 13°03′ | 31 | 0.3 | − 0.4 | 12.6 | 12.1 | 13.0 |
| Svolvær | 68°14′ | 14°37′ | 1 | − 1.1 | − 1.8 | 13.9 | 13.4 | 15.7 |
| Narvik | 68°28′ | 17°30′ | 32 | − 3.7 | − 4.2 | 14.4 | 12.6 | 18.6 |
| Björnfjell | 68°26′ | 18°05′ | 512 | −10.6 | −11.2 | 11.7 | 9.6 | 22.9 |
| Riksgränsen | 68°26′ | 18°08′ | 508 | −10.3 | −10.8 | 11.4 | 9.6 | 22.2 |
| Kiruna | 67°51′ | 20°14′ | 505 | −12.2 | −12.4 | 12.9 | 10.5 | 25.3 |
| Muonio | 67°57′ | 23°41′ | 242 | −13.0 | −12.5 | 14.1 | 11.9 | 27.1 |
| Sodankylä Vuotso | 68°06′ | 27°11′ | 246 | −14.6 | −14.6 | 13.9 | 11.2 | 28.5 |
| *Variation of temperature from south to north along the west coast* | | | | | | | | |
| Borris | 55°57′ | 8°38′ | | 0.0 | − 0.3 | 16.1 | 15.8 | 16.1 |
| Lindesnes | 58°00′ | 7°06′ | 4 | 0.7 | 0.2 | 16.2 | 16.0 | 16.0 |
| Bergen (Fr.berg) | 60°24′ | 5°19′ | 43 | 1.5 | 1.3 | 15.0 | 14.7 | 13.7 |
| Bodö | 67°17′ | 14°25′ | 10 | − 2.1 | − 2.4 | 13.6 | 12.7 | 16.0 |
| Gamvik | 71°04′ | 28°15′ | 6 | − 3.6 | − 4.6 | 9.5 | 9.8 | 14.0 |
| *Variation of temperature from south to north inland* | | | | | | | | |
| Växjö | 56°53′ | 14°49′ | 172 | − 3.2 | − 3.0 | 16.5 | 15.2 | 19.7 |
| Karasjok | 69°28′ | 25°31′ | 129 | −14.8 | −14.6 | 13.9 | 11.5 | 28.7 |
| *Variation of temperature from south to north along the coast of Sweden* | | | | | | | | |
| Karlshamn | 56°10′ | 14°52′ | 7 | − 0.9 | − 0.9 | 17.3 | 16.4 | 18.2 |
| Västervik | 57°47′ | 16°36′ | 12 | − 2.0 | − 2.2 | 17.4 | 16.4 | 19.6 |
| Härnösand | 62°38′ | 17°57′ | 9 | − 6.4 | − 5.8 | 16.4 | 15.1 | 22.8 |
| Haparanda | 62°50′ | 24°09′ | 9 | −10.7 | −10.9 | 16.3 | 14.0 | 27.2 |

for example, (see Table VI) is −10.9°C, and only −0.9°C at Karlshamn further to the south on the east coast of Sweden. The mean temperature of July, on the other hand, decreases only by 1.0°C from Karlshamn to Haparanda.

TABLE VII

MEAN UPPER AIR TEMPERATURES IN °C IN THE PERIOD 1951–1960

|  | Jan. | Apr. | July | Oct. | Year | | Jan. | Apr. | July | Oct. | Year |
|---|---|---|---|---|---|---|---|---|---|---|---|
| | *Sodankylä: 67°22′N 26°39′E* | | | | | | *Frösön: 63°11′N 14°37′E* | | | | |
| 850 mbar | −11.1 | − 7.2 | 6.7 | − 3.5 | − 3.8 | | − 8.8 | − 4.9 | 7.0 | − 0.4 | − 1.9 |
| 700 mbar | −18.9 | −15.2 | − 2.1 | −10.7 | −11.7 | | −16.1 | −13.1 | − 1.8 | − 7.8 | − 9.7 |
| 500 mbar | −34.6 | −30.5 | −17.2 | −25.8 | −27.0 | | −32.3 | −28.7 | −17.0 | −23.3 | −25.3 |
| 400 mbar | −45.0 | −41.3 | −28.6 | −37.4 | −37.9 | | −42.3 | −39.9 | −28.9 | −34.7 | −36.6 |
| 300 mbar | – | – | – | – | – | | −55.6 | −51.7 | −43.6 | −48.4 | −49.8 |
| 200 mbar | – | – | – | – | – | | −61.2 | −53.4 | −49.0 | −56.3 | −54.5 |
| | *Ilmala: 60°12′N 24°55′E* | | | | | | *Bromma: 59°21′N 17°57′E* | | | | |
| 850 mbar | − 8.3 | − 4.0 | 8.4 | 0.3 | − 1.0 | | − 7.3 | − 3.7 | 8.0 | 1.0 | − 0.6 |
| 700 mbar | −15.7 | −11.9 | − 0.5 | − 6.8 | − 8.8 | | −15.0 | −11.7 | − 0.8 | − 6.1 | − 8.5 |
| 500 mbar | −31.6 | −27.6 | −16.4 | −22.1 | −24.4 | | −31.2 | −27.5 | −16.3 | −21.7 | −24.2 |
| 400 mbar | −42.0 | −38.4 | −27.3 | −33.1 | −35.1 | | −42.2 | −38.8 | −28.2 | −33.1 | −35.6 |
| 300 mbar | – | – | – | – | – | | −54.4 | −51.3 | −43.4 | −47.3 | −49.1 |
| 200 mbar | – | – | – | – | – | | −60.1 | −54.1 | −50.2 | −56.8 | −55.1 |
| | *Torslanda: 57°43′N 11°47′E* | | | | | | *Kastrup: 55°46′N 12°31′E* | | | | |
| 850 mbar | − 5.7 | − 2.6 | 8.0 | 2.0 | 0.4 | | − 5.0 | − 1.7 | 8.5 | 3.0 | 1.2 |
| 700 mbar | −13.7 | −10.8 | − 0.4 | − 5.2 | − 7.6 | | −12.7 | − 9.7 | 0.2 | − 4.2 | − 6.6 |
| 500 mbar | −29.8 | −26.7 | −16.1 | −20.8 | −23.3 | | −29.0 | −25.8 | −15.4 | −20.1 | −22.6 |
| 400 mbar | −41.0 | −38.0 | −28.0 | −32.5 | −34.8 | | −40.4 | −37.4 | −27.2 | −31.8 | −34.2 |
| 300 mbar | −53.4 | −50.8 | −42.9 | −46.6 | −48.5 | | −52.9 | −50.5 | −42.4 | −46.1 | −48.1 |
| 200 mbar | −59.5 | −54.8 | −50.4 | −56.3 | −55.2 | | −59.0 | −54.8 | −50.8 | −56.7 | −55.2 |

In Denmark and Finland, where the terrain is flat, the air temperature does not vary much from point to point. But in Norway and in the mountainous sections of Sweden large differences in the air temperature may occur even over very short horizontal and vertical distances making the temperature range very complex. The temperature differences due to these terrestrial influences, however, are smoothed out with elevation. At higher levels the longitudinal effect on the temperature distribution discussed above, is therefore almost negligible. At these levels the temperature is mainly subjected to a latitudinal effect. This also stands out very clearly in Table VII showing the mean temperatures (1951–1960) at six points in the upper air over Scandinavia.

In Scandinavia there is no dry or wet season so that the beginning and end of the four seasons of the year mainly depend on the air temperature. For practical reasons winter, therefore, is defined as the time of the year when the daily mean air temperature is <0°C and summer as the time when it is ≥10°C.

**The winter**

The winter normally begins between the 5th and the 10th October in the mountains of interior arctic Scandinavia, and about ten days later in the high mountain plateaus of southern Norway. Between 15th and 25th November the winter also begins in southern and southwestern Finland and about one month later it arrives in the coastal districts of southern Sweden and southern Norway too. But it is not until 15–20 January that the winter reaches the southern sections of Denmark.

The 0°C isotherm of the coldest month runs parallel to the west coast of Jutland, crosses the Skagerrak and continues parallel to the Norwegian west coast some kilometers inland to the Trondheimfjord. From there it runs almost parallel to the coast, but at some kilometers off the coast to the Lofoten Islands. Thus these coastal regions have no real winter. Winter is marked by some days with minimum temperatures below zero. But the high relative humidity and the great wind velocities make the mild temperatures seem colder than they really are. Within these regions we even find the ilex growing. The longest winter occurs on Finnmarksvidda and in Lapland where the length ranges from 214 to 201 days at Siccajavre and Sodankylä, respectively. The lowest mean monthly air temperatures are also experienced in these areas, that of January being −14.8°C and −14.6°C at Karasjok and Sodankylä, respectively. But low winter temperatures also occur in East Finland, in Bergslagen, Sweden, and in eastern Norway near the Swedish border where Tynset (62°18′N 10°45′E) has a mean temperature of −12.8°C in January. In Denmark, however, the mean monthly temperature of the coldest month seldom drops much below 0°C.

A typical phenomenon on the Scandinavian Peninsula in winter is the drainage of the cold mountain air down-valley and down-fjord towards the sea. In arctic Norway the air temperatures at inland points are normally much lower than at coastal points in winter. The down-fjord, bora-like, winter winds of arctic Norway are, therefore, generally much colder and stronger than in the rest of Scandinavia.

Stations situated near the inland end of fjords exposed to down-fjord winds have much lower temperature than stations near the mouth of the fjords. The down-fjord winds, therefore, greatly influence the temperature of the fjord and coastal districts of arctic Norway in winter. They are also the primary cause of the 0°C isotherm of January being displaced many kilometres off the coastline on the stretch between the Trondheimfjord and the Lofoten Islands.

The coldest and strongest down-fjord winds are experienced in the Ranafjord immediately north off the 60°N, and in the Lyngenfjord, Tanafjord and Varangerfjord in northernmost arctic Norway. At Mo i Rana near the inland end and at Nord-Solvær in the skerries near the mouth of this fjord, the mean temperatures of January, for instance, are −6.5° and −0.1°C, respectively. Further to the north, however, the differences in temperature between the inland end and the mouth of the fjords, are even larger. From the inland end to the mouth of the Tanafjord, for instance, the mean temperature of February decreases by no less than 6.6°C.

Even the winter temperatures in the deep fjords of western Norway are considerably influenced by the cold mountain air. This stands out very clearly in Table VI when comparing the mean temperature of February at Bulandet and at Fortun located near the mouth and the inland end, respectively, of the Sognefjord.

In southeastern Norway the cold inland air flows down the Oslofjord and the north–south orientated river valleys in the southernmost part of the country and makes the mean temperature of the coldest month fall 1°–4° below 0°C along the Skagerrak coast.

Another large drainage area of cold mountain air occurs in Norrland in northern Sweden. There the air from the mountains near the Norwegian border and from the nearby plateau occupied by the great lakes, streams down the many river valleys and overflows the large forested area between the lake region in the west and the coastal districts in the east with rather cold air in winter.

In high-pressure situations with much calm weather over the Scandinavian Peninsula cold air from the mountains and the hills also has a tendency to gather in depressions in the terrain where it is subjected to continued cooling by radiation. Because of its very irregular topography the temperature regime of many districts of interior Scandinavia, is therefore, characterized by rather strong temperature inversions in winter. This is the case in many river valleys, lowland districts, forested regions and even at many points in the central plateaus and mountains in eastern and arctic Norway, in northern and central Sweden and in Finland. Compared to the low temperatures within these inversion areas those experienced in the high mountain regions in winter are relatively warm. Thus the mean temperature of the coldest month at many points with elevations of 1,000–1,300 m above M.S.L. in the central plateaus of southern Norway ranges from −7° to −10°C while that at Røros (628 m) and Tynset (483 m) is −11.2° and −12.8°C, respectively.

The actual temperature regime of Scandinavia is subjected to large variations in winter, and the air temperatures frequently drop far below 0°C, especially in the interior of the countries. The lowest temperatures ever observed in winter are −51.4°C at Karasjok in Norway, −53.3°C at Laxbäcken in Sweden and −31.0°C at Silkeborg in Denmark. Temperatures of −45° and −50°C have also been observed in eastern and northern Finland.

The lowest (min.) and the highest (max.) mean temperatures of January experienced at some stations in the period 1931–1960 are tabulated in Table VIII, together with the standard deviation and the normal temperature (mean) for the same period. As will be seen, the mean temperature of January is subjected to large variations from year to year, and the same also applies to the mean temperatures of the other winter months. The variations, however, are much larger at inland than at coastal points. At Lille-hammer, for example, the difference between the extremes is 21.4°C, and only 8.5°C at

TABLE VIII

THE LOWEST (MIN.) AND HIGHEST (MAX.) MEAN MONTHLY AIR TEMPERATURES OF JANUARY AND JULY EX-
PERIENCED IN THE PERIOD 1931–1960 TOGETHER WITH THE NORMAL TEMPERATURE (MEAN) AND THE STANDARD
DEVIATIONS OF THE MEAN MONTHLY AIR TEMPERATURES IN THE SAME PERIOD OF TIME

| Station | Lat. (N) | Long. (E) | H (m) | January temperature (°C) | | | July temperature (°C) | | | Standard deviation (°C) | |
|---|---|---|---|---|---|---|---|---|---|---|---|
| | | | | min. | mean | max. | min. | mean | max. | Jan. | July |
| Borris | 55°57′ | 8°38′ | 25 | − 7.1 | 0.0 | 3.9 | 14.3 | 16.1 | 18.7 | 2.58 | 1.01 |
| Bergen (Fr.berg) | 60°24′ | 5°19′ | 43 | − 3.7 | 1.5 | 4.8 | 12.9 | 15.0 | 17.4 | 1.93 | 1.01 |
| Lillehammer | 61°07′ | 10°28′ | 266 | −17.0 | − 8.6 | −4.4 | 11.2 | 16.2 | 18.3 | 3.15 | 1.15 |
| Uppsala | 59°53′ | 17°36′ | 24 | −12.6 | − 4.2 | 0.6 | 15.3 | 17.3 | 20.0 | 2.95 | 1.22 |
| Helsinki | 60°12′ | 24°55′ | 45 | −16.6 | − 6.1 | −1.0 | 15.0 | 17.2 | 20.1 | 3.65 | 1.28 |
| Vallersund | 63°51′ | 9°44′ | 4 | − 6.1 | − 0.4 | 3.6 | 11.5 | 13.9 | 16.4 | 2.15 | 1.18 |
| Gisselås | 63°42′ | 15°22′ | 320 | −19.3 | −11.2 | −4.0 | 11.0 | 14.2 | 17.2 | 3.96 | 1.41 |
| Vaasa | 63°05′ | 21°37′ | 4 | −16.6 | − 6.7 | −1.0 | 14.0 | 16.5 | 18.3 | 3.54 | 1.14 |
| Maaninka | 63°09′ | 27°19′ | 88 | −19.1 | − 9.9 | −3.0 | 14.3 | 16.7 | 19.6 | 3.85 | 1.40 |
| Andenes | 69°19′ | 16°07′ | 5 | − 3.1 | − 0.9 | 1.7 | 9.1 | 11.3 | 13.8 | 1.48 | 1.26 |
| Kvikkjokk | 66°57′ | 17°45′ | 330 | −20.2 | −13.5 | −4.6 | 10.5 | 13.7 | 16.8 | 3.96 | 1.50 |
| Karasjok | 69°28′ | 25°31′ | 129 | −20.9 | −14.8 | −5.8 | 10.7 | 13.9 | 17.9 | 4.14 | 1.93 |

Bergen. The contrasting influence on the temperature conditions in winter caused by the ice-covered Gulf of Bothnia and by the ice-free, warm Atlantic also stands out very clearly in Table VIII. The lowest mean temperature in January at Bergen, for example, is only $-3.7°C$, while that at Vaasa on the west coast of Finland is $-16.6°C$.

**The spring**

The spring makes its entrance in the southernmost districts of Denmark as early as about the 10th of February and about one month later in extreme southern Sweden and southern Norway. In southern Finland the spring begins between the 5th and the 10th of April, and by April 20–25 it arrives at the coast of Finnmark too. Between the 1st and the 10th of May it also appears in the mountains and mountain plateaus of southern Norway and arctic Scandinavia.

Frequent blockings occur over northern Russia in spring, and the Scandinavian countries, therefore, are very often exposed to cold and dry easterly winds in this season of the year. Hence, only slight cloudiness is usual, and the incoming and outgoing radiation is large. The temperature regime of Scandinavia in spring, therefore, is generally characterized by great daily amplitudes and by rapid fluctuations. Much heat is, of course, used to melt the snow cover and to thaw the soil in spring. But nevertheless the strong and increasing global radiation provides for a rapid rise of the daily mean temperatures from day to day and makes the spring cover a shorter period of time than does the autumn. In April and May, however, an extensive blocking anticyclone very often occurs over the northeastern Atlantic and causes large outbreaks of cold polar or even arctic air from northerly directions in over Scandinavia. If these air masses are sufficiently cold, the spring may be considerably prolonged and severe damage caused to plants and fruit trees.

**The summer**

The summer begins about the 5th of May in southern Denmark and about 10–15 May in extreme southern Sweden. About 15–20 May it makes its entrance on the southern coast of Norway and some five days later in southern Finland. In the coastal districts of western Finland and in the archipelago, however, it is delayed until about June 1–10. Between 15 May and 1 June the summer also makes its entrance in the inner fjord districts of western Norway, and the areas along the Hardangerfjord and Sognefjord are famous for their early-ripening fruit. But it is not until about 15–20 July that the summer appears along the coast of Finnmark. In the high mountain plateaus of southern Norway and arctic Scandinavia, however, it begins about July 1–10.

The sea surface is relatively cold in summer both along the west coast of Scandinavia and in the Baltic Sea and its gulfs. The mean air temperature of the warmest months, therefore, does not rise very much above 16°C either along the west coast of Jutland or in the coastal districts around the Gulf of Bothnia in summer. It is about 15°C along the west coast of Norway, about 14°C on the coastal stretch between Stad and the Lofoten Islands, and about 10°C along the north coast of Finnmark. Some kilometres inland from the coast line and near the inland end of the great fjords of western and northern Norway, however, the air temperatures are much higher. A similar increase in temperature inland from the coastline is also experienced in the coastal districts bounded by the Gulf of Bothnia and the Baltic Sea. But there the increase is not as perceptible as in the fjords on the Norwegian coasts.

The warmest regions in summer occur on Sjelland, in central and southern Sweden, in the Oslo area and in southern Finland where the mean temperature of the warmest month is about 17°C. In the surroundings of Copenhagen, Helsinki, Stockholm and Oslo the mean temperature of July is 18°C or even more.

On Finnmarksvidda and in Lapland the mean temperature of July is about 14°C and daily maximum temperatures of 32°C have occurred at Karasjok. But in spite of the midnight sun, chilly nights characterize the summer in arctic Scandinavia and frost may occur even during the summer nights.

The coldest summer temperatures are experienced in the central plateaus and mountains in southern Norway and northern Sweden. The mean temperature of the warmest month within these regions thus ranges from 7° to 11°C, at elevations ranging between 900 m and 1,300 m above M.S.L., and frost is not unusual even in June and July.

In stable high pressure situations the daily maximum temperatures may rise very high in the interior regions throughout Scandinavia in summer. The warmest temperatures ever observed are 35.8°C at Slagelse in Denmark, 38.0°C at Ultuna in central Sweden, and 35.0°C at Oslo and Trondheim. Similar temperatures have also been observed in interior Finland.

As will be seen from Table VIII, the mean temperatures of the individual summer months are subjected to variations from year to year. But the fluctuations about the monthly normal is much smaller in summer than in winter.

**The autumn**

The autumn normally begins about 15th August on the coast of Finnmark and about 25th August or even earlier in the central plateaus of arctic Scandinavia and southern Norway. About 20th September it reaches southern Finland and between 1st and 10th October autumn makes its entrance in the coastal regions of southernmost Sweden and Norway and in Denmark.

In summer large amounts of heat have been stored in the soil and lakes and serve to keep the air temperature relatively high for a longer time in autumn. The increased advection of mild Atlantic air masses and the increasing precipitation also act in the same way. Autumn, therefore, is generally comparatively mild in the Scandinavian countries and lasts for a longer interval of time than does the spring.

**The heating season**

Heating of houses is necessary in all Scandinavian countries every winter and at some points in arctic Scandinavia and in the high mountain regions throughout the year. The heating season normally begins when the mean daily air temperature falls below 11°C in autumn and ends when it rises above 9°C in spring. The beginning and end of the heating season, therefore, very closely coincides with the beginning of autumn and the end of spring.

At Copenhagen, Bergen, Oslo, Stockholm and Helsinki the heating season normally (1931/1960) lasts for 209, 229, 232, 231 and 237 days, respectively; and for 254 and 291 at Lillehammer and Karasjok. Further, the degree-days of the heating season at the five first stations average 2,831, 2,882, 3,759, 3,551 and 4,015, respectively, and at Lillehammer and Karasjok 4,647 and 6,404.

Comparing the degree-days of the heating season at inland and coastal points, those at the latter points are relatively small considering the length of the heating season. This does not mean, however, that the consumption of fuel is much less in the coastal districts than inland. Even though the air temperatures in winter does not fall much below 0°C neither in the coastal districts of western Norway, Sweden nor in Denmark, they are usually associated with high wind velocities. Hence, houses are exposed to a much larger loss of heat through natural ventilation than those at much colder inland points with almost calm weather. This, therefore, accounts for the relatively large consumption of fuel, even in the mild regions of Scandinavia.

**Sea ice**

Sea ice of any importance to navigation does not form along either the west coast of Jutland or along the coasts of Norway. Even the fjords of arctic Norway are free of ice of any importance in winter. It is only in abnormally cold winters that ice may form in the Danish Sounds, along the west coast of Sweden, in the Oslofjord and along the Skagerrak coast.

In the Baltic Sea and its gulfs, however, where the salinity of the water is low and the air temperature is comparatively low in winter, ice forms every winter. Normally fast ice begins to form early in November, in the bays, inlets and harbours along the northern coast of the Gulf of Bothnia. In the middle of the winter the whole gulf is usually covered by continuous ice. Around the turn of the year fast and continuous ice forms also at Helsinki and at Stockholm and far into the Bothnian Sea, although the central area of this sea is normally covered by drift ice or pack ice only. But it is not until the beginning of February that ice forms in the southernmost harbours of Sweden and in the ports of the Danish islands of the Baltic Sea.

The northernmost harbours in Sweden and Finland are normally free of ice at the end of May and those in southernmost Sweden in the middle of March or even earlier. The surface water thus remains frozen from one to six months in the southern and northern waters along the east coast of Sweden. The use of ice breakers is, therefore, necessary. It must be emphasized that some winters may deviate considerably from these average conditions. In severe winters the formation of ice may start about one month earlier and in mild winters may be delayed two or three months. It also happens in abnormally cold winters with persistent easterly winds that ice in the Baltic Sea is pushed out through the Danish sounds into Kattegat and Skagerrak. In such ice winters, even the Oslofjord and the ports along the coasts of Kattegat and Skagerrak may be blocked by ice for a longer period of time.

**Humidity of the air**

As a result of the great frequency of moisture-bearing Atlantic air masses over Scandinavia, the humidity of the air is comparatively high. But because of the annual evaporation from the sea and the large area covered by Scandinavia, the humidity of the air is subjected to pronounced seasonal variations and large terrestrial influences.

The lowest vapour pressure normally occurs in February and the highest in July or August. The mean monthly vapour pressure at Bergen averages 4.1 mm and 9.6 mm in February and July, respectively, and 1.5 mm and 7.6 mm in the same months at Karasjok. The corresponding figures for Stockholm are 3.1 mm and 10.2 mm, and for Helsinki 2.8 mm and 10.8 mm. The annual vapour pressure at Copenhagen, Göteborg, Haparanda and Tromsö averages 6.5 mm, 6.2 mm, 4.1 mm and 4.2 mm, respectively, while at Bergen Mariehamn and Helsinki it is 6.3 mm, 6.2 mm and 5.9 mm respectively.

In autumn and winter the mild Atlantic winds bring much moisture to Scandinavia. In spring and early summer, however, the Atlantic air masses are relatively dry because of the smaller evaporation from the sea surface in this season. On the other hand, Scandinavia is frequently invaded by dry air from the east in spring. The relative humidity of the air is, therefore, on an average greatest in autumn and winter and smallest in spring and early summer, ranging between 90 and 95% in winter, and between 65 and 75% in spring and early summer.

**Cloudiness**

Scandinavia is one of the cloudiest regions of the world with considerable cloudiness throughout the year. The greatest cloudiness occurs in the high mountain regions of Norway and Sweden and along the western, northwestern and northern coasts of Norway. Denmark and the coasts along the Baltic Sea and the Gulf of Bothnia on the other hand, have a relatively low cloudiness.

Autumn and winter are the seasons of the greatest cloudiness. Thus the mean monthly cloudiness in winter averages 65–80% on the coasts and 14–15 cloudy days. This high degree of cloudiness, however, does not arise only from frontal and orographical lifting alone, but from turbulent mixing with the moist air near the surface as well. Hence stratocumulus and low stratus types of clouds are most frequent in autumn and winter. Spring in most of Norway, Sweden and Denmark, and summer in Finland have less cloudiness than the remaining seasons of year. In May, for instance, cloudiness averages 60–75% in the high mountain regions of Norway and Sweden and along the west and north coasts of Norway, and only 45–55% in the remainder of Scandinavia. The greatest contrast between winter and summer cloudiness averages about 20–25% in Denmark, interior Finland, in Sweden and along the southeastern coast of Norway.

The highest number of cloudy days is experienced on the western and northwestern slopes of the Scandinavian mountain chain, about 200 days annually. Because of frequent föhn winds, however, southeastern Norway, northernmost Sweden and Finland as well as the east coast of Sweden are favoured by a very small frequency of cloudy days averaging 120–150 days annually.

In a belt along the Scandinavian western, northwestern and northern coasts the number of clear days is small, averaging 25–35 days/year, apart from in the Varangerfjord where it averages only 12 days. The high mountain regions of the Scandinavian Peninsula experience an average of 35–45 clear days/year and so do Denmark and Finland. The highest annual frequency of clear days, however, occurs in eastern Sweden near the Gulf of Bothnia and the Baltic Sea where it averages 75–100 days. A secondary area of rather high frequency of clear days is experienced in a belt along the southeastern and southern coast of Norway, the mean annual frequency being about 55 days.

**Fog**

The foggiest areas are the high mountain regions of the Scandinavian Peninsula and the west coast of Jutland. Within these areas the annual frequency of fog averages 20–25% and even more. Local areas of high annual fog frequency (20–45%) are also experienced in the hill land north of Oslo, the uplands of south Sweden (15–20%) and in the coastal waters east of Stockholm. On the other hand, the Norwegian coast and the fjord districts between Bergen and Vardö have a very small annual amount of fog. Similarly the interior of Finland, the forested regions of Norrland (northern Sweden) and the interior of eastern Norway experience little fog, less than 5% annually.

In general autumn and winter are the seasons of most frequent fog in the Scandinavian area. But because of the temperature extremes of the ocean surface being delayed 2–3 months in time in relation to those of the earth's surface, some coastal waters and coastal districts are more exposed to fog in spring and summer than in autumn and winter. Thus spring and early summer are the seasons of maximum fog along the southern coast of Norway between the Swedish border and Bergen. The same also applies to the coasts of Finland and Sweden along the Baltic Sea and its gulfs. Summer is the foggiest season along the Norwegian coast between Bergen and Kirkenes.

Radiation fog is most frequent in August when the humidity of the air is highest and the nights become cooler and longer. It is experienced in late evening and early morning near lakes, in river valleys, at spots in the plains and in the hill lands and may spread over large areas. It usually tends to thin out in late morning or early afternoon.

Advection fog is common throughout the year and accounts for the high frequency of fog along the western coast of Denmark and in the high mountain regions of the Scandinavian Peninsula. This so-called upslope or "hill" fog is in reality low-lying clouds and accounts for the great number of cloudy days in the Norwegian and Swedish mountains exposed to the mild, moisture-bearing Atlantic winds. The sea fog experienced along the southern coasts of Norway and in the coastal waters of eastern Sweden and Finland in spring and early summer is also due to advection of mild, moist air. The same also applies to the sea fog occurring along the western and northern coasts of Norway in summer. But this fog is caused mainly by cooling of the lowest layers of warm continental air coming from northern Finland or Russia and taking part in the summer monsoon. During the day this sea fog is experienced as a dense fog bank lying off the coast in spite of the moderate fresh breeze. In the evening, however, when the wind weakens and the air temperature falls, the fog is carried over land, but seldom penetrates into the deep fjords. The inner fjord districts between Bergen and Vardö are, therefore, favoured by less cloudy and foggy days and by higher air temperatures than are the islands and coastal waters off the coast.

When a blocking high-pressure ridge from east or northeast or a blocking anticyclone occurs over Scandinavia in winter, cold air from the Norwegian mountain plateaus pours down-valley and down-fjord like a bora. A layer of "steam fog" is usually then caused near the sea surface. This evaporation fog or "arctic sea smoke", is of frequent occurrence in the fjords of arctic Norway, especially in the Varangerfjord where it very often is rather dense and deep. Even in the Oslofjord the sea smoke frequently occurs when the cold drainage air from southeastern Norway streams down-fjord in late autumn and winter.

## Precipitation

Moisture-bearing air currents other than the westerlies invade Scandinavia or parts of it and produce rain or snow at different times of the year. But none of these air currents are as moisture-bearing, strong or persistent as the westerlies and southwesterlies. The precipitation regime of Scandinavia is, therefore, to a great extent determined by the amounts of water released within these air currents under the influence of the topography. Particularly, the latter one plays an important role in Norway and Sweden. The heaviest precipitation occurs on the slopes of the Norwegian coastal mountains which are exposed to the moist Atlantic winds.

After depositing moisture in the form of rain or snow on the west slopes of the Scandinavian mountain chain the westerly winds become relatively dry. Besides, the other moisture-bearing winds invading Scandinavia are generally of small intensity and frequency during the greatest part of the year. On the whole, the mean annual amounts of precipitation, therefore, decrease eastwards in Scandinavia. Thus, while about 3,500–3,000 mm occur in a zone of maximum precipitation on the western slopes of the Norwegian coastal mountains between Bergen and Bodö, they are about 600 mm in southern and central Finland.

Only small geographical differences occur in the mean annual precipitation in Denmark and Finland where the influence of topography is negligible. That of Denmark ranges between about 650 mm in the western and 550 mm in the eastern sections. In Norway and Sweden, however, for topographical reasons the geographical differences are rather large in some regions. In Sweden where the mean annual precipitation per year is about 600 mm, there are three well marked areas of high annual precipitation caused by orographic upglide. One area with annual amounts of 850–900 mm runs parallel to the west coast in a narrow zone some kilometres inland. Another area with amounts of about 1,000 mm/year occurs on the western slopes of the Småland uplands in southern Sweden, and the third one lies on the western slopes of the high mountains of Norrland near the Norwegian border where the Sarek Massif receive about 2,000 mm annually. Large local differences in mean annual precipitation also occur in the zone of maximum precipitation on the western slopes of the coastal mountains in western and northern Norway. The heaviest falls occur south of Nordfjord and south of Bodö where the mean annual precipitation is about 5,000 mm and 4,000 mm respectively. In eastern Norway, which lies in the lee of the westerly and northerly winds, however, the annual precipitation is fairly evenly distributed, the mean areal amount being about 600 mm. In some districts and localities on the Scandinavian Peninsula the screening effect of mountains upon the moisture-bearing winds causes very small amounts of precipitation over larger areas. This is the case, for example, in the northwestern districts of eastern Norway where the annual precipitation averages less than 400 mm. Skjåk (61°53′N 8°28′E) and Lom (61°50′N 8°34′E) in the Otta Valley, a tributary valley of the Gudbrand Valley have a mean annual precipitation of only 251 mm and 274 mm, respectively. Screening effects are also obvious in many valleys, even in the zone of maximum precipitation in western Norway. Lærdal (61°6′N 7°29′E) and Ljøsne (61°3′N 7°37′E) at the inland end of the Sognefjord have a mean annual precipitation of 444 mm and 375 mm, respectively, while the corresponding amounts in the surrounding mountains average 1,800–2,000 mm. Similarly the high mountains near the Norwegian border in northern Sweden account for the small annual amounts of precipitation (300–400 mm) experienced in the forested regions of Norrland.

Screening effects also account to some extent for the small precipitation in interior Finnmark ($\leqslant 400$ mm annually) and in Finnish Lapland where the annual amounts of precipitation average 450–500 mm. But within these regions the distance from the sea and the relative low water vapour content of the air in high latitudes also serve to minimize the precipitation.

The geographical distribution and the amounts of precipitation in Scandinavia in the different seasons of the year are closely related to the prevailing type of atmospheric circulation. The zonal or westerly type of circulation is the most frequent throughout the year. But it is strongest in winter and so is the cyclonic activity. In spring, particularly in March and April, dry easterly winds are as persistent and strong in Scandinavia as are the westerlies because of frequent blocking over northern and northwestern Russia. On the other hand, Scandinavia is often for longer times subjected to cold and dry northwesterly and northerly winds in May and June because of the frequent blockings in the Norwegian Sea in these months. In summer, however, when the different types of atmospheric circulations are weak and the solar radiation is strong, precipitation is mainly caused by vertical convection and not so much by advection.

The typical maritime regions of Norway, west of the Scandinavian mountain chain, experience most precipitation in autumn and winter, from October through January. The maximum rainfall ever recorded in a 24-h period (230 mm) in Scandinavia occurred within these regions and was recorded at Indre Matre (59°51′N 6°00′E) on the 26th of November, 1940, in a typical westerly situation with deep cyclones. The remainder of Norway, practically all parts of Sweden, Finland and Denmark experience most precipitation in summer. It is mainly caused by vertical convection and the largest amounts of precipitation inland occur normally in July, and in August in the coastal districts. The months of least precipitation are February and March, except in western and northern Norway where March and April have the least precipitation.

The annual frequency of days with precipitation $> 0.1$ mm, like the annual amounts of precipitation, is greatest on the western coast of Norway. But the differences between the stations in the frequency of precipitation are not as great as differences in amount. Thus at Stadlandet (62°09′N 5°13′E) there are 234 days annually, while at Mehamn (71°02′N 27°51′E) with less than half as much precipitation, there are 218 days annually with precipitation. In interior Norway the average frequency of days with precipitation ranges between 95 and 180, the lowest frequency occurring in the dry area south of Dovre and in interior Finnmark. In Denmark, Sweden and Finland the corresponding frequency ranges from about 120 days to 200 days annually, and in all Scandinavian countries the greatest frequency of days with precipitation occurs from July through December.

Rain is the usual form of precipitation in summer and snow in winter, although on the southern and western coasts of Scandinavia and in the Lofoten Islands rain or sleet alternate with snow. Showers of hail, though rare, occur as well. They are most frequent in spring and summer at inland points, while in winter at coastal points. The districts most exposed to hail are the western and northern coasts of Norway, the western coasts of Jutland and southern Sweden where hail occurs on an average of about 15–20 days annually. In the plains of the interior of the countries the average is 2–4 days. Similarly thunderstorms are usually infrequent in the Scandinavian countries. In the most exposed districts, they occur on an average of 8–16 days/year, and of 2–3 days in the mountains and in arctic Scandinavia. Thunderstorms are most likely to occur in July and in August. They may occur, although this is rare, in conjunction with front passages in all months of the year. Finally, tornadoes may occur from time to time in summer. But on the whole this phenomenon is rare in Scandinavia.

## State of ground

Normally the temperature of the earth's surface falls below 0°C just before the air temperature does it. As long as the ground is bare in autumn and in early winter, glazed ice is a common phenomenon when warm air masses invade Scandinavia and frost starts to penetrate into the soil. The depth of ice in the soil ranges greatly from place to place depending on the soil cover and water content of the soil. But on an average the depth of ice in the soil is about 2.5 m in northern Scandinavia, 1.5 m in central and 1 m or less in southernmost Norway and Sweden. The conditions may considerably vary from year to year depending upon how early the continuous snow cover begins, as well as upon the air temperature.

The first snowfall generally occurs in early October and the last one in May with some local snowfall in September and June. The frequency of snowfall in arctic Scandinavia ranges from an average of 74 days with snowfall annually at Glomfjord (66°49′N 13°59′ E) to 130 such days annually at Vardö and still higher frequencies at some mountain stations. Southwards from the Arctic Circle the difference between inland and coastal points is marked. For example, at Bergen there are on the average 47 days with snowfall annually while at Lillehammer there are 69 such days and at Slirå (1,300 m above M.S.L.) 173. In Sweden the area of least frequent snowfall is along the western and southern coasts where Göteborg and Lund have 37 days with snowfall annually. Stockholm has 60 days and Östersund 104. A similar increase in the frequency of snowfall from coastal to inland points and with latitude is experienced in Finland too. Mariehamn on the Åland Islands, for example has an average of about 50 days with snowfall annually, Sartavala has 80 such days and Sodankylä has 95. In Denmark, on the other hand, the average annual frequency of days with snowfall ranges between 19 and 39 days. Snow lies on the ground for more than half the year in most parts of Scandinavia. Thus the snow cover begins to form in the high mountains of Norway and in northern Sweden as early as the end of September and persists until the end of May or early June. In arctic Scandinavia snow cover generally lasts from the end of October until the middle of May, while in southern Norrland, central Finland and southeastern Norway, from the beginning of November until the middle or end of April. The duration of the snow cover, however, varies considerably with exposure and elevation, generally being shorter along the coasts where the full effects of the mild Atlantic air flow may be felt and also at lower elevations, e.g., in valleys. Because of the thawing effect of the mild Atlantic air, snow cover is an irregular phenomenon in Denmark, southwestern Finland, and along the western and southernmost coast of Sweden and Norway. On the Åland Islands, in Skåne and Denmark it lasts, on the average, from the beginning of January to the middle of March, while in the uplands of southern Sweden it lasts almost as long as it does in southern Norrland.

Except at coastal points where frequent thawing usually occurs, the snow cover builds up during the winter and the snow generally attains its greatest depth in March. Snow depth normally reaches an average of 100 cm and more in northern Norway and northern Sweden, and averages less than about 15 cm in the snowiest month in southern Sweden, extreme southern Norway and at some coastal points in western Norway. In Finland the average snow depths range from 20 cm at the Åland Islands to about 80 cm in the northern and eastern districts, and observations show that as much as about 3 m snow accumulates on the ground at a number of stations in the Scandinavian mountain chain.

Above certain levels there is a snow cover that lasts throughout the year. These levels—the height of the so-called "snow-line"—vary from year to year. But on an average the snow-line in southern Norway occurs at about 1,450 m and 2,000 m above M.S.L. in the western and eastern districts, respectively, and in northern Norway and northern Sweden at about 800 m above M.S.L.

The permanent snow fields and glaciers above the snow-line in Scandinavia are the most extensive ones on the European mainland and have runners extending far down into many valleys and fjords. The largest glaciers are the Jostedal Glacier (815.0 km²) in southern Norway and the Svartisen Glacier (476.5 km²) in northern Norway.

**Drifting snow**

Drifting snow is often experienced in the high mountains and mountain plateaus of Norway and Sweden when strong wind forces occur in winter. Snow-fences are, therefore, necessary on the most exposed stretches along the railways and roads crossing the high mountains from west to east and from south to north in Norway. But drifting snow also occurs over the plains and often causes large snow-drifts, especially in the three northern Scandinavian countries.

**Avalanches**

Avalanches often occur in the high mountains and in valleys leading down from these mountains both in Norway and Sweden. They are most frequent during the thawing period in spring and may cause much damage to houses, roads, animals and human beings. Nearly every year people perish in Norway because of avalanches, especially during the Easter holidays when there are many people skiing in the mountains. But avalanches sometimes also occur in winter in the narrow, steep valleys dissecting the western slopes of the coastal mountains in western and northern Norway. This happens when very warm air from southern latitudes comes in over Norway from the west and causes heavy orographical rainfalls and great increase in the air temperatures in the high mountains on both sides of the valleys.

**Hoar frost**

Hoar frost is a frequent phenomenon in the winter half year in the high mountain regions exposed to the moisture-bearing, warm Atlantic winds. The deposit of hoar frost on the snow cover may be rather deep and electrical power lines crossing these regions may be weighed down. But hoar frost is often also experienced over low-lying regions, especially in spring when warm air comes in over Scandinavia. It even happens that the deposit of hoar frost on the trees becomes so thick that the branches break off.

**Flash floods**

Flash floods occur in all the Scandinavian countries and sometimes cause much damage. Normally they are most frequent in autumn and early winter when the precipitation is generally greatest.

**Hydrology**

The variations of the amount of precipitation and temperature during the year, the snow accumulation during the winter and the snow melting in spring highly influence the hydrology of the rivers and lakes in the Scandinavian countries. The influence is best characterized by the march of the discharge during the year. In this respect the mountain rivers of Norway and Sweden—in general—show similar conditions having a *minimum* discharge in late summer and winter and a *maximum* discharge in spring and early summer. In western Norway, however, the large amount of precipitation throughout the year gives rise to a comparatively even hydrographic curve for most of the rivers. Maxima may occur almost in any season, though the most general feature is high water in spring and autumn, the latter one due to the increasing precipitation in autumn.

In many districts the discharge is increased because of ice in the ground, which prevents the water from penetrating into the soil. Similarly the ice and ice drift in many mountain rivers start breaking up just before the increase in discharge in spring.

The maximum discharge in the mountain rivers in spring is often characterized by two or three peaks. The first one is due to the melting of snow in low forested regions and corresponds to the so-called "home flood". The next peak is due to the snow melting in the hill lands and lower mountains and indicates the "mountain flood". The third peak—if any—is caused by melting from glaciers in the highest mountains where snow first starts melting in summer and is called "highsummer flood".

In the lowland rivers of eastern Norway, central Sweden and Finland the maximum discharge in spring shows one, sometimes two, peaks only. But owing to the regulating effect of the lakes—particularly in central Sweden and Finland—these peaks are much less pronounced than those in the mountain rivers. Rivers in southern Norway, Denmark and southwestern Sweden may also show a secondary maximum discharge in autumn due to the comparatively heavy precipitation in the westerly maritime regions during this period. Similarly the rivers of southern Sweden and Denmark have a lower discharge in summer than in winter because of the comparatively large evaporation during the summer in these southern latitudes.

The time at which the fresh-water lakes freeze and thaw is a complex problem, depending on the depth of the lake and the air temperature. Lakes less than 25 m deep or more than 100 m deep freeze for about 2 weeks and 11 weeks, respectively, after the daily mean air temperature has fallen below 0°C. On the other hand all lakes thaw about 3–5 weeks after the daily mean air temperature has risen above 0°C in spring. Some lakes in Scandinavia begin to freeze as late as January and thaw as early as April.

### Evapotranspiration and water balance

Because of the relatively large deficit in radiation balance of the earth's surface in Scandinavia in winter, energy available for evaporation in this season is small. The evaporation from the snow fields and ice-covered lakes is, therefore, negligible in winter and the evaporation is mainly restricted to the summer half year.

On an average the annual potential evapotranspiration from grass surfaces is about 90% of the evaporation from free water surfaces. The geographical distribution of the mean annual potential evapotranspiration from grass surfaces in Scandinavia shown in Fig.3, is therefore also representative of the mean annual evaporation from water surfaces in Scandinavia.

Low cloudiness, much solar radiation, moderate fresh wind velocities and great water vapour deficit in the air are factors that greatly support and further the evaporation and evapotranspiration. All or most of these conditions are present in the coastal districts of eastern Sweden, in southeasternmost Norway and in Denmark. These regions, therefore, have the greatest potential evapotranspiration in Scandinavia, ranging from 500 mm to about 600 mm/year. In the high mountain regions of southern Norway and northwestern Sweden, in Finnmark and northernmost Finland, however, the energy available for evaporation is small. Within these regions is also the least annual potential evapotranspiration, ranging from about 100 mm to 250 mm.

The mean annual amount of precipitation within all sections of Scandinavia is larger than the mean annual evaporation and potential evapotranspiration experienced within them. Taken over the whole year, Scandinavia has a positive water balance. But in Denmark and Finland this is very small, and both countries represent sub-arid areas. Similar areas also occur on the Scandinavian Peninsula east of the Scandinavian mountain chain. Thus dry areas occur in northern Sweden, near the Finnish border, on Öland and Gotland, in a broad zone in eastern Sweden south of Stockholm and in the lake districts of central Sweden. Likewise most of eastern Norway and interior Finnmark are more or less sub-arid areas and in the northern parts of the Gudbrand Valley even *arid* areas occur. Dry areas are also experienced in the inner districts of many large fjords of western Norway, e.g., in Lærdal at one of the inland ends of the Sognefjord.

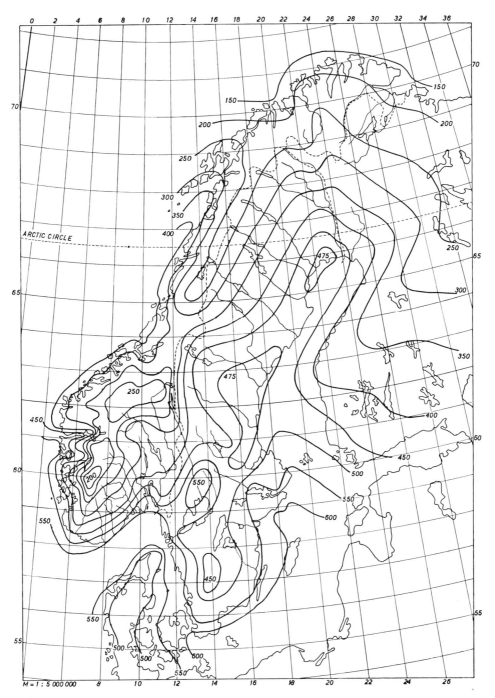

Fig.3. The mean annual potential evapotranspiration from grass surfaces (mm). The chart is based on data computed from the Penman formula for 25 Swedish and 70 Norwegian stations by Wallén (1966) and the author, respectively. Further, data computed from Turc's formula by Mohrmann and Kessler (1959) for 6 Finnish and 5 Danish stations have been used.

## The vegetation period

In some sections of Scandinavia grass begins to grow at a mean daily air temperature of 3°C. But most plants, fruits and crops first start growing at higher air temperatures, and

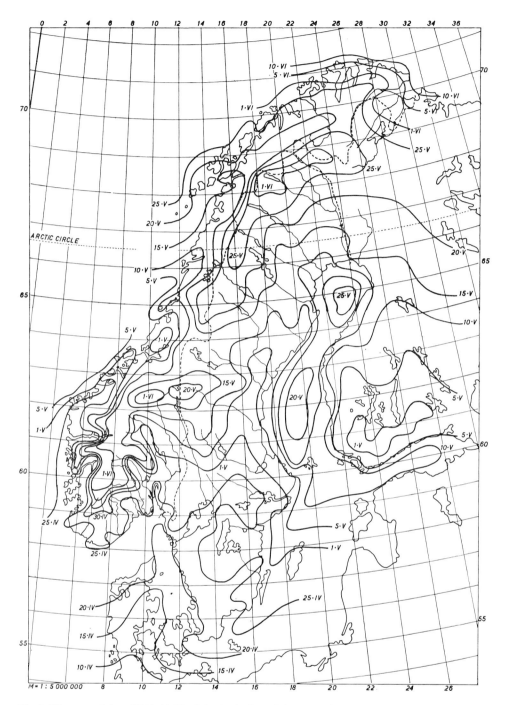

Fig.4. The normal date (1931–1960) of the beginning of the vegetation period (the date when the daily mean air temperature rises above 6°C in spring).

the vegetation period is usually assumed to cover the time of the year when the mean daily air temperature is 6°C or higher.

In southern Denmark the vegetation period, as defined above, begins on the average about 10th April, and between 20th and 25th April in southern Sweden, in the coastal districts of eastern and southern Norway and at the inner fjord districts of western

Norway. In southern Finland the vegetation period begins during the first days of May, and at the end of May or beginning of June in the mountains of Norway and Sweden, in interior Finnmark and in the northernmost part of Finnish Lapland. Because of the sea being cooler than the continent in spring, however, the beginning of the vegetation period in the coastal districts along the Gulf of Bothnia is delayed 5–10 days. The same is also the case along the western and northwestern coasts of Norway. On the west coast the vegetation period begins at the end of April while in the Lofoten Islands about 17th May (see Fig.4).

Except in the mountain plateaus and in arctic Scandinavia, frost is very rare in June and July. But in May, when Scandinavia from time to time is invaded by cold air masses from the north or northwest, advection frost is experienced on some days in almost all Scandinavia. On the other hand, most of the air masses invading Scandinavia in May–June are relatively dry. The begin of the vegetation period is characterized by clear skies, strong solar radiation, great daily temperature variations and little precipitation. Hence evapotranspiration is great and water deficit occurs in many sections of Scandinavia during the first time of the vegetation period. But during the "highsummer" frequent convective showers gradually diminish the water deficit and in August it is negligible or even positive in most parts of Scandinavia.

Because of the outstanding topography and the decrease in length of the vegetation period with altitude great areal differences are experienced in the length of the vegetation period even over short distances both in Norway and Sweden. In Norway the length of the vegetation period ranges from 207 days in the outer coastal districts of western Norway to only 61 days at Slirå (1,300 m above M.S.L.) in the central mountain plateau of southern Norway; and from 206 days in southernmost Sweden to 97 days at Riksgränsen (508 m above M.S.L.) in northwestern Sweden. The corresponding figures for Finland are: 170 days at Kaisaniemi (near Helsinki) and 108 days at Utsjok, and for Denmark, 212 days and 195 days, respectively. At the outer and inner fjord districts of Finnmark the vegetation period lasts for about 105 days and 125 days, respectively. Due to the midnight sun, however, the vegetation period is very intensive both there and in interior Finnmark and development rapidly takes place (Fig.5).

In Denmark and Finland precipitation is scarce during most of the vegetation period. But even if irrigation is necessary in many rural districts, the air temperature and not the precipitation is the limiting factor to agriculture in most of Scandinavia. Daily maximum temperatures of say 25°C and more are infrequent and frost occurs both in spring and autumn. But the long, light summer nights north of 60°N with their good light are of great importance for the development of vegetation and the possibilities of cultivation. Advection frost is most frequent in spring because of outbreaks of cold polar or arctic air from northwest and north and may sometimes causes large damages to agriculture and horticulture. In autumn radiation frost is most common. It usually begins towards the end of August and may be especially strong in Finland.

The normal number of degree-days in the vegetation period in Norway ranges from 1,382 at Fornebu near Oslo to 62 at Slirå and in Sweden from 1,494 at Hälsingborg to 308 at Riksgränsen. The corresponding figures for Finland are 1,256 at Kaisaniemi and 415 at Utsjok in northern Finland, and for Denmark 1,590 and 1,251, respectively. Agriculture in Denmark, accordingly, is much more fortunate with respect to air temperature during the vegetation period than in any of the other Scandinavian countries.

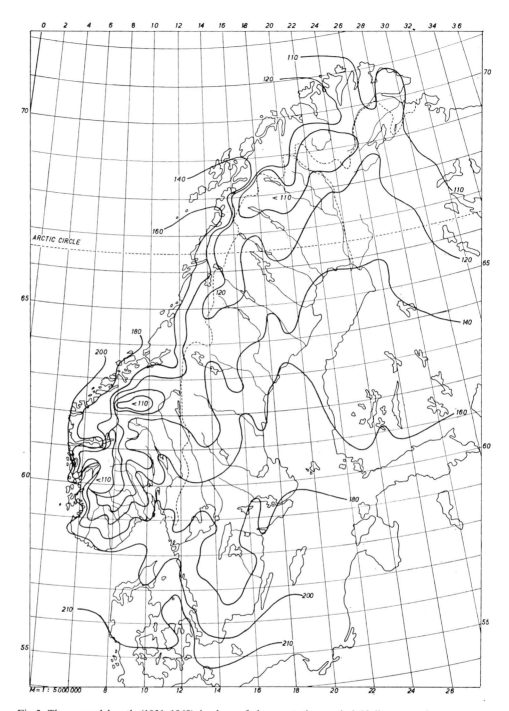

Fig.5. The normal length (1931–1960) in days of the vegetation period (daily mean air temperature ⩾ 6°C).

In the mountain plateaus of southern Norway and northern Sweden the vegetation period lasts, on an average (see Fig.6), until about 15th September. In southern Finland and Sweden it lasts until about 10th to 25–30th October, respectively; and in southern Norway, until about 15–30th October. Along the west coast of Norway and in Denmark, it lasts until about 10–15th November.

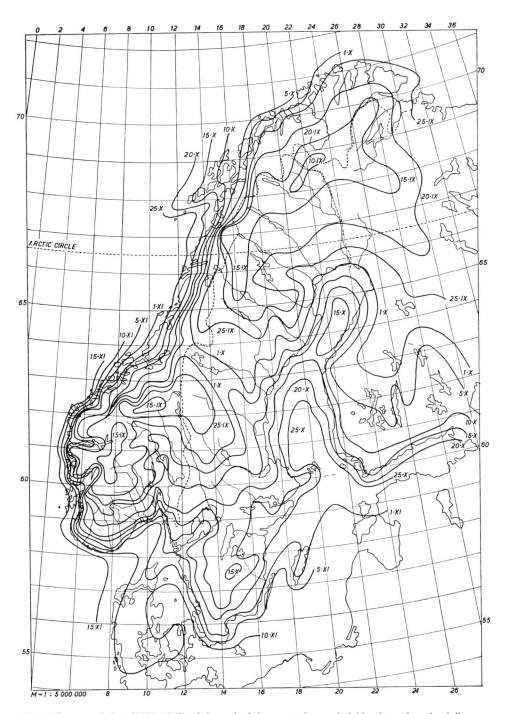

Fig.6. The normal date (1931–1960) of the end of the vegetation period (the date when the daily mean air temperature drops below 6°C in autumn).

## Climatic fluctuations

There is considerable evidence of climatic fluctuations in Scandinavia, in geological, prehistorical and historical times. Pronounced fluctuations have even been observed during the short period of meteorological records since the last half of the nineteenth century.

The last Glacial Period in Scandinavia finished about 7500 B.C. At that time the remains of the large ice-sheet that had covered northwestern Europe for millions of years had retreated to the high mountain plateaus of Norway and Sweden. The melting of these glaciers continued during the following postglacial period and they probably disappeared completely during the "Climatic Optimum" (4000–2000 B.C.) of the Atlantic time (5000–2000 B.C.), or in the beginning of the subsequent warm, dry subboreal time (2000–500 B.C.).

But towards the end of the subboreal time the climate of Scandinavia deteriorated again, and the following sub-Atlantic time (from about 500 B.C.) was characterized by cool summers and increased precipitation. The firn line dropped, and about 600 B.C. new glaciers were formed in the mountains of Norway and Sweden. The Svartisen, Jostedal and Folgefonn glaciers are assumed to have originated from that time. The new glaciers advanced considerably during the first half of the Celtic Iron Age (500–0 B.C.) and the height of the firn line is assumed to have been about 1,500 m in western Norway.

In the Roman Iron Age (0–400 A.D.), there is evidence of an improvement of the climate in Scandinavia with a consequent increase in the height of the firn line and recession of the glaciers. But from the end of this period the firn line begins to drop again, and we know from historical records that the Baltic Sea was very often covered by ice during the Middle Ages (about 1000–1350 A.D.). A general advance of the glaciers in Norway and Sweden took place during this period. But the advance was small from about 1300 to 1550 when the climate of Scandinavia seems to have been relatively warm, with few severe winters. From about 1550, however, the firn line dropped further and from about 1740–1750 the glaciers of Norway and Sweden reached their maximum extension since the last Ice Age.

Since 1750 the height of the firn line has increased more or less continuously up to about 1955. Between 1850 and 1955 there was a general recession of the glaciers in Scandinavia as well as in Spitsbergen and Greenland. Small, temporary advances, however, occurred between 1904 and 1908 and in the years 1923–1925. Likewise, the pack ice in the arctic waters has diminished considerably during the last century.

The improvement of the climate experienced with regard to the glaciers since 1850 has also manifested itself in a pronounced temperature increase in all the Scandinavian countries. This stands out very clearly in Table IX, showing the successive 30 year means of the air temperature in the individual seasons of the year between 1876 and 1960. According to the table the increase in the mean annual temperature averages no less than 0.9°C for the Scandinavian Peninsula. The increase is by far greatest in autumn and spring, and it is greatest and most pronounced in arctic Scandinavia. At Karasjok the rise in the mean annual temperature was 1.2°C from the period 1876–1905 to the period 1931–1960.

The general rise in the winter temperatures (December–February) started about 1850, and though severe ice winters occurred in 1870/1871, 1874/1875, 1879/1880, 1880/1881, 1887/1888, 1892/1893, 1916/1917 and 1923/1924, the rise continued until about 1933. The mean temperatures of the summers (June–August), on the other hand, first began to increase about 1900 in arctic Scandinavia and about 1915 in the remainder of the area. The climate of Scandinavia consequently has become more and more maritime since 1850. Until the beginning of the 1920's the increase of the temperature in spring (March–May) and autumn (September–November) was small. But later on both these seasons have

TABLE IX

SUCCESSIVE 30-YEARS MEANS OF THE AIR TEMPERATURES (°C) OF THE FOUR SEASONS
OF THE YEAR AT FIVE SELECTED STATIONS IN SCANDINAVIA

| Station | Period | Winter | Spring | Summer | Autumn |
|---|---|---|---|---|---|
| Skudenes | 1876–1905 | − 1.4 | 1.1 | 9.4 | 4.0 |
| $\varphi = 69°19'N$ | 1901–1930 | − 1.3 | 1.5 | 9.6 | 4.1 |
| $\lambda = 16°07'E$ | 1931–1960 | − 0.6 | 2.0 | 10.2 | 5.0 |
| Karasjok | 1876–1905 | −15.0 | −4.0 | 10.7 | −2.3 |
| $\varphi = 69°28'N$ | 1901–1930 | −14.2 | −3.4 | 11.1 | −2.1 |
| $\lambda = 25°31'E$ | 1931–1960 | −13.8 | −3.0 | 11.8 | −0.8 |
| Stockholm | 1876–1905 | − 2.7 | 3.6 | 15.5 | 6.4 |
| $\varphi = 59°21'N$ | 1901–1930 | − 2.1 | 4.1 | 15.3 | 6.4 |
| $\lambda = 18°03,5'E$ | 1931–1960 | − 2.0 | 4.7 | 16.4 | 7.3 |
| Hellisöy Fyr | 1876–1905 | 2.3 | 5.0 | 12.4 | 8.2 |
| $\varphi = 60°45'N$ | 1901–1930 | 2.8 | 5.3 | 12.4 | 8.2 |
| $\lambda = 4°43'E$ | 1931–1960 | 2.8 | 5.6 | 13.0 | 9.2 |
| Ås | 1876–1905 | − 4.2 | 3.8 | 14.9 | 4.9 |
| $\varphi = 59°40'N$ | 1901–1930 | − 3.6 | 4.2 | 14.8 | 5.2 |
| $\lambda = 10°47'E$ | 1931–1960 | − 4.0 | 4.4 | 15.6 | 5.8 |

been a good deal warmer, and this especially applies to autumn which is at present the warmest season of the year.

This large scale improvement of the temperature conditions has resulted in better possibilities for agriculture, horticulture and forestry than in earlier times. The vegetation period has become longer and frost damages to cereal crops and potatoes have diminished. This particularly applies to the agriculture in the mountain communities and in the northern sections but also to the remainder of Scandinavia. The forest limit has moved upwards in many mountain areas and the setting of seed in the mountain forests has improved. Because of the increased maritimity, seabirds are now observed much further inland than earlier and some migratory birds now spend the winter in Scandinavia. As a result of less sea ice in the Spitsbergen waters the navigation between Svalbard and Norway has also been considerably prolonged in autumn during the last 50 years.

But there is evidence indicating that the general rise in air temperature has reached its peak. This stands out very clearly in Fig.7 and 8, showing the variation in time of the departures of the successive 5 years means of the temperatures from the mean temperatures in the period 1876–1960. As will be seen from the figures, the winters have gradually become cooler since the period 1931–1935 in all Scandinavia. The same also applies to the summers that have been colder again since the period 1936–1940. But autumn is still getting warmer and the increase is greatest in arctic Scandinavia. The decrease in temperature has also resulted in small advances of some glaciers after 1961. Likewise more drift ice has been observed in the arctic waters than earlier and severe ice winters occurred in 1939/1940, 1941/1942, 1946/1947, 1962/1963 and 1965/1966. The winters from 1940 to 1942 were the coldest ever observed in Scandinavia, and the winter 1965/1966 was also very cold and long.

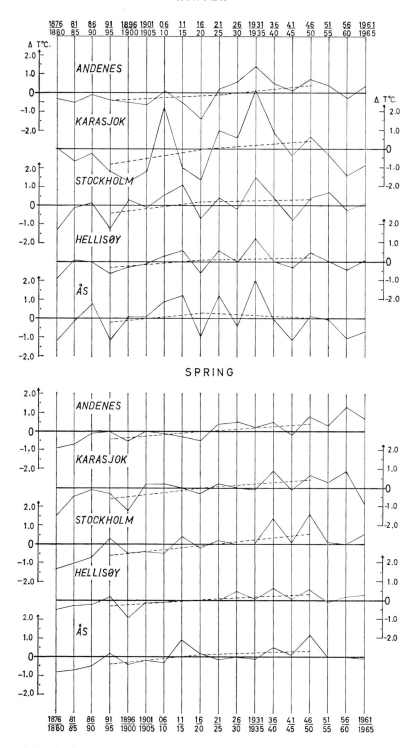

Fig.7. The departure of the successive 5-years means of the air temperature in winter and spring from the mean temperatures (1876–1960) of these seasons. The dotted curve shows the corresponding departure of the 30-year means.

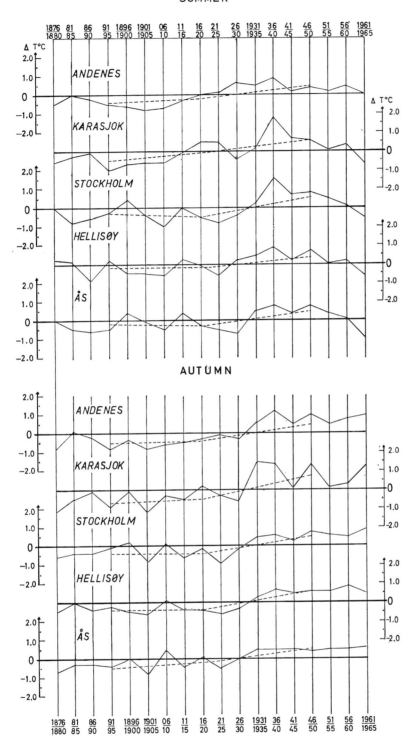

Fig.8. The departure of the successive 5-years means of the air temperature in summer and autumn from the mean temperatures (1876–1960) of these seasons. The dotted curve shows the corresponding departure of the 30-year means.

The fluctuations of the air temperature in Scandinavia during the last hundred years are closely connected with corresponding variations of the circulation of the atmosphere. As shown by HESSELBERG and BIRKELAND (1940, 1943), the circulation over the northeast Atlantic was mainly zonal until about 1915. The frequency of westerly and southwesterly winds was, therefore, great, resulting in the pronounced maritimity with warm winters and cool summers discussed above. From about 1915 on, however, the circulation of the atmosphere over northwestern Europe, became more and more meridional. As shown by the two above-mentioned authors this change in circulation was followed by a pronounced increase in the frequency of southerly winds over Scandinavia all the year and by rising temperatures both in winter and summer. But from about 1940 the meridional type of circulation has been more complicated and considerably influenced by the Siberian high-pressure cell in winter and by the northeast Atlantic blocking in late spring and summer. Easterly and northeasterly winds in winter and northwesterly and northerly winds in summer have, therefore, become more frequent in Scandinavia than before 1940. This also explains the lower air temperatures experienced in Scandinavia both in winter and summer during the last 2–3 decades of this century.

## References

HESSELBERG, TH. und BIRKELAND, B. J., 1940. Säkulare Schwankungen des Klimas von Norwegen— Die Lufttemperatur. *Geofys. Publikasjoner, Norske Videnskapsakad. Oslo*, 14 (4): 106 pp.

HESSELBERG, TH. und BIRKELAND, B. J., 1943. Säkulare Schwankungen des Klimas von Norwegen— Die Lufttemperatur. *Geofys. Publikasjoner, Norske Videnskapsakad. Oslo*, 14 (6): 76 pp.

MOHRMANN, J. C. J. and KESSLER, J., 1959. Water deficiencies in European agriculture. A climatological survey. *Intern. Inst. Land Reclamation Improvement, Publ.*, 5: 60 pp.

WALLÉN, C. C., 1966. Global solar radiation and potential evapotranspiration in Sweden. *Tellus*, 18 (4): 786–800.

TABLE X

CLIMATIC TABLE FOR TROMSÖ
Latitude 69°39′N, longitude 18°57′E, elevation 114.5 m

| Month | Mean sta. press. (mbar) | Mean daily temp. (°C) | Mean daily temp. range (°C) | Temp. extremes (°C) | | Mean vapor press. (mbar) | Mean precip. (mm) | Max. precip., 24 h |
|---|---|---|---|---|---|---|---|---|
| | | | | highest | lowest | | | |
| Jan. | 990.8 | − 3.5 | 4.2 | 7.4 | −15.8 | 4.1 | 96 | 38 |
| Feb. | 991.0 | − 4.0 | 4.4 | 8.2 | −15.0 | 4.0 | 79 | 30 |
| Mar. | 995.0 | − 2.7 | 4.8 | 8.6 | −15.5 | 4.1 | 91 | 38 |
| Apr. | 995.0 | 0.3 | 5.3 | 13.9 | −11.2 | 4.8 | 65 | 22 |
| May | 1001.2 | 4.1 | 5.5 | 22.6 | − 6.2 | 6.1 | 61 | 29 |
| June | 998.1 | 8.8 | 6.5 | 27.7 | − 2.0 | 8.3 | 59 | 22 |
| July | 997.9 | 12.4 | 7.0 | 28.5 | 1.7 | 10.5 | 56 | 30 |
| Aug. | 997.1 | 11.0 | 5.9 | 26.6 | 1.1 | 10.1 | 80 | 29 |
| Sept. | 994.8 | 7.2 | 4.9 | 22.4 | − 3.2 | 8.1 | 109 | 45 |
| Oct. | 992.9 | 3.0 | 3.8 | 15.4 | − 8.0 | 5.9 | 115 | 34 |
| Nov. | 993.4 | − 0.1 | 3.7 | 11.3 | −12.5 | 4.8 | 88 | 36 |
| Dec. | 990.7 | − 1.9 | 4.0 | 8.8 | −14.9 | 4.3 | 95 | 42 |
| Annual | 994.8 | 2.9 | 5.0 | 28.5 | −15.8 | 6.3 | 994 | 45 |

| Month | Number of days | | | Mean cloudiness (%) | Mean sunshine[1] (h) | Most freq. wind direction[2] | Mean wind speed[2] (m/sec) |
|---|---|---|---|---|---|---|---|
| | precip. ⩾0.1 mm | thunderstorm | fog | | | | |
| Jan. | 18 | 0.3 | 0.1 | 66 | 2 | SSW | 3.9 |
| Feb. | 17 | 0.1 | 0.2 | 68 | 35 | SSW | 3.6 |
| Mar. | 19 | 0.0 | 0.1 | 70 | 101 | SSW | 3.5 |
| Apr. | 18 | 0.1 | 0.2 | 71 | 176 | SW | 3.0 |
| May | 18 | 0.0 | 0.7 | 74 | 172 | SW | 2.9 |
| June | 18 | 0.2 | 0.7 | 72 | 202 | NE | 2.5 |
| July | 15 | 0.5 | 1.8 | 67 | 226 | NE | 2.1 |
| Aug. | 19 | 0.4 | 2.0 | 73 | 176 | NE | 2.1 |
| Sept. | 21 | 0.0 | 1.2 | 75 | 97 | SW | 2.6 |
| Oct. | 21 | 0.0 | 0.4 | 73 | 52 | SSW | 3.1 |
| Nov. | 18 | 0.0 | 0.3 | 68 | 7 | SSW | 3.1 |
| Dec. | 19 | 0.2 | 0.3 | 68 | – | SSW | 3.7 |
| Annual | 221 | 2 | 8 | 70 | 1246 | SW | 3.0 |

[1] Period: 1956–1965.
[2] Period: 1941–1950.

TABLE XI

CLIMATIC TABLE FOR KARESUANDO
Latitude 68°27′N, longitude 22°30′E, elevation 327 m

| Month | Mean sta. press. (mbar) | Mean daily temp. (°C) | Mean daily temp. range (°C) | Temp. extremes (°C) | | Mean vapor press. (mbar) | Mean precip. (mm) | Max. precip., 24 h |
|---|---|---|---|---|---|---|---|---|
| | | | | highest | lowest | | | |
| Jan. | 1009.6 | −14.0 | 9.0 | 7.2 | −34.0 | 2.3 | 19 | 13.8 |
| Feb. | 1009.7 | −13.9 | 9.3 | 6.0 | −33.0 | 2.1 | 18 | 11.6 |
| Mar. | 1012.5 | − 9.9 | 11.4 | 10.0 | −31.3 | 2.8 | 17 | 11.6 |
| Apr. | 1010.9 | − 3.6 | 10.5 | 15.5 | −22.4 | 4.1 | 19 | 14.5 |
| May | 1015.7 | 3.0 | 8.7 | 25.0 | − 8.0 | 5.9 | 26 | 23.3 |
| June | 1011.5 | 9.8 | 8.7 | 32.2 | − 0.4 | 8.5 | 46 | 28.0 |
| July | 1011.6 | 13.7 | 9.9 | 32.2 | 3.7 | 11.6 | 63 | 49.0 |
| Aug. | 1011.6 | 11.2 | 9.4 | 29.0 | 0.2 | 10.8 | 57 | 31.7 |
| Sept. | 1010.0 | 5.4 | 7.7 | 22.5 | − 5.2 | 7.7 | 41 | 30.3 |
| Oct. | 1009.3 | − 1.6 | 6.3 | 16.0 | −16.3 | 5.1 | 26 | 27.8 |
| Nov. | 1011.3 | − 7.3 | 7.1 | 7.0 | −25.7 | 3.5 | 26 | 17.6 |
| Dec. | 1009.0 | −11.2 | 7.9 | 6.0 | −30.2 | 2.7 | 22 | 13.8 |
| Annual | 1011.0 | − 1.5 | 8.6 | 32.2 | −34.0 | 5.6 | 380 | 49.0 |

| Month | Number of days | | | | Mean cloudiness (%) | Mean sunshine[1] (h) | Most freq. wind direction | Mean wind speed (m/sec) |
|---|---|---|---|---|---|---|---|---|
| | precipitation | | thunderstorm | fog | | | | |
| | ⩾0.1 mm | ⩾1.0 mm | | | | | | |
| Jan. | 10 | 5.7 | 0 | 1.7 | 65 | 0 | SW | 1.1 |
| Feb. | 10 | 5.6 | 0 | 1.1 | 65 | 36 | SW | 1.2 |
| Mar. | 9 | 5.4 | 0 | 0.9 | 62 | 120 | SW | 1.3 |
| Apr. | 9 | 5.1 | 0 | 0.8 | 64 | 170 | SW | 1.5 |
| May | 9 | 5.8 | 0.2 | 0.4 | 68 | 195 | SW | 2.0 |
| June | 11 | 8.6 | 1.2 | 0.1 | 71 | 225 | SW | 2.3 |
| July | 12 | 9.1 | 2.9 | 0.4 | 67 | 240 | SW | 1.8 |
| Aug. | 13 | 10.2 | 1.1 | 1.6 | 70 | 160 | SW | 1.7 |
| Sept. | 12 | 8.4 | 0.1 | 2.0 | 73 | 115 | SW | 1.8 |
| Oct. | 10 | 6.5 | 0 | 2.9 | 69 | 75 | SW | 1.3 |
| Nov. | 10 | 6.7 | 0 | 2.9 | 73 | 5 | SW | 1.2 |
| Dec. | 9 | 6.6 | 0 | 2.4 | 70 | 0 | SW | 1.2 |
| Annual | 124 | 84 | 6 | 17 | 68 | 1341 | SW | 1.5 |

[1] These values refer to Abisko (68°20′N, 18°50′E, 388 m).

TABLE XII

CLIMATIC TABLE FOR SODANKYLÄ
Latitude 67°22′N, longitude 26°39′E, elevation 178 m

| Month | Mean sta. press. (mbar) | Mean daily temp. (°C) | Mean daily temp. range (°C) | Temp. extremes (°C) | | Mean vapor press. (mbar) | Mean precip. (mm) | Max. precip., 24 h |
|-------|---------|---------|---------|---------|---------|---------|---------|---------|
| | | | | highest | lowest | | | |
| Jan. | 1010.6 | −13.5 | 9.9 | 6.0 | −45.6 | 1.7 | 27 | 13.1 |
| Feb. | 1010.4 | −13.0 | 10.0 | 5.6 | −41.6 | 1.7 | 26 | 12.4 |
| Mar. | 1012.7 | − 9.0 | 12.4 | 9.0 | −38.9 | 2.2 | 20 | 13.8 |
| Apr. | 1011.2 | − 2.1 | 10.6 | 16.4 | −31.8 | 2.9 | 31 | 20.8 |
| May | 1015.3 | 4.9 | 9.6 | 25.4 | −13.7 | 4.3 | 31 | 31.8 |
| June | 1015.3 | 11.3 | 10.4 | 30.7 | − 5.0 | 6.4 | 56 | 32.6 |
| July | 1011.7 | 14.7 | 11.4 | 31.5 | − 1.1 | 8.4 | 74 | 40.1 |
| Aug. | 1011.6 | 12.0 | 10.9 | 28.6 | − 5.1 | 8.2 | 71 | 48.2 |
| Sept. | 1010.1 | 6.2 | 8.3 | 24.0 | −10.6 | 5.9 | 57 | 35.6 |
| Oct. | 1009.5 | − 0.5 | 6.5 | 14.5 | −25.9 | 4.0 | 43 | 26.1 |
| Nov. | 1012.4 | − 5.8 | 7.3 | 7.9 | −40.3 | 2.7 | 39 | 17.3 |
| Dec. | 1010.6 | − 9.8 | 8.6 | 5.5 | −43.1 | 2.2 | 31 | 16.7 |
| Annual | 1011.4 | − 0.4 | 9.7 | 31.5 | −45.6 | 4.2 | 508 | 48.2 |

| Month | Mean snow depth on the 15th (cm) | Number of days | | | Mean cloudiness (%) | Mean sunshine (h) | Most. freq. wind direction | Mean wind speed (m/sec) |
|-------|---------|---------|---------|---------|---------|---------|---------|---------|
| | | precip. ⩾0.1 mm | thunderstorm | fog | | | | |
| Jan. | 47 | 18.8 | 0.0 | 2.7 | 74 | 8.5 | S | 2.8 |
| Feb. | 61 | 17.1 | 0.0 | 2.2 | 75 | 49.0 | S | 2.9 |
| Mar. | 68 | 14.4 | 0.0 | 0.9 | 67 | 140.2 | S | 3.3 |
| Apr. | 65 | 14.9 | 0.0 | 1.7 | 71 | 191.0 | S | 3.4 |
| May | | 13.2 | 0.3 | 0.7 | 71 | 222.4 | N | 3.2 |
| June | | 15.5 | 1.9 | 1.1 | 71 | 277.4 | S | 3.2 |
| July | | 15.1 | 4.2 | 4.8 | 66 | 278.3 | NNE | 2.9 |
| Aug. | | 16.5 | 2.5 | 10.4 | 71 | 184.4 | S | 2.6 |
| Sept. | | 16.8 | 0.2 | 6.6 | 77 | 101.6 | S | 3.1 |
| Oct. | | 17.3 | 0.0 | 5.0 | 78 | 60.4 | S | 2.9 |
| Nov. | 13 | 19.4 | 0.0 | 4.6 | 81 | 19.2 | S | 2.4 |
| Dec. | 30 | 19.0 | 0.0 | 3.1 | 79 | 0.3 | S | 2.8 |
| Annual | | 198 | 9.1 | 4 | 73 | 1533.7 | S | 3.0 |

TABLE XIII

CLIMATIC TABLE FOR KAJAANI
Latitude 64°17′N, longitude 27°41′E, elevation 134 m

| Month | Mean sta. press. (mbar) | Mean daily temp. (°C) | Mean daily temp. range (°C) | Temp. extremes (°C) | | Mean vapor press. (mbar) | Mean precip. (mm) | Max. precip., 24 h |
|---|---|---|---|---|---|---|---|---|
| | | | | highest | lowest | | | |
| Jan. | 1012.2 | −10.6 | 6.8 | 6.4 | −40.1 | 2.1 | 34 | 14.4 |
| Feb. | 1011.3 | −10.5 | 7.2 | 5.5 | −39.0 | 2.0 | 27 | 19.0 |
| Mar. | 1013.2 | − 6.7 | 9.5 | 11.8 | −35.6 | 2.6 | 24 | 17.0 |
| Apr. | 1011.4 | 0.3 | 8.9 | 21.2 | −25.8 | 3.6 | 35 | 26.3 |
| May | 1014.8 | 6.9 | 10.3 | 27.1 | −10.9 | 5.2 | 38 | 41.0 |
| June | 1010.9 | 12.9 | 10.1 | 30.6 | − 2.8 | 7.6 | 67 | 48.6 |
| July | 1010.6 | 16.1 | 10.2 | 31.6 | 0.2 | 9.8 | 72 | 34.1 |
| Aug. | 1011.6 | 14.0 | 9.2 | 29.4 | − 3.5 | 9.7 | 72 | 43.1 |
| Sept. | 1011.6 | 8.3 | 7.0 | 25.2 | − 7.6 | 7.0 | 63 | 58.0 |
| Oct. | 1010.6 | 2.1 | 4.8 | 15.0 | −21.5 | 4.9 | 53 | 25.4 |
| Nov. | 1013.6 | − 2.6 | 5.6 | 9.7 | −25.8 | 3.5 | 43 | 30.0 |
| Dec. | 1011.9 | − 7.0 | 5.7 | 6.0 | −36.8 | 2.7 | 36 | 22.0 |
| Annual | 1011.9 | 1.9 | 7.9 | 31.6 | −40.1 | 5.1 | 563 | 58.0 |

| Month | Mean snow depth on the 15th (cm) | Number of days | | | Mean cloudi-ness (%) | Mean sun-shine (h) | Most freq. wind direction | Mean wind speed (m/sec) |
|---|---|---|---|---|---|---|---|---|
| | | precip. ⩾0.1 mm | thunder-storm | fog | | | | |
| Jan. | 35 | 19.0 | 0.0 | – | 77 | – | – | – |
| Feb. | 48 | 16.8 | 0.0 | – | 72 | – | – | – |
| Mar. | 56 | 13.7 | 0.0 | – | 62 | – | – | – |
| Apr. | 40 | 14.8 | 0.0 | – | 63 | – | – | – |
| May | | 12.5 | 0.6 | – | 58 | – | – | – |
| June | | 14.8 | 2.6 | – | 60 | – | – | – |
| July | | 14.7 | 4.1 | – | 57 | – | – | – |
| Aug. | | 15.0 | 2.1 | – | 61 | – | – | – |
| Sept. | | 16.1 | 0.2 | – | 69 | – | – | – |
| Oct. | | 17.5 | 0.0 | – | 78 | – | – | – |
| Nov. | 6 | 18.9 | 0.0 | – | 84 | – | – | – |
| Dec. | 17 | 19.3 | 0.0 | – | 82 | – | – | – |
| Annual | | 193 | 10 | – | 69 | – | – | – |

TABLE XIV

CLIMATIC TABLE FOR TRONDHEIM (VOLL)
Latitude 63°25′N, longitude 10°27′E, elevation 133.0 m

| Month | Mean sta. press. (mbar) | Mean daily temp. (°C) | Mean daily temp. range (°C) | Temp. extremes (°C) | | Mean vapor press. (mbar) | Mean precip. (mm) | Max. precip., 24 h |
|---|---|---|---|---|---|---|---|---|
| | | | | highest | lowest | | | |
| Jan. | 992.1 | − 3.4 | 5.7 | 10.8 | −25.2 | 4.0 | 68 | 51 |
| Feb. | 992.9 | − 2.9 | 6.0 | 10.4 | −25.0 | 4.0 | 67 | 51 |
| Mar. | 996.7 | − 0.7 | 6.9 | 12.0 | −21.5 | 4.3 | 67 | 65 |
| Apr. | 994.8 | 3.2 | 6.9 | 16.0 | −15.4 | 5.6 | 60 | 40 |
| May | 1000.1 | 7.9 | 7.6 | 23.8 | − 4.6 | 7.5 | 48 | 24 |
| June | 997.0 | 11.3 | 7.6 | 29.2 | − 0.8 | 9.6 | 66 | 31 |
| July | 995.6 | 14.4 | 7.5 | 32.9 | 4.8 | 11.9 | 70 | 38 |
| Aug. | 995.8 | 13.3 | 7.2 | 27.8 | 1.4 | 11.1 | 78 | 88 |
| Sept. | 995.0 | 9.5 | 6.4 | 24.0 | − 1.8 | 8.8 | 92 | 53 |
| Oct. | 993.9 | 5.1 | 5.2 | 18.0 | −11.8 | 6.5 | 98 | 81 |
| Nov. | 993.4 | 1.5 | 4.9 | 15.0 | −12.5 | 5.6 | 67 | 43 |
| Dec. | 991.5 | − 1.0 | 5.0 | 10.8 | −20.2 | 4.1 | 76 | 39 |
| Annual | 994.9 | 4.9 | 6.4 | 32.9 | −25.2 | 6.9 | 857 | 88 |

| Month | Number of days | | | Mean cloudi-ness (%) | Mean sun-shine[1] (h) | Most freq. wind direction[2] | Mean wind speed[2] (m/sec) |
|---|---|---|---|---|---|---|---|
| | precip. ⩾0.1 mm | thunder-storm | fog | | | | |
| Jan. | 15 | 0.1 | 1.3 | 62 | 19 | S | 3.4 |
| Feb. | 15 | 0.2 | 0.8 | 63 | 63 | SSW | 4.0 |
| Mar. | 15 | 0.2 | 1.3 | 62 | 138 | SW | 3.6 |
| Apr. | 16 | 0.0 | 0.9 | 66 | 162 | W | 3.3 |
| May | 15 | 0.2 | 0.8 | 61 | 193 | W | 3.1 |
| June | 18 | 1.2 | 0.9 | 68 | 181 | W | 2.8 |
| July | 17 | 1.7 | 1.2 | 66 | 182 | W | 2.2 |
| Aug. | 16 | 0.7 | 2.0 | 65 | 178 | W | 2.2 |
| Sept. | 19 | 0.2 | 2.0 | 69 | 116 | SSW | 3.0 |
| Oct. | 19 | 0.0 | 2.0 | 66 | 72 | SW | 3.3 |
| Nov. | 14 | 0.0 | 1.6 | 63 | 31 | S | 3.5 |
| Dec. | 16 | 0.2 | 1.9 | 61 | 11 | S | 3.6 |
| Annual | 195 | 5 | 17 | 64 | 1346 | S | 3.2 |

[1] Period: 1956–1965.
[2] Period: 1941–1950.

TABLE XV

CLIMATIC TABLE FOR ÖSTERSUND
Latitude 63°10′N, longitude 14°40′E, elevation 328 m

| Month | Mean sta. press. (mbar) | Mean daily temp. (°C) | Mean daily temp. range (°C) | Temp. extremes (°C) | | Mean vapor press. (mbar) | Mean precip. (mm) | Max. precip., 24 h |
|---|---|---|---|---|---|---|---|---|
| | | | | highest | lowest | | | |
| Jan. | 1011.1 | − 8.4 | 6.7 | 9.8 | −33.8 | 3.2 | 34 | 19.0 |
| Feb. | 1011.5 | − 7.1 | 7.2 | 9.2 | −31.4 | 3.3 | 23 | 12.8 |
| Mar. | 1014.7 | − 4.1 | 8.9 | 16.0 | −30.0 | 4.0 | 23 | 17.4 |
| Apr. | 1011.7 | 1.2 | 8.0 | 18.8 | −18.0 | 5.3 | 29 | 31.7 |
| May | 1016.3 | 6.7 | 10.3 | 26.5 | − 6.8 | 6.9 | 31 | 22.1 |
| June | 1012.2 | 11.3 | 9.8 | 33.5 | − 1.5 | 9.9 | 69 | 38.7 |
| July | 1011.2 | 14.7 | 9.7 | 32.5 | 2.5 | 12.4 | 77 | 71.8 |
| Aug. | 1011.8 | 13.4 | 9.3 | 31.3 | 0.2 | 11.7 | 74 | 66.8 |
| Sept. | 1011.4 | 8.9 | 7.1 | 25.2 | − 5.2 | 9.3 | 51 | 53.9 |
| Oct. | 1011.2 | 3.8 | 5.2 | 16.4 | −15.2 | 6.7 | 43 | 19.2 |
| Nov. | 1011.9 | − 0.8 | 4.3 | 12.2 | −20.4 | 5.1 | 42 | 22.9 |
| Dec. | 1010.3 | − 4.5 | 5.4 | 8.5 | −31.0 | 4.0 | 36 | 31.5 |
| Annual | 1012.1 | 2.9 | 7.6 | 33.5 | −33.8 | 6.8 | 532 | 71.8 |

| Month | Number of days | | | | Mean cloudi-ness (%) | Mean sun-shine[1] (h) | Most freq. wind direction | Mean wind speed (m/sec) |
|---|---|---|---|---|---|---|---|---|
| | precipitation | | thunder-storm | fog | | | | |
| | ⩾0.1 mm | ⩾1.0 mm | | | | | | |
| Jan. | 15 | 8.9 | 0 | 3.1 | 66 | 26 | SE | 2.0 |
| Feb. | 13 | 6.3 | 0 | 0.9 | 65 | 61 | SE | 2.1 |
| Mar. | 12 | 6.5 | 0 | 1.3 | 61 | 121 | NW | 2.0 |
| Apr. | 12 | 6.2 | 0.1 | 0.4 | 60 | 180 | NW | 2.1 |
| May | 11 | 5.9 | 0.3 | 0.3 | 57 | 252 | NW | 2.3 |
| June | 14 | 9.9 | 1.4 | 0.2 | 62 | 240 | NW | 2.7 |
| July | 15 | 10.9 | 2.4 | 0.3 | 60 | 249 | NW | 2.4 |
| Aug. | 14 | 9.7 | 1.1 | 0.6 | 61 | 204 | NW | 2.1 |
| Sept. | 14 | 9.0 | 0.2 | 1.9 | 66 | 126 | NW | 2.4 |
| Oct. | 14 | 8.4 | * | 2.0 | 68 | 74 | SE | 2.3 |
| Nov. | 15 | 8.9 | 0 | 2.5 | 77 | 29 | SE | 2.1 |
| Dec. | 14 | 8.5 | 0 | 2.4 | 70 | 10 | SE | 2.0 |
| Annual | 163 | 99 | 6 | 15 | 64 | 1572 | NW | 2.2 |

[1] These values refer to Gisselås (63°42′N, 15°22′E, 320 m).

* Indicates a frequency >0 but <0.05.

TABLE XVI

CLIMATIC TABLE FOR VAASA
Latitude 63°03′N, longitude 21°46′E, elevation 6 m

| Month | Mean sta. press. (mbar) | Mean daily temp. (°C) | Mean daily temp. range (°C) | Temp. extremes (°C) | | Mean vapor press. (mbar) | Mean precip. (mm) | Max. precip., 24 h |
|---|---|---|---|---|---|---|---|---|
| | | | | highest | lowest | | | |
| Jan. | 1010.4 | − 7.3 | 6.7 | 7.5 | −32.1 | 2.8 | 35 | 18.5 |
| Feb. | 1010.5 | − 7.5 | 7.0 | 7.0 | −33.4 | 2.5 | 21 | 34.2 |
| Mar. | 1013.5 | − 4.7 | 9.5 | 11.4 | −34.1 | 3.0 | 20 | 13.7 |
| Apr. | 1011.4 | 1.3 | 8.8 | 19.7 | −19.1 | 4.2 | 31 | 21.1 |
| May | 1015.2 | 7.5 | 10.3 | 29.0 | − 7.1 | 5.7 | 30 | 25.8 |
| June | 1011.4 | 12.8 | 10.1 | 31.8 | − 1.9 | 8.1 | 45 | 32.5 |
| July | 1010.4 | 16.2 | 10.5 | 30.6 | 1.8 | 10.3 | 63 | 50.4 |
| Aug. | 1011.2 | 14.6 | 9.9 | 29.3 | 0.5 | 10.3 | 65 | 40.1 |
| Sept. | 1010.6 | 9.6 | 8.1 | 27.5 | − 5.0 | 7.9 | 67 | 50.5 |
| Oct. | 1010.9 | 3.8 | 6.2 | 15.4 | −14.7 | 5.7 | 65 | 37.9 |
| Nov. | 1012.4 | − 0.5 | 5.0 | 10.5 | −19.2 | 4.1 | 53 | 27.3 |
| Dec. | 1010.6 | − 3.7 | 7.9 | 7.9 | −30.2 | 3.4 | 39 | 24.0 |
| Annual | 1011.6 | 3.5 | 8.3 | 31.8 | −34.1 | 5.7 | 532 | 50.5 |

| Month | Mean snow depth on the 15th (cm) | Number of days | | | Mean cloudi-ness (%) | Mean sun-shine (h) | Most freq. wind direction | Mean wind speed (m/sec) |
|---|---|---|---|---|---|---|---|---|
| | | precip. ⩾0.1 mm | thunder-storm | fog | | | | |
| Jan. | 20 | 14.7 | 0.0 | 4.7 | 71 | − | SW | 3.8 |
| Feb. | 26 | 12.7 | 0.0 | 3.2 | 68 | − | SW | 3.9 |
| Mar. | 27 | 10.5 | 0.0 | 2.6 | 64 | − | SW | 3.9 |
| Apr. | 9 | 10.7 | 0.1 | 2.8 | 63 | − | SW | 4.3 |
| May | | 9.0 | 0.6 | 1.7 | 57 | − | N | 3.8 |
| June | | 10.9 | 1.8 | 2.6 | 58 | − | W | 3.9 |
| July | | 11.5 | 3.5 | 3.0 | 57 | − | N | 3.4 |
| Aug. | | 12.7 | 2.1 | 6.5 | 62 | − | N | 3.2 |
| Sept. | | 14.2 | 0.6 | 7.0 | 66 | − | SSW | 4.2 |
| Oct. | | 15.1 | 0.0 | 4.6 | 71 | − | SW | 3.9 |
| Nov. | 4 | 16.2 | 0.0 | 5.9 | 76 | − | S | 3.9 |
| Dec. | 9 | 16.5 | 0.0 | 3.4 | 76 | − | E | 4.0 |
| Annual | | 155 | 9 | 48 | 66 | − | SW | 3.8 |

TABLE XVII

CLIMATIC TABLE FOR HÄRNÖSAND
Latitude 62°28′N, longitude 17°57′E, elevation 8 m

| Month | Mean sta. press. (mbar) | Mean daily temp. (°C) | Mean daily temp. range (°C) | Temp. extremes (°C) | | Mean vapor press. (mbar) | Mean precip. (mm) | Max. precip., 24 h |
|---|---|---|---|---|---|---|---|---|
| | | | | highest | lowest | | | |
| Jan. | 1010.5 | − 6.4 | 7.3 | 10.2 | −30.0 | 3.6 | 62 | 46.6 |
| Feb. | 1010.9 | − 5.8 | 8.2 | 10.4 | −30.6 | 3.6 | 37 | 29.5 |
| Mar. | 1013.9 | − 2.9 | 9.3 | 17.0 | −31.0 | 4.3 | 32 | 23.2 |
| Apr. | 1011.5 | 2.1 | 9.0 | 20.0 | −17.5 | 5.6 | 46 | 78.0 |
| May | 1015.9 | 7.7 | 10.4 | 27.4 | − 5.6 | 7.3 | 34 | 28.2 |
| June | 1011.8 | 12.8 | 10.1 | 31.6 | − 2.7 | 10.4 | 52 | 49.4 |
| July | 1010.8 | 16.4 | 10.0 | 31.6 | 4.8 | 13.9 | 58 | 73.6 |
| Aug. | 1011.3 | 15.1 | 9.6 | 30.0 | 0.2 | 13.3 | 78 | 107.5 |
| Sept. | 1011.1 | 10.4 | 8.8 | 26.0 | − 5.8 | 10.4 | 68 | 58.8 |
| Oct. | 1011.1 | 5.0 | 6.9 | 19.6 | −11.2 | 7.2 | 63 | 38.5 |
| Nov. | 1012.0 | 0.7 | 5.1 | 11.5 | −19.0 | 5.6 | 87 | 50.5 |
| Dec. | 1010.1 | − 2.8 | 5.8 | 10.3 | −27.0 | 4.5 | 80 | 42.7 |
| Annual | 1011.7 | 4.4 | 8.3 | 31.6 | −31.0 | 7.5 | 697 | 107.5 |

| Month | Number of days | | thunder-storm | fog | Mean cloudi-ness (%) | Mean sun-shine (h) | Most freq. wind direction | Mean wind speed (m/sec) |
|---|---|---|---|---|---|---|---|---|
| | precipitation | | | | | | | |
| | ⩾0.1 mm | ⩾1.0 mm | | | | | | |
| Jan. | 15 | 10.7 | 0 | 2.1 | 65 | 54 | N | 2.0 |
| Feb. | 12 | 7.5 | 0 | 2.1 | 65 | 70 | N | 1.9 |
| Mar. | 10 | 6.7 | 0 | 2.6 | 62 | 144 | N | 2.0 |
| Apr. | 10 | 6.6 | * | 3.2 | 61 | 195 | S | 2.4 |
| May | 8 | 5.4 | 0.3 | 1.6 | 55 | 259 | S | 2.8 |
| June | 10 | 7.1 | 1.1 | 2.3 | 58 | 251 | S | 3.6 |
| July | 11 | 7.4 | 2.1 | 2.0 | 56 | 267 | S | 2.9 |
| Aug. | 12 | 8.2 | 1.4 | 2.5 | 59 | 217 | S | 2.6 |
| Sept. | 11 | 8.2 | 0.5 | 3.5 | 63 | 145 | S | 2.6 |
| Oct. | 12 | 8.7 | * | 3.4 | 64 | 90 | S | 2.1 |
| Nov. | 15 | 11.2 | 0 | 2.5 | 72 | 52 | S | 2.0 |
| Dec. | 16 | 12.0 | 0 | 2.1 | 69 | 27 | S | 2.1 |
| Annual | 142 | 100 | 6 | 30 | 63 | 1771 | S | 2.4 |

* Indicates a frequency >0 but <0.05.

TABLE XVIII

CLIMATIC TABLE FOR BERGEN (FREDRIKSBERG)
Latitude 60°24′N, longitude 5°19′E, elevation 44.4 m

| Month | Mean sta. press. (mbar) | Mean daily temp. (°C) | Mean daily temp. range (°C) | Temp. extremes (°C) | | Mean vapor press. (mbar) | Mean precip. (mm) | Max. precip., 24 h |
|---|---|---|---|---|---|---|---|---|
| | | | | highest | lowest | | | |
| Jan. | 1004.0 | 1.5 | 4.0 | 13.3 | −13.5 | 5.7 | 179 | 60 |
| Feb. | 1005.5 | 1.3 | 4.3 | 11.2 | −10.9 | 5.5 | 139 | 54 |
| Mar. | 1008.8 | 3.1 | 5.6 | 19.8 | −10.0 | 5.7 | 109 | 66 |
| Apr. | 1006.6 | 5.8 | 6.0 | 22.1 | − 5.6 | 6.5 | 140 | 55 |
| May | 1010.4 | 10.2 | 7.5 | 27.1 | − 0.1 | 8.5 | 83 | 44 |
| June | 1007.8 | 12.6 | 6.9 | 31.8 | 0.6 | 10.7 | 126 | 67 |
| July | 1006.0 | 15.0 | 6.7 | 30.5 | 5.2 | 12.8 | 141 | 79 |
| Aug. | 1006.3 | 14.7 | 6.5 | 29.7 | 5.4 | 12.5 | 167 | 58 |
| Sept. | 1006.4 | 12.0 | 5.6 | 26.0 | 1.3 | 10.8 | 228 | 61 |
| Oct. | 1005.9 | 8.3 | 4.8 | 19.5 | − 3.1 | 8.5 | 236 | 91 |
| Nov. | 1004.6 | 5.5 | 4.1 | 15.4 | − 5.6 | 6.7 | 207 | 76 |
| Dec. | 1003.4 | 3.3 | 3.8 | 16.4 | − 8.4 | 6.1 | 203 | 99 |
| Annual | 1006.3 | 7.8 | 5.5 | 31.8 | −13.5 | 8.4 | 1958 | 99 |

| Month | Number of days | | | Mean cloudi- ness (%) | Mean sun- shine[1] (h) | Most freq. wind direction[2] | Mean wind speed[2] (m/sec) |
|---|---|---|---|---|---|---|---|
| | precip. ⩾0.1 mm | thunder- storm | fog | | | | |
| Jan. | 21 | 0.4 | 0.9 | 71 | 22 | S | 3.2 |
| Feb. | 18 | 0.7 | 0.8 | 72 | 54 | S | 3.5 |
| Mar. | 16 | 0.4 | 1.2 | 67 | 127 | S | 3.1 |
| Apr. | 19 | 0.1 | 0.8 | 72 | 142 | S | 3.6 |
| May | 15 | 0.6 | 1.3 | 63 | 191 | NNW | 3.1 |
| June | 18 | 1.2 | 0.5 | 68 | 175 | NNW | 2.9 |
| July | 21 | 1.7 | 1.0 | 74 | 167 | NNW | 2.4 |
| Aug. | 20 | 1.4 | 1.4 | 71 | 132 | NNW | 2.6 |
| Sept. | 22 | 1.0 | 0.9 | 73 | 102 | S | 3.2 |
| Oct. | 24 | 1.3 | 0.6 | 73 | 69 | S | 3.3 |
| Nov. | 22 | 0.9 | 1.2 | 73 | 29 | S | 3.3 |
| Dec. | 23 | 1.2 | 1.0 | 73 | 13 | S | 3.6 |
| Annual | 239 | 11 | 12 | 71 | 1223 | S | 3.2 |

[1] Period: 1956–1965.
[2] Period: 1941–1950.

TABLE XIX

CLIMATIC TABLE FOR ILMALA
Latitude 60°12′N, longitude 24°55′E, elevation 45 m

| Month | Mean sta. press. (mbar) | Mean daily temp. (°C) | Mean daily temp. range (°C) | Temp. extremes (°C) | | Mean vapor press. (mbar) | Mean precip. (mm) | Max. precip., 24 h |
|-------|------|------|------|------|------|------|------|------|
| | | | | highest | lowest | | | |
| Jan. | 1012.6 | − 6.1 | 5.1 | 6.8 | −33.2 | 3.0 | 57 | 21.4 |
| Feb. | 1012.2 | − 6.6 | 5.4 | 11.8 | −30.2 | 2.6 | 42 | 18.4 |
| Mar. | 1014.3 | − 3.4 | 6.9 | 15.0 | −26.0 | 2.9 | 36 | 22.4 |
| Apr. | 1013.1 | 2.6 | 7.1 | 20.5 | −13.4 | 4.0 | 44 | 34.7 |
| May | 1015.0 | 8.8 | 9.2 | 26.1 | − 5.5 | 5.6 | 41 | 31.1 |
| June | 1012.0 | 14.0 | 9.4 | 31.2 | − 0.3 | 8.4 | 51 | 56.3 |
| July | 1010.6 | 17.2 | 9.1 | 33.1 | 5.4 | 10.6 | 68 | 88.4 |
| Aug. | 1011.9 | 16.0 | 8.4 | 30.1 | 3.5 | 10.5 | 72 | 59.8 |
| Sept. | 1012.2 | 11.1 | 7.5 | 24.3 | − 4.1 | 8.1 | 71 | 56.6 |
| Oct. | 1012.8 | 5.4 | 5.4 | 17.8 | −10.0 | 6.2 | 73 | 50.2 |
| Nov. | 1014.4 | 1.0 | 3.9 | 10.7 | −16.3 | 4.4 | 68 | 35.2 |
| Dec. | 1013.0 | − 2.6 | 4.3 | 9.4 | −27.8 | 3.8 | 66 | 39.6 |
| Annual | 1012.8 | 4.8 | 6.8 | 33.1 | −33.2 | 5.8 | 692 | 88.4 |

| Month | Mean snow depth on the 15th (cm) | Number of days | | | Mean cloudiness (%) | Mean sunshine (h) | Most freq. wind direction | Mean wind speed (m/sec) |
|-------|------|------|------|------|------|------|------|------|
| | | precip. ⩾0.1 mm | thunderstorm | fog | | | | |
| Jan. | 23 | 20.1 | 0.0 | 5.4 | 83 | 30.6 | S | 4.3 |
| Feb. | 34 | 17.5 | 0.0 | 6.1 | 78 | 62.8 | N | 4.1 |
| Mar. | 40 | 14.0 | 0.0 | 5.7 | 69 | 135.5 | WNW | 3.9 |
| Apr. | 15 | 13.4 | 0.2 | 4.2 | 70 | 183.9 | S | 3.8 |
| May | | 11.6 | 0.8 | 4.8 | 65 | 269.6 | SSW | 3.8 |
| June | | 13.1 | 2.2 | 3.5 | 63 | 294.3 | SSW | 3.7 |
| July | | 14.4 | 3.9 | 6.1 | 63 | 294.6 | S | 3.3 |
| Aug. | | 14.6 | 2.7 | 7.8 | 64 | 251.0 | S | 3.4 |
| Sept. | | 15.3 | 1.9 | 6.6 | 72 | 152.0 | WNW | 3.8 |
| Oct. | | 17.6 | 0.2 | 6.8 | 80 | 76.3 | S | 3.9 |
| Nov. | 2 | 19.1 | 0.0 | 5.1 | 87 | 30.2 | S | 4.2 |
| Dec. | 9 | 19.7 | 0.0 | 6.7 | 88 | 18.1 | S | 4.0 |
| Annual | | 190 | 12 | 79 | 74 | 150 | S | 3.8 |

TABLE XX

CLIMATIC TABLE FOR OSLO (BLINDERN)
Latitude 59°56′N, longitude 10°44′E, elevation 95.6 m

| Month | Mean sta. press. (mbar) | Mean daily temp. (°C) | Mean daily temp. range (°C) | Temp. extremes (°C) | | Mean vapor press. (mbar) | Mean precip. (mm) | Max. precip., 24 h |
|---|---|---|---|---|---|---|---|---|
| | | | | highest | lowest | | | |
| Jan. | 1000.2 | − 4.7 | 5.4 | 9.9 | −26.0 | 4.1 | 49 | 33 |
| Feb. | 1000.7 | − 4.0 | 6.5 | 13.8 | −21.9 | 4.4 | 35 | 21 |
| Mar. | 1003.4 | − 0.5 | 8.0 | 15.5 | −21.3 | 5.2 | 26 | 16 |
| Apr. | 1000.4 | 4.8 | 8.8 | 21.5 | −14.9 | 6.1 | 44 | 29 |
| May | 1003.8 | 10.7 | 10.4 | 28.4 | − 3.4 | 8.3 | 44 | 43 |
| June | 1000.4 | 14.7 | 10.0 | 33.7 | 1.8 | 10.9 | 71 | 43 |
| July | 999.1 | 17.3 | 9.5 | 32.8 | 3.7 | 13.7 | 84 | 46 |
| Aug. | 999.9 | 15.9 | 9.0 | 30.9 | 3.9 | 12.7 | 96 | 52 |
| Sept. | 1000.6 | 11.3 | 7.9 | 25.5 | − 2.0 | 10.4 | 83 | 38 |
| Oct. | 1000.8 | 5.9 | 6.3 | 20.1 | − 7.6 | 7.6 | 76 | 50 |
| Nov. | 1000.8 | 1.1 | 4.4 | 11.7 | −15.7 | 5.6 | 69 | 32 |
| Dec. | 999.5 | − 2.0 | 4.6 | 10.8 | −20.2 | 4.7 | 63 | 24 |
| Annual | 1000.8 | 5.9 | 7.5 | 33.7 | −26.0 | 7.9 | 740 | 52 |

| Month | Number of days | | | Mean cloudiness (%) | Mean sunshine[1] (h) | Most freq. wind direction[2] | Mean wind speed[2] (m/sec) |
|---|---|---|---|---|---|---|---|
| | precip. ⩾0.1 mm | thunderstorm | fog | | | | |
| Jan. | 15 | 0.0 | 10.1 | 72 | 45 | NNE | 2.1 |
| Feb. | 13 | 0.0 | 7.3 | 68 | 83 | NNE | 1.9 |
| Mar. | 9 | 0.0 | 6.0 | 64 | 152 | NNE | 2.0 |
| Apr. | 11 | 0.1 | 2.8 | 65 | 182 | S | 2.5 |
| May | 10 | 1.5 | 1.0 | 59 | 233 | S | 2.5 |
| June | 13 | 3.8 | 0.2 | 63 | 244 | S | 2.7 |
| July | 15 | 5.3 | 0.2 | 65 | 219 | S | 2.5 |
| Aug. | 14 | 4.6 | 0.8 | 64 | 183 | S | 2.3 |
| Sept. | 14 | 0.9 | 2.3 | 67 | 138 | S | 2.2 |
| Oct. | 14 | 0.2 | 3.8 | 69 | 87 | NNE | 2.1 |
| Nov. | 16 | 0.3 | 7.5 | 78 | 41 | NNE | 2.4 |
| Dec. | 17 | 0.0 | 11.1 | 77 | 25 | NNE | 1.8 |
| Annual | 161 | 17 | 53 | 68 | 1632 | S | 2.2 |

[1] Period: 1956–1965.
[2] Period: 1941–1950.

TABLE XXI

Latitude 59°21′N, longitude 18°4′E, elevation 44 m

| Month | Mean sta. press. (mbar) | Mean daily temp. (°C) | Mean daily temp. range (°C) | Temp. extremes (°C) | | Mean vapor press. (mbar) | Mean precip. (mm) | Max. precip., 24 h |
|---|---|---|---|---|---|---|---|---|
| | | | | highest | lowest | | | |
| Jan. | 1011.7 | − 2.9 | 3.7 | 9.6 | −28.2 | 4.4 | 43 | 26.1 |
| Feb. | 1012.1 | − 3.1 | 4.3 | 11.8 | −24.9 | 4.3 | 30 | 23.0 |
| Mar. | 1014.8 | − 0.7 | 5.5 | 15.2 | −22.0 | 4.7 | 26 | 17.0 |
| Apr. | 1012.6 | 4.4 | 7.6 | 20.2 | −11.5 | 6.1 | 31 | 20.3 |
| May | 1015.9 | 10.1 | 8.9 | 28.0 | − 3.3 | 8.0 | 34 | 36.1 |
| June | 1012.7 | 14.9 | 8.8 | 32.2 | 1.0 | 11.2 | 45 | 29.5 |
| July | 1011.1 | 17.8 | 7.8 | 34.6 | 8.0 | 14.3 | 61 | 42.3 |
| Aug. | 1011.8 | 16.6 | 6.9 | 31.0 | 4.8 | 14.3 | 76 | 47.4 |
| Sept. | 1012.4 | 12.2 | 5.9 | 25.7 | 0.1 | 11.6 | 60 | 44.4 |
| Oct. | 1012.9 | 7.1 | 4.2 | 17.4 | − 6.5 | 8.7 | 48 | 38.4 |
| Nov. | 1013.2 | 2.8 | 3.5 | 12.4 | −11.0 | 6.7 | 53 | 39.1 |
| Dec. | 1011.7 | 0.1 | 3.8 | 12.2 | −16.3 | 5.6 | 48 | 32.5 |
| Annual | 1012.7 | 6.6 | 5.9 | 34.6 | −28.2 | 8.3 | 555 | 47.4 |

| Month | Number of days | | | | Mean cloudi- ness (%) | Mean sun- shine (h) | Most freq. wind direction | Mean wind speed (m/sec) |
|---|---|---|---|---|---|---|---|---|
| | precipitation | | thunder- storm | fog | | | | |
| | ≥0.1 mm | ≥1.0 mm | | | | | | |
| Jan. | 16 | 9.6 | 0 | 3.9 | 76 | 41 | WSW | 3.9 |
| Feb. | 13 | 7.4 | 0 | 3.2 | 74 | 76 | W | 3.6 |
| Mar. | 10 | 6.2 | 0 | 3.1 | 62 | 151 | WSW | 3.6 |
| Apr. | 11 | 6.7 | 0.1 | 1.9 | 60 | 208 | SW | 3.8 |
| May | 11 | 6.7 | 0.5 | 0.3 | 53 | 292 | NNE | 4.0 |
| June | 13 | 8.3 | 1.7 | 0.2 | 55 | 318 | WSW | 4.3 |
| July | 13 | 8.9 | 2.5 | 0.2 | 54 | 295 | SW | 3.7 |
| Aug. | 14 | 9.7 | 1.8 | 0.7 | 56 | 248 | SW | 3.8 |
| Sept. | 14 | 9.2 | 0.4 | 2.7 | 61 | 174 | WSW | 3.9 |
| Oct. | 15 | 8.7 | 0.1 | 3.6 | 66 | 103 | SW | 4.1 |
| Nov. | 16 | 10.3 | 0 | 5.2 | 80 | 41 | SW | 3.9 |
| Dec. | 17 | 10.7 | * | 5.3 | 81 | 26 | SW | 3.9 |
| Annual | 163 | 102 | 7 | 30 | 65 | 1973 | WSW | 3.8 |

* Indicates a frequency >0 but <0.05.

## TABLE XXII

CLIMATIC TABLE FOR GÖTEBORG
Latitude 57°42′N, longitude 11°58′E, elevation 31 m

| Month | Mean sta. press (mbar) | Mean daily temp. (°C) | Mean daily temp. range (°C) | Temp. extremes (°C) | | Mean vapor press. (mbar) | Mean precip. (mm) | Max. precip., 24 h |
|---|---|---|---|---|---|---|---|---|
| | | | | highest | lowest | | | |
| Jan. | 1012.1 | − 1.1 | 4.1 | 8.0 | −26.0 | 5.1 | 51 | 27.7 |
| Feb. | 1012.6 | − 1.2 | 4.9 | 9.0 | −20.0 | 4.9 | 34 | 14.9 |
| Mar. | 1015.4 | 1.0 | 5.8 | 17.0 | −19.2 | 5.2 | 29 | 29.0 |
| Apr. | 1012.7 | 5.6 | 6.7 | 22.0 | −11.0 | 6.5 | 39 | 22.4 |
| May | 1015.9 | 11.0 | 8.3 | 28.3 | − 3.0 | 8.7 | 34 | 28.1 |
| June | 1013.2 | 14.5 | 7.4 | 32.0 | 3.0 | 11.6 | 54 | 33.2 |
| July | 1011.4 | 17.0 | 7.1 | 32.0 | 8.0 | 14.0 | 86 | 66.1 |
| Aug. | 1012.0 | 16.3 | 6.7 | 30.0 | 5.2 | 14.0 | 84 | 40.6 |
| Sept. | 1013.0 | 12.9 | 6.0 | 25.0 | − 0.5 | 11.7 | 75 | 45.2 |
| Oct. | 1013.1 | 8.8 | 4.7 | 20.0 | − 6.0 | 9.1 | 65 | 48.3 |
| Nov. | 1012.6 | 4.2 | 3.6 | 13.0 | − 7.8 | 7.2 | 62 | 29.8 |
| Dec. | 1011.7 | 1.6 | 3.6 | 10.9 | −15.8 | 6.1 | 57 | 22.8 |
| Annual | 1013.0 | 7.6 | 5.7 | 32.0 | −26.0 | 8.7 | 670 | 66.1 |

| Month | Number of days | | | | Mean cloudi-ness (%) | Mean sun-shine (h) | Most freq. wind direction | Mean wind speed (m/sec) |
|---|---|---|---|---|---|---|---|---|
| | precipitation | | thunder-storm | fog | | | | |
| | ⩾0.1 mm | ⩾ 1.0 mm | | | | | | |
| Jan. | 15 | 9.6 | 0.1 | 5.4 | 73 | 48 | ESE | 4.1 |
| Feb. | 12 | 7.5 | 0 | 5.3 | 69 | 76 | E | 4.0 |
| Mar. | 10 | 6.5 | 0 | 5.5 | 61 | 151 | ESE | 3.7 |
| Apr. | 12 | 7.7 | 0.2 | 2.5 | 60 | 201 | WSW | 4.2 |
| May | 10 | 6.8 | 0.7 | 0.9 | 55 | 274 | SW | 3.8 |
| June | 12 | 8.6 | 1.9 | 0.3 | 58 | 286 | W | 4.3 |
| July | 14 | 10.9 | 2.9 | 0.1 | 60 | 285 | W | 3.9 |
| Aug. | 14 | 11.2 | 2.9 | 0.8 | 61 | 245 | W | 3.8 |
| Sept. | 16 | 11.1 | 1.1 | 1.7 | 64 | 178 | W | 3.9 |
| Oct. | 15 | 10.8 | 0.2 | 3.0 | 69 | 108 | WSW | 4.3 |
| Nov. | 16 | 11.3 | 0.1 | 4.0 | 80 | 47 | SE | 4.0 |
| Dec. | 17 | 11.6 | * | 4.9 | 79 | 29 | SE | 4.1 |
| Annual | 163 | 114 | 10 | 34 | 66 | 1928 | WSW | 4.0 |

* Indicates a frequency >0 but <0.05.

TABLE XXIII

CLIMATIC TABLE FOR TYLSTRUP
Latitude 57°11′N, longitude 9°57′E, elevation 13 m

| Month | Mean sta. press. (mbar) | Mean daily temp. (°C) | Mean daily temp. range (°C) | Temp. extremes (°C) | | Mean press. vapor (mbar) | Mean precip. (mm) |
|---|---|---|---|---|---|---|---|
| | | | | highest | lowest | | |
| Jan. | 1010.0 | − 0.5 | 5.1 | 10.6 | −27.2 | 5.24 | 48.5 |
| Feb. | 1010.7 | − 0.9 | 6.1 | 13.4 | −23.5 | 4.96 | 34.1 |
| Mar. | 1013.2 | 1.1 | 7.1 | 17.5 | −21.2 | 5.56 | 26.7 |
| Apr. | 1010.8 | 5.8 | 8.9 | 23.1 | − 9.6 | 7.28 | 38.1 |
| May | 1013.6 | 10.8 | 11.1 | 31.1 | − 5.6 | 9.32 | 34.2 |
| June | 1011.3 | 14.2 | 10.8 | 31.9 | 1.0 | 11.96 | 51.1 |
| July | 1009.6 | 16.4 | 10.6 | 34.0 | 2.1 | 14.36 | 76.3 |
| Aug. | 1010.0 | 15.9 | 9.9 | 33.6 | 1.0 | 14.44 | 72.1 |
| Sept. | 1011.2 | 12.6 | 8.9 | 27.8 | − 1.3 | 12.12 | 73.3 |
| Oct. | 1011.1 | 8.0 | 7.2 | 19.8 | − 7.0 | 9.24 | 70.0 |
| Nov. | 1010.1 | 4.4 | 5.4 | 14.5 | − 7.8 | 7.52 | 64.6 |
| Dec. | 1009.5 | 1.8 | 4.7 | 12.7 | −18.2 | 6.32 | 47.7 |
| Annual | 1010.9 | 7.5 | 8.0 | 34.0 | −27.2 | 8.60 | 636.7 |

| Month | Max. precip., 24 h | Number of days | | | Mean cloudi- ness (%) | Mean sun- shine (h) | Mean wind speed (m/sec) |
|---|---|---|---|---|---|---|---|
| | | precip. ⩾1.0 mm | thunder- storm | fog | | | |
| Jan. | 25.5 | 14.5 | 0.13 | 8.6 | 77 | 40.2 | 4.4 |
| Feb. | 43.9 | 11.3 | 0 | 6.8 | 72 | 71.0 | 4.4 |
| Mar. | 17.4 | 9.9 | 0.03 | 7.1 | 65 | 134.0 | 4.4 |
| Apr. | 23.1 | 11.4 | 0.33 | 3.5 | 61 | 189.7 | 4.6 |
| May | 19.9 | 9.5 | 1.50 | 2.8 | 55 | 265.9 | 4.2 |
| June | 39.8 | 11.1 | 2.53 | 1.4 | 58 | 272.7 | 4.4 |
| July | 61.8 | 12.7 | 4.03 | 1.6 | 59 | 264.0 | 4.1 |
| Aug. | 56.3 | 13.7 | 3.67 | 3.7 | 60 | 229.1 | 3.9 |
| Sept. | 60.9 | 14.7 | 1.77 | 4.2 | 60 | 165.2 | 4.1 |
| Oct. | 26.7 | 15.0 | 0.60 | 5.4 | 69 | 101.5 | 4.1 |
| Nov. | 22.8 | 16.8 | 0.23 | 6.5 | 77 | 46.3 | 4.1 |
| Dec. | 17.3 | 16.0 | 0.10 | 8.5 | 77 | 27.7 | 4.2 |
| Annual | 61.8 | 157 | 15 | 60 | 66 | 1807 | 4.2 |

TABLE XXIV

CLIMATIC TABLE FOR VIBORG (FOLKEKUREN HALD)
Latitude 56°25′N, longitude 9°21′E, elevation 25 m

| Month | Mean sta. press. (mbar) | Mean daily temp. (°C) | Mean daily temp. range (°C) | Temp. extremes (°C) | | Mean vapor press. (mbar) | Mean precip. (mm) |
|---|---|---|---|---|---|---|---|
| | | | | highest | lowest | | |
| Jan. | 1008.8 | − 0.3 | 5.6 | 10.0 | −30.3 | 5.32 | 62.4 |
| Feb. | 1009.5 | − 0.6 | 6.7 | 14.5 | −25.4 | 5.08 | 41.3 |
| Mar. | 1011.9 | 1.5 | 8.0 | 18.2 | −23.9 | 5.64 | 34.8 |
| Apr. | 1009.7 | 6.1 | 9.5 | 23.5 | −10.4 | 7.08 | 42.9 |
| May | 1009.6 | 11.3 | 11.8 | 30.9 | − 6.5 | 8.96 | 37.6 |
| June | 1010.7 | 14.5 | 11.1 | 33.7 | − 2.1 | 11.40 | 48.0 |
| July | 1008.8 | 16.3 | 10.1 | 33.9 | 2.0 | 13.72 | 86.8 |
| Aug. | 1009.2 | 15.8 | 9.9 | 33.0 | 1.6 | 14.00 | 90.6 |
| Sept. | 1010.3 | 12.5 | 9.1 | 27.1 | − 3.1 | 11.88 | 79.8 |
| Oct. | 1009.9 | 8.1 | 7.6 | 20.0 | − 9.7 | 9.20 | 77.8 |
| Nov. | 1008.8 | 4.7 | 5.3 | 15.4 | − 9.3 | 7.60 | 65.0 |
| Dec. | 1008.4 | 2.1 | 4.8 | 12.5 | −17.6 | 6.48 | 55.6 |
| Annual | 1009.9 | 7.7 | 8.3 | 33.9 | −30.3 | 9.84 | 722.6 |

| Month | Max. precip., 24 h | Number of days | | | Mean cloudi- ness (%) | Mean sun- shine (h) |
|---|---|---|---|---|---|---|
| | | precip. ⩾0.1 mm | thunder- storm | fog | | |
| Jan. | 23.8 | 16.9 | 0.10 | 7.5 | 76 | 41.1 |
| Feb. | 20.2 | 14.0 | 0.07 | 5.7 | 72 | 67.3 |
| Mar. | 22.2 | 12.0 | 0.13 | 6.6 | 63 | 133.8 |
| Apr. | 36.2 | 12.6 | 0.40 | 3.2 | 61 | 180.2 |
| May | 31.9 | 10.6 | 1.67 | 1.8 | 54 | 248.0 |
| June | 28.2 | 12.3 | 2.10 | 1.0 | 59 | 247.4 |
| July | 61.6 | 14.8 | 3.67 | 1.7 | 61 | 237.6 |
| Aug. | 66.5 | 15.0 | 3.50 | 3.4 | 61 | 210.5 |
| Sept. | 37.3 | 15.5 | 1.93 | 4.6 | 59 | 158.7 |
| Oct. | 39.1 | 16.4 | 0.43 | 5.5 | 69 | 96.5 |
| Nov. | 20.4 | 18.2 | 0.10 | 6.6 | 79 | 40.7 |
| Dec. | 27.0 | 18.4 | 0.07 | 7.1 | 79 | 29.8 |
| Annual | 66.5 | 177 | 14 | 55 | 66 | 1692 |

TABLE XXV

CLIMATIC TABLE FOR COPENHAGEN (AGRICULTURAL UNIVERSITY)
Latitude 55°41′N, longitude 12°33′E, elevation 9 m

| Month | Mean sta. press. (mbar) | Mean daily temp. (°C) | Mean daily temp. range (°C) | Temp. extremes (°C) | | Mean vapor press. (mbar) | Mean precip. (mm) |
|---|---|---|---|---|---|---|---|
| | | | | highest | lowest | | |
| Jan. | 1013.2 | 0.1 | 4.0 | 9.9 | −24.2 | 5.36 | 48.8 |
| Feb. | 1013.5 | − 0.1 | 4.6 | 14.0 | −19.6 | 5.16 | 38.5 |
| Mar. | 1015.7 | 1.9 | 5.9 | 18.5 | −17.8 | 5.84 | 32.2 |
| Apr. | 1013.6 | 6.6 | 7.3 | 22.1 | − 8.8 | 7.40 | 38.1 |
| May | 1016.0 | 11.8 | 8.6 | 27.7 | − 1.5 | 9.40 | 42.4 |
| June | 1014.3 | 15.6 | 8.6 | 32.7 | 3.0 | 12.04 | 47.1 |
| July | 1012.5 | 17.8 | 8.2 | 30.7 | 7.5 | 14.48 | 71.0 |
| Aug. | 1012.9 | 17.3 | 7.8 | 30.5 | 5.6 | 14.60 | 66.1 |
| Sept. | 1014.5 | 13.9 | 7.0 | 26.7 | 0.9 | 12.40 | 61.9 |
| Oct. | 1014.4 | 9.3 | 5.4 | 19.9 | − 4.0 | 9.72 | 58.6 |
| Nov. | 1013.6 | 5.4 | 3.9 | 14.4 | − 6.8 | 7.80 | 47.8 |
| Dec. | 1012.9 | 2.5 | 3.5 | 12.3 | −11.4 | 6.44 | 49.3 |
| Annual | 1013.9 | 8.5 | 6.2 | 32.7 | −24.2 | 8.76 | 601.7 |

| Month | Max. precip., 24 h | Number of days | | | Mean cloudi-ness (%) | Mean sun-shine (h) | Mean wind speed (m/sec) |
|---|---|---|---|---|---|---|---|
| | | precip. ⩾0.1 mm | thunder-storm | fog | | | |
| Jan. | 24.0 | 16.7 | 0.03 | 5.7 | 73 | 35.9 | 2.7 |
| Feb. | 24.0 | 13.5 | 0.03 | 4.1 | 71 | 55.5 | 2.7 |
| Mar. | 25.2 | 11.8 | 0.03 | 4.5 | 59 | 118.3 | 2.5 |
| Apr. | 38.0 | 13.0 | 0.27 | 1.2 | 55 | 161.2 | 2.3 |
| May | 38.2 | 10.9 | 0.83 | 0.5 | 47 | 244.7 | 2.3 |
| June | 37.0 | 12.7 | 1.33 | 0.1 | 50 | 245.3 | 2.2 |
| July | 76.8 | 14.2 | 2.00 | 0.3 | 51 | 239.2 | 2.0 |
| Aug. | 41.0 | 14.3 | 2.10 | 0.9 | 52 | 204.5 | 2.0 |
| Sept. | 50.2 | 15.2 | 0.77 | 2.0 | 52 | 156.6 | 2.0 |
| Oct. | 47.0 | 15.7 | 0.27 | 3.6 | 64 | 86.5 | 2.3 |
| Nov. | 34.2 | 16.3 | 0.03 | 3.1 | 75 | 34.0 | 2.3 |
| Dec. | 30.0 | 17.2 | 0.00 | 5.0 | 79 | 18.8 | 2.3 |
| Annual | 76.8 | 172 | 8 | 31.0 | 61 | 1600 | 2.3 |

TABLE XXVI

CLIMATIC TABLE FOR FANÖ
Latitude 55°27′N, longitude 8°24′E, elevation 3 m

| Month | Mean sta. press. (mbar) | Mean daily temp. (°C) | Mean daily temp. range (°C) | Temp. extremes (°C) | | Mean vapor press. (mbar) | Mean precip. (mm) |
|-------|------|------|------|---------|--------|------|------|
| | | | | highest | lowest | | |
| Jan. | 1011.9 | 0.6 | 4.4 | 8.5 | −22.0 | 5.68 | 60.3 |
| Feb. | 1012.5 | 0.2 | 5.1 | 9.6 | −21.2 | 5.52 | 44.7 |
| Mar. | 1014.5 | 2.3 | 6.0 | 18.0 | −18.0 | 6.28 | 37.7 |
| Apr. | 1012.8 | 6.5 | 7.4 | 25.0 | − 5.2 | 8.04 | 38.5 |
| May | 1014.9 | 11.3 | 8.7 | 30.0 | − 2.9 | 10.44 | 43.4 |
| June | 1013.2 | 14.5 | 8.2 | 33.0 | 1.4 | 12.56 | 42.8 |
| July | 1012.0 | 16.7 | 7.6 | 35.0 | 5.1 | 14.84 | 73.9 |
| Aug. | 1012.1 | 16.6 | 7.7 | 32.5 | 3.5 | 14.85 | 85.2 |
| Sept. | 1013.3 | 13.9 | 7.1 | 29.0 | 1.1 | 13.04 | 87.4 |
| Oct. | 1012.9 | 9.6 | 5.8 | 19.6 | − 6.0 | 10.28 | 82.3 |
| Nov. | 1011.7 | 5.6 | 4.7 | 14.0 | − 6.7 | 8.20 | 67.5 |
| Dec. | 1011.5 | 2.9 | 4.1 | 14.5 | −13.2 | 7.08 | 63.6 |
| Annual | 1012.8 | 8.4 | 6.4 | 35.0 | −22.0 | 9.28 | 727.3 |

| Month | Max. precip., 24 h | Number of days | | | Mean cloudiness (%) | Mean wind speed (m/sec) |
|-------|------|--------------|-------------|------|------|------|
| | | precip. ≥0.1 mm | thunderstorm | fog | | |
| Jan. | 21.3 | 17.3 | 0.03 | 4.8 | 70 | 4.0 |
| Feb. | 23.4 | 13.2 | 0.07 | 4.3 | 66 | 4.0 |
| Mar. | 26.3 | 11.8 | 0.07 | 5.2 | 59 | 3.8 |
| Apr. | 21.7 | 12.4 | 0.40 | 2.4 | 56 | 4.0 |
| May | 27.1 | 11.4 | 1.30 | 1.3 | 52 | 3.8 |
| June | 33.4 | 11.8 | 1.53 | 0.2 | 54 | 3.7 |
| July | 94.4 | 14.3 | 2.37 | 0.4 | 58 | 3.8 |
| Aug. | 43.4 | 15.0 | 3.00 | 0.7 | 57 | 3.4 |
| Sept. | 44.8 | 16.2 | 1.93 | 1.4 | 57 | 3.7 |
| Oct. | 50.7 | 17.5 | 1.23 | 2.7 | 65 | 3.8 |
| Nov. | 20.7 | 18.9 | 0.50 | 3.7 | 76 | 3.8 |
| Dec. | 25.0 | 19.0 | 0.40 | 4.6 | 76 | 3.8 |
| Annual | 94.4 | 178 | 13 | 32 | 62 | 3.8 |

# Chapter 3

# The Climate of the British Isles

GORDON MANLEY

## Historical introduction

Lying throughout within the domain of the westerlies, beneath the sinuosities of that upper wave-pattern that girdles the Northern Hemisphere, the moist and temperate climate of the British Isles has been described by many authors. Indeed, no climate in the world has been more provocative of comment. By some, it has been described as the best in the world. For others, the astonishing variety of small-scale effects that it can and does repeatedly provide, induces resigned tolerance, irritation, or downright dislike, accompanied by declarations that the British Isles have no climate, but merely weather. This paradoxical British climate can now be given statistical expression (BILHAM, 1938; METEOROLOGICAL OFFICE, 1952).

The earliest known comments are those of Tacitus, who, very appropriately for a Roman, recognised the dampness of a wild undrained island, the mild winter, and the fact that summers were too cool for satisfactory cultivation of the vine. At the same time he did not overlook the fact that crops were quick to grow, although slow to ripen.

Bede in his *Ecclesiastical History*, written in 731, emphasised the advantage of the mild winters of Ireland. Without doubt this was the most noteworthy characteristic of the lowlands throughout Britain that would appeal to those invading Northern peoples under whose sway these European islands gradually fell. In Bede's time the Anglian and Saxon folk-memory of the greater cold of winter in the lands from which they had so recently migrated would be vivid; and the like would be true of the Danish and Norwegian invaders that were to follow. To the vast majority of Frenchmen it was forever illogical to embark on the conquest of lands across the Channel that from experience they must suppose to be even more cloudy, damp and windy than those green shores of Normandy that they already knew. In such fundamentals, we may seek the reasons for the dominance throughout the British Isles of a northern rather than a Latin tongue.

It is fitting to recall that one of the most recent French geographical texts stresses that the affinities of Britain lie far more with Scandinavia than with her neighbour to the south. Yet the author continues to marvel at the magnificent wheat crop that is grown under what he describes as the sad grey skies of Yorkshire.

It has been for the northern peoples to appreciate the advantage of the milder winter and the longer growing season, accompanying the west wind and the variable sky to whose lively behaviour they were already accustomed. For the Latins, little benefit beyond that of a dubious liberty was to be gained by moving northwestward towards the Icelandic low.

In medieval times, the early 13th century appears to have brought a minor climatic optimum (LAMB, 1965). Crusaders brought back vegetables and fruits, monastic gardens flourished and vineyards were found widely scattered through the drier and sunnier southeast. But the vicissitudes of the weather were always present in men's minds; and it is noteworthy that the earliest meteorological record we know of north of the Alps, was kept by an Englishman between 1337 and 1344.

Throughout the centuries men have written for and against the British climate. In the 16th century Venetian ambassadors and others commented that it was healthy, and favourable to bodily activity; that it produced abundance of grass; that in a stock-rearing land the Englishman liked his meat and was given to good feeding. Of Scotland and Ireland we hear less, but the more limited production of grain crops in the cooler summers of the north and west was recognised.

With the gradual improvement of the means and opportunity to journey abroad in the 17th century, travellers became less immediately concerned with the practical objects and needs, less confined to the military-minded and the merchants. Thoughtful and observant men began to embark on philosophical discussion, mingled with subjective opinions that often owed much to prejudice, jealousy or uncritical enthusiasm. Even a Frenchman crossing Kent on a fine June morning on his way to London might be led to unwonted praise of what he saw. The age of science, of instrumental measurement and objective comparison was dawning; by 1660, men were beginning to count and measure those qualities of the atmosphere that, when assembled, enabled us to make a truer assessment of climate.

In the same century, with the rapid increase in knowledge of the world's vegetation and its possible uses, experiments in the cultivation of exotics were begun on a ever-increasing scale. The monastic achievements of the 13th century were revived and extended. Surprising efforts were made to introduce many trees, foodstuffs and fruits to southern England in particular. Potatoes and pine-trees, tobacco, mulberries and oranges, the Mediterranean evergreens such as the laurel, were followed in succeeding centuries by a variety of trees, shrubs, flowers and crops. Eucalyptus and Araucaria, Lonicera and Fuchsia, tulip and Chrysanthemum, Lucerne and Broccoli can all thrive. To this day, there are a number of plants whose cultivation lies just within the limits of possibility, given care; outdoor tomatoes, maize, soya beans for example. And Frenchmen who hold traditional views will find it noteworthy that for a number of years a light white wine has been regularly produced from a small open-air vineyard on a favourable slope in Hampshire (LAMB, 1964).

Experience indeed shows that the climate of Britain is marginal in very many respects (MANLEY, 1952). So much so, that the diversity of the land surface arising from small-scale local relief, from soil, from proximity to water bodies, and from the many local modifications that can be ascribed to the work of man, plays an unusually large part.

This is a result of the range of mean temperature at sea level. Over a large part of the lowlands the mean for the coldest month lies sufficiently above the freezing point to ensure that the mean minimum temperature at night for at least three months of the year lies very close to 0°C. Hence quite small differences of aspect, relief, soil and exposure make for wide variations in the number of nights on which the minimum temperature will fall effectively below freezing point (MANLEY, 1966; Table XI).

Secondly, the mean for the warmest months lies just below that which is found most effective for the ripening of a number of crops; but with a little care, such as the provision of local shelter, very rewarding results can be obtained.

Suffice it also to point out that as a result of the rather "flat" curve of the annual march of temperature, over the agricultural English Midlands, the length of the period during which the mean temperature exceeds 6°C is about eight months. But the slow rise of mean temperature through the spring months implies that over open country there is a considerable risk of frost, given a clear radiation night, until far on into May. Moreover the incidence of frost varies so greatly in different locations that it can readily be shown that, over the English Midlands, the average length of the frost-free period ranges from about 2½ months only, on the sandy heathlands, to six months on favourable slopes and to upward of seven months within the built-up areas of the larger cities. "Frost" here refers to "air frost", that is, a screen minimum temperature 0°C or below.

Likewise it is demonstrable that the rate of change with altitude in the length and effective warmth of the growing season is probably more rapid than in any other country in the world. One result of this can be seen in the remarkably low altitude of the treeline, between 600 and 700 m (MANLEY, 1945).

Again, so great are the variations in the amount and frequency of rainfall, that the ratio between the number of sunny hours and the number of rainy hours during the daytime is eight times greater on the Sussex coast than that which is measured among the mountains of the west Highlands in Scotland.

Contrasts such as these go far to explain why, within what appears from the textbooks to be the proverbial cool temperate maritime climate of widespread uniformity and limited range, the actual experience, and in particular the sensations and impressions of the visitor can be so very different from what he might have been led to expect.

Lastly, as a result of the peculiarly close relationship between evaporation and rainfall, over the greater part of agricultural England, artificial irrigation can not only be shown to be desirable but is frequently used. Such a fact comes as a great surprise to many who have been influenced by the conventional attitudes that have long been fashionable (PENMAN, 1948).

That there are biological advantages in such diversity of climatic stress within a small area can be recognised. It will be the purpose of this chapter to demonstrate the reasons for the remarkably varied characteristics that the climate of Britain presents; and to provide quantitative assessments of the range of behaviour and of possibilities both in space and time.

**The observational material**

In a country in which the vicissitudes of weather have always played so large a part in the expenditure of energy, in welfare and in economic gain it is appropriate that the earliest meteorological record that we know of, in Europe, was that kept by an Englishman from 1337–1344, giving daily notes of weather and wind direction. Scattered records of floods and droughts are many; but records of the freezing of rivers and lakes are relatively few, on account of the milder winters compared with those on the continent. All records of this type, moreover, are distinctly unreliable and must be treated with care.

There is evidence that in common with other European countries, systematic daily records of some kind were being made in the 16th century, but only fragmentary notes remain. Instrumental measurements of temperature and pressure, rainfall, humidity and wind strength began in the 17th century. Soon after the foundation of the Royal Society in 1660, the need for the keeping of a "history of the weather" was recognised. Robert Hooke, the "Curator of Experiments", can be regarded as the founder of English scientific meteorology. In 1663 he was commissioned to draw up a plan for the recording of daily observations. In 1664 he invented a simple swinging-plate anemometer, and in 1665 he constructed a thermometer of a type which for long served as a standard. Daily thermometer readings were kept by John Locke, the philosopher, at Oxford in 1666; these are the earliest series known. For the London area, daily observations of the weather can be assembled in a continuous sequence from 1668 onward to the present day, including for example days of rain and of snow (MANLEY, 1961). Wind direction and estimates of force begin in 1673.

Early instrumental observations suffer from many imperfections and are very difficult to interpret. Not only were instruments relatively crude and scales unstandardised; exposures were highly unsatisfactory and techniques of observation varied so that the risks of error are great. About 100 years elapsed before the improved Fahrenheit thermometer, out-of-doors, came into general use. Early barometers tended to read too low on account of the imperfection of the Torricellian vacuum. From about 1755 onward instruments and techniques became more trustworthy, and fairly reliable reductions can be made from the growing number of fixed-hour out-of-door observations.

The longest series for pressure and temperature in the same location throughout, comes from the Radcliffe Observatory at Oxford, beginning in 1815. Using the method of overlapping series from neighbouring stations, it is possible to take London and Edinburgh observations of temperature back into the 18th century (BUCHAN, 1893; MOSSMAN, 1897, 1900; BRUNT, 1925; BIRKELAND, 1957). By integrating the results of series of daily observations kept in several parts of the country, and regarding the mean value derived from them as likely to represent the central parts of England, the present writer has essayed the assemblage of monthly means of temperature since 1670 (MANLEY, 1964). The results derived from this "central England" series provide a very good indication of the overall fluctuations of temperature for England as a whole (Fig.1, see also Table X). Temperature records for various parts of Scotland are available from the mid-18th century onward and for Edinburgh itself since 1764. For the far north however the records are more interrupted, and for the Shetlands they are not available before 1858.

Scattered rainfall measurements are known from 1677 onward and estimates of the monthly fall for England and Wales as a whole, as a percentage of normal, are available since 1727 (GLASSPOOLE, 1928). These can be compared with Dutch records that begin in 1715. For individual localities records are shorter; they have been compiled for Edinburgh from 1785. For the London area scattered records from 1695 onward have been assembled (BRAZEELL, 1968).

In Ireland, meteorological observations have been maintained in Dublin since the early 18th century (DIXON, 1953), but there are some breaks and the record does not become continuous until the early 19th century; soon afterwards, Armagh becomes available from 1833 onward.

It is characteristic of the British Isles that almost all these earlier observations were kept by amateurs. Official meteorology can only be said to have begun very tardily, with the establishment of the observations at Greenwich Observatory late in 1840; this was largely

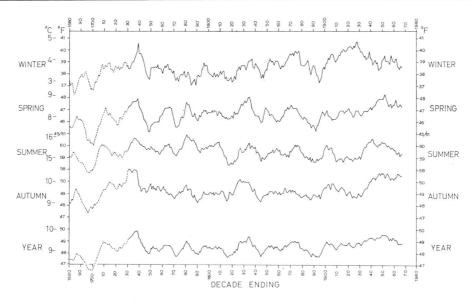

Fig.1. Decadal running averages of seasonal and annual mean temperatures in central England for the years 1670–1968.

in order to provide basic data in relation to public health and welfare. The founding of the Meteorological Office in 1855 led to the setting up of a number of "telegraphic stations", while climatological stations, still very often kept by amateurs, came under the aegis of the Royal Meteorological Society, founded in 1850, and the Scottish Meteorological Society, founded in 1855. Since the adoption of standardised exposure and techniques of observation, in 1881, the number of reliable sources of data upon which our climatic statistics must be founded has become large. In recent years, the data from upwards of 400 stations have been summarised in the *Monthly and Annual Weather Reports* of the Meteorological Office. The British Rainfall Organisation, founded in 1860, supervises the collection of data from over 5,000 rain gauges; originally a mainly amateur body, it is now associated with the Meteorological Office.

The Meteorological Office publishes numerous tables of averages. Among the latest are those for mean temperature, 1931–1960 (M.O. 635) at about 100 stations; of rainfall (M.O. 635), 1916–1950, at over 700 stations; and of sunshine duration (M.O.743), relative humidity, and other elements such as visibility and the incidence of fog. The important series of handbooks for seamen, *Weather in Home Fleet Waters* (M.O.446, published in 1940), and the volumes of the *Admiralty Pilot* (1957–1968), give much statistical information for coastal stations. The *Climatological Atlas of the British Isles* (M.O.488, 1952) makes an additional reference of the greatest value for those wanting more detail than this chapter can give. An extensive summary of official records, with detailed tables, is now in preparation and is to appear in the new edition of *Weather in Home Fleet Waters* as Volume 2, *The British Coasts* from the Meteorological Office; at the time of writing (1966) it is not yet ready for publication. The tables that it will contain will greatly supplement and extend those that are attached to this article. For the Irish Republic monthly reports of the Irish Meteorological Service begin in 1941; but for Northern Ireland, like Britain, those of the Meteorological Office should be consulted.

**Descriptive climatology**

Sixteenth-century descriptions of the characteristics of the counties of England commonly included some estimate of the climate, using adjectives such as "mild" or "piercing". This is a sufficient reminder that throughout the British Isles the sensations experienced by anyone out-of-doors depend much more on the kind of wind that is blowing rather than on the temperature. The cooling power of a strong wind at $+5°C$ is about equal to that

TABLE I

GEOGRAPHICAL LOCATIONS OF STATIONS USED IN
CLIMATOLOGICAL TABLES

| Station | Position | Altitude (m) |
|---|---|---|
| Lerwick | 60°08'N 01°11'W | 83 |
| Stornoway | 58°13'N 06°20'W | 3 |
| Aberdeen (Dyce) | 57°12'N 02°12'W | 58 |
| Eskdalemuir | 55°19'N 03°12'W | 242 |
| Manchester Airport | 53°21'N 02°16'W | 76 |
| Waddington | 53°10'N 00°31'W | 71 |
| Yarmouth (Gorleston) | 52°35'N 01°43'E | 2 |
| London (Kew Observatory) | 51°28'N 00°19'W | 5 |
| Plymouth (Mount Batten) | 50°21'N 04°07'W | 26 |
| Aldergrove (N. Ireland) | 54°39'N 06°13'W | 67 |
| Dublin (Ireland) | 53°22'N 06°21'W | 48 |
| Valentia (Ireland) | 51°56'N 10°15'W | 9 |
| *Additional stations:* | | |
| Braemar | 57°00'N 03°24'W | 340 |
| Renfrew | 55°52'N 04°24'W | 6 |
| Durham | 54°46'N 01°35'W | 102 |
| Cambridge | 52°12'N 00°08'E | 13 |
| Oxford | 51°46'N 01°16'W | 63 |
| Aberystwyth | 52°25'N 04°04'W | 4 |

of still air at −10°C; so that exposure to wind can make one locality feel very different from another nearby.

Comparison of climates by other than qualitative judgments was not practicable until enough observations had been gathered. The appearance of Père Cotte's treatise in France (1773), was followed by Kirwan's work in 1787, by Dalton's first attempts to compare the amount of rainfall received in different places, and by Howard's work on the climate of London (1833).

Later in the 19th century Buchan and Mill, in Scotland, made their outstanding contributions to climatology. From the Meteorological Office since 1920 the outstanding contributors have been Brooks, Glasspoole, Bilham and Lamb. In particular, BROOKS (1949) became renowned for his work on past changes of climate. More recently these have been extended by LAMB (1963, 1964, 1965), as well as by the present writer (MANLEY, 1939, 1964). Of the older descriptions, the essay on the *Climate of Britain* by MILL (1928) is essentially very simple, written for geographers and unadorned by any excess of meteorological comment, but it remains one of the best. Of the newer, more interpretative treatments, that by MILLER (1962) in his well-known *Climatology* is widely appreciated in educational centres.

The abundance of observational material with regard to two of the climatic elements to which the public is sensitive, namely duration of sunshine and amount, frequency and duration of rainfall, has given rise to many studies of interest. In recent years much attention has been given to the assessment and measurement of evaporation, developing from the extensive studies by PENMAN (1948). A summary by F. J. Holland in the annual volume *British Rainfall, 1961*, published from the METEOROLOGICAL OFFICE (1967) is very useful.

Viewed in the most general terms, the climate of the British Isles is fundamentally an expression of increasing proximity to that principal route of travel of active frontal depressions which is represented throughout the year on charts of mean surface pressure by the Icelandic low. Northwestward, there is always the trend towards more wind and cloud, increased rainfall, more frequent rain, diminished annual range of temperature; the cooler and more windy summer, the mild but very stormy winter; the greater predominance of air masses of maritime-polar origin, variously modified. But the uplands towards the north and west of the British Isles also play their part; rainfall diminishes rapidly to leeward.

In essence, the climate of the agricultural lowlands south and east of a line from Exeter to Middlesbrough, does not differ in any serious degree from that of the neighbouring countries of continental Europe. Throughout the growing season, Cornwall and Brittany, Sussex and Normandy, Lincolnshire and The Netherlands, Fife and Denmark are very similar. From the standpoint of quantity of rain measured, London is distinctly drier than Lille; Edinburgh is drier than Hamburg; even Dublin has less rain than Cologne, while Manchester has little more rainfall than Brussels.

In the characteristic short sharp spells of frosty weather, that occur when in winter pressure becomes high towards the Baltic, and the whole of western Europe comes within the dominance of the outflowing continental air from the interior, the severity of cold is never so great, nor so prolonged, in Britain, and still less so in Ireland. The open waters of the North Sea and the English Channel have an appreciable effect, to which that of the Irish Sea can be added. It is, therefore, quite rare to see the larger rivers and lakes frozen over, and very rarely, even in England, does the frost penetrate into the ground to a depth exceeding 50 cm. Accordingly, in a normal year the pastures remain green, and over much of the Midlands the grass begins to grow soon after the beginning of March.

In the spring, however, the risk of frost is as great, if not greater than on the continent, especially when clear calm nights succeed the characteristic springtime outbursts of direct arctic air from the north; at the season when the pack-ice limit lies not far north of Iceland, arctic outbursts can still reach the north coast of Scotland with much the same temperature as in January. In summer, the slightly lower temperatures in Britain imply that the evaporation-losses are not quite so rapid as they can become in France or Germany; hence, while lengthy dry spells can occur, throughout the agricultural lowlands, drought is rarely catastrophic.

But the incidence of dry and wet spells varies so widely from year to year, and their effects on farming can differ so much, depending on the season and on the weather just past, that an unusual alertness of mind and capacity for judgment, estimation and adaptation are called for. Moreover in a country where the evaporation-loss throughout the winter is too small for comfort, water can accumulate on the land and drain away very slowly indeed unless it is assisted. Hence it has for long paid the English farmer, tilling the lower ground throughout the southern and eastern lowlands, to drain his fields. In the narrow but agriculturally favoured coastal lowlands of eastern Scotland the same is true; and one of the major difficulties of Ireland arises from the fact that, while the rainfall is but little greater, evaporation is a little lower and the physical build of the country renders the broad central lowland difficult to drain.

Clearance of the forest followed by drainage of the land was for centuries the English objective, and the high standards and yields of English farming are the product of prolonged care in this respect. The rainfall–evaporation ratio, in these islands lying on the margin of the principal track of so many Atlantic depressions, and beset by many others diverted from time to time to follow the less common tracks, can be regarded as a fundamental variable upon which agricultural prosperity largely depends.

Lastly, while maps of the mean distribution of pressure remind us that the west-south-westerly flow between an Azores high and an Icelandic low is not only characteristic but dominant, migrant anticyclones take up their station and intensify around the British Isles with considerable, though very variable frequency. In particular, the "blocking anticyclone" interrupting the normal westerly flow may persist, occasionally for a month at a time, as far north as the Norwegian Sea (SUMNER, 1954). Such developments are most common in spring, but are known at all seasons. As a result, in August 1947 no rain whatever fell throughout a large part of the normally very wet west Highlands; in February 1932, April 1938 and September 1959, large areas towards the west remained dry throughout; in March 1938 persistence of an anticyclone over northeast France ensured record-breaking warmth and dryness in eastern England, exceptional warmth and very heavy rains in western Scotland. No part of the British Isles is wholly free from the effects of such persistent stagnation of the circulation off northwestern Europe; even the normally disturbed, stormy and cloudy Hebrides and northwest Ireland enjoy, occasionally, remarkable spells of dryness when blocking anticyclones become established over the northeastern Atlantic.

The winter continental anticyclone from time to time extends westward, or builds up a separate cell over Scandinavia; and, more rarely, as in January 1963, a vigorous anticyclone repeatedly spreads southward from Greenland. Should such developments occur and persist, a protracted spell of cold over the British Isles may occur; although, thanks to the encircling seas, it can never produce the degree of severity that occurs on the adjacent continent.

## The meteorological background

Dr. C. C. Wallén in his introductory chapter (pp.1–21) has discussed the principal characteristics of the atmospheric circulation as it affects western Europe. It will not be necessary here to do more than remind the reader of the outstanding features of the average surface pressure distribution; the Icelandic low, the Azores high, the prevailing trend of the isobars across the British Isles, the slight eastward rise of pressure towards the continent in winter. The mean wind direction therefore tends to be westerly in the summer, with a slightly greater southerly component giving a southwesterly resultant in the winter. The prevailing characteristics of the climate are very simply explained.

Such pressure distributions when mapped are merely the integration of a wide variety of actual conditions. It can be agreed that over a long period the centres of the majority of active low-pressure systems will pass eastward between Iceland and Scotland, and that a majority of the accompanying frontal systems crossing Britain will have become more or less occluded. But the speed with which depressions develop and decay, the routes that they follow, and the location of the areas in which they commonly become more or less stationary, varies very widely.

In Fig.2, illustrations are given of the extent to which the surface pressure distribution departed from normal in several outstanding months, of which two fell in the same year, 1947. This will serve to show the extent to which pressure distributions very different from the average are capable of persistence over considerable periods. It will thus be evident that those climatic characteristics that might be expected to prevail as a result of the average pressure distribution, presenting an Icelandic low with an Azores high and accompanying westerly winds are often very far from being fulfilled.

The existence of such marked deviations of the pressure distribution from the normal, serves to show how widely the characteristics of one month can differ from the average. Almost every district in Britain, even among the mountains, has at one time or another experienced a whole month without any measurable rain. In the brilliant anticyclonic June of 1925 nearly 80% of the possible duration of sunshine was recorded in Cornwall, a figure comparable with the expectation that we find on the drier Mediterranean islands. By contrast, in August 1912 the persistence of low stratus cloud near the North Sea coast was so great that at Durham less than 14% of the possible duration of sunshine was recorded, even in a summer month.

The persistence, or otherwise, of unusual pressure distribution around Britain depends greatly on the wave pattern assumed by the upper westerly flow, characteristically shown by the 500 mbar contours. We have now sufficient material to recognise that in general, a warm ridge extends northeastward from the region of the Azores high, with a cold trough somewhere in mid-Atlantic. But if this pattern develops further, with a decreased wave length accompanied by increased meridional flow, the warm ridge extending over Britain and on towards Scandinavia may be enlarged, and develop into the "blocking

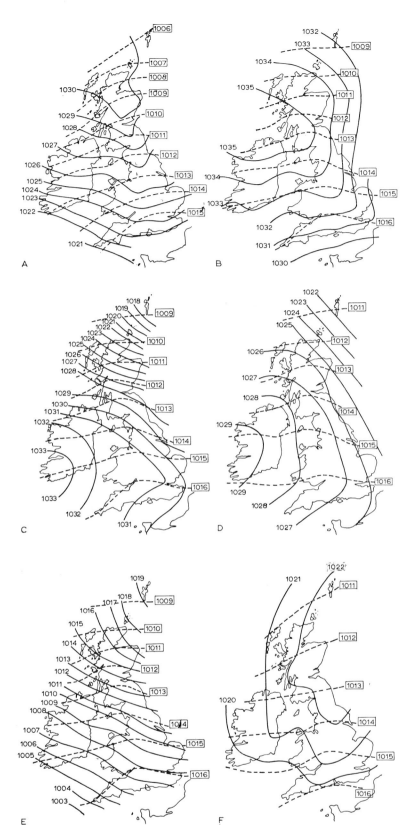

Fig.2. Wind directions and mean pressure for exceptional months. A. January 1963, extremely cold; B. February 1932, and C. February 1934, anticyclonic, very dry; D. April 1938, extremely dry; E. February 1947, and F. August 1947, persistent blocking anticyclones, very cold and very hot respectively. The normal distribution of pressure is shown by dashed lines.

(Reproduced from the *Monthly Weather Report* of the Meteorological Office, by permission of the Controller, Her Britannic Majesty's Stationery Office.)

anticyclone" that, we now know, is from time to time a necessary feature of the westerly flow through the temperate zone.

The position of the mid-Atlantic trough, the western European ridge and the accompanying troughs and ridges over America and Russia determine to a large extent the frequency of development and route of travel of the frontal systems that affect us. To this must be added the as yet unexplained deviations of the location of the principal centre of high pressure over the Arctic. For example, in the severe winters of 1891 and 1917 pressure remained very high for long periods over the Baltic region; in 1947 and 1963 the persistence of high pressure over the Iceland–Greenland region became noteworthy. It has been suggested that one of the "preferred patterns" into which the pressure distribution around Britain now and then tends to fall is characterised by very persistent low pressure off the coasts of France and Portugal, associated with a "cold pool" at higher levels; at the same time, the Azores anticyclone and its warm ridge is displaced southwestward.

There is no doubt that the location of upper troughs and ridges over the Atlantic sector can become displaced for considerable periods, and indeed, as LAMB (1963) has shown, the known vicissitudes of the climate of Europe over several centuries can perhaps be related to such displacements tending to prevail on some larger time-scale.

What we have to explain, in regard to the climate of these relatively small islands, is the remarkably wide and effective range of possibilities, and the extent of the differences that can prevail in unusual seasons. For example, the year 1937 in northwestern Scotland was phenomenally dry, whereas in parts of the south of England it was one of the wettest years on record. The winter of 1963 was not only the most severe for over 200 years; it was very much more severe in the south than in northern Scotland; whereas exactly the opposite can be said of 1959.

### Characteristics of the prevailing air-masses

To an overwhelming extent the climate of Britain is the product of the predominance of the mobile air from the Atlantic, with widely varying results. It is particularly in Britain that air-masses must be treated on their merits, and not categorised too closely. For it will be evident that, pressure distribution and the frequency of depressions being what they are, with the greater number becoming occluded by the time they have crossed the country, the predominant air-mass must be described as transitional maritime-polar, capable of rapid modification if the vigour of the circulation is suddenly checked (BELASCO, 1952).

Air derived from the broad stretches of the cool northwestern Atlantic, approaches the British Isles across the warmer waters lying towards the eastern side of the ocean. Therefore it is always liable to develop some tendency to instability; but this becomes very much more marked in autumn, winter, and early spring. Plotting of the trajectories however shows to how great an extent such air may or may not be further modified as it approaches. Much of the air that enters Britain from the southwest can be called "returning maritime-polar". Having travelled far down into the central Atlantic with its warmer surface waters, it then returns over rather cooler water towards Britain so that in the lower layers it becomes a little more stable; at the same time, very extensive stratus cloud is characteristic of such air. Especially in the cooler months this stratus persists over the land; in the summer, the cloud sheet may break up, as the air moves into the Midlands.

Much of the maritime-polar air approaches the west of Scotland over a shorter ocean track, and hence at all seasons tends to be more unstable than further south. This goes far to explain the greater frequency with which cloud and passing showers beset the Hebrides and the western Highlands, by comparison with, say, Cornwall. One of the most fundamental characteristics of the climate of Britain is the greater frequency of rain towards the northwest. London has about 167 days a year with measurable rain (0.2 mm or more); at Falmouth in Cornwall and also at Cardiff and Glasgow this figure passes 200, and Stornoway in the Outer Hebrides reports 263 (see Table XIV).

The effects of the uplands and mountains lying athwart the flow of air from the south-southwest are, however, considerable; Liverpool to leeward of Wales, Edinburgh and Inverness to leeward of the Scottish uplands record fewer days of rain than would be the case if the mountain barriers did not exist. Dublin is similarly favoured.

Maritime-polar air coming by different tracks, with varying amounts of moisture and cloud in it, becomes further modified as it moves over the British Isles. If it stagnates overnight, a very common result in the autumn and early winter will be morning fog. In those broad valleys that are occupied by the greater cities, such fog acquires the unpleasant characteristics and the greater density that result from the presence of smoke held down beneath an inversion 300 m or so above the surface. In the country the sun will break through later in the morning; but in and around the city, the fog may persist all day and, as the depression fills up and the air remains quiet on the margin of a continental anticyclone, it may last into a second or third day. In December 1962, in such a situation, fog developed inland and persisted in the London Basin for the greater part of five days; 80 km away on the south coast, there was continuous sunshine beneath a cloudless sky. This exceptional spell brought the result that December 1962 on the south coast ranked as the most sunny midwinter month that has been recorded for upwards of seventy years.

Should this type of air approach from the southwest, bringing with it extensive low stratus cloud, the effects of the mountains and hills become noteworthy. The cloud sheet breaks up to leeward and, in winter, pleasantly mild sunny days occur to the east and north of the uplands. Föhn effects are quite well developed on occasion and will be mentioned again.

Within the circulation of large and slow-moving depressions centred to the north, minor fronts develop from time to time between long-track and short-track maritime-polar air, and may give protracted rainfall here and there, especially where such fronts may be held up for a time against the uplands, in place of the expected passing showers. The forecaster in the British Isles must always be on the alert for such localised developments over the Irish Sea and elsewhere, for example in the Bristol Channel, and for the modifying effects of the relatively low but important ranges of mountains. Convergence ahead of a stationary front, for example, has been known to produce remarkably heavy localised snowfalls in winter on the west coast, between a mild southerly airstream and the much colder continental air arriving from the southeast. Such air, if it reaches England in winter by the short sea route, undergoes relatively little modification.

But from the climatic standpoint, the principal result of the predominance of transitional maritime-polar air that has recently undergone a long Atlantic travel lies in the very rapid change of effective climate with altitude. Such air streams can be excessively unstable when, in winter and spring, they arrive directly from the Arctic; but as we have seen, in the summer months they are not more than conditionally unstable as a rule, and if they are "returning" from the southwest after a passage over the warmer waters further south, they may be moderately stable but with extensive stratus cloud. In every case however, the cloudiness will increase upon any uplands that they must surmount.

We therefore find that the increased cloud, frequency of precipitation, stronger wind, relatively low maximum temperatures by day, diminished evaporation, and deficiency of bright sunshine are all very conspicuous features of the uplands, especially as most of the higher ranges lie towards the west.

Moreover, in this type of air the average temperature of the lowlands, summer and winter, is such that the fall of temperature with altitude has most conspicuous effects upon the vegetation. Based on records from mountain summits, the lapse rate averages nearly 0.7°C/100 m in April–May, falling below 0.6°C in November. Hence the mean temperature of the winter months, about 4° at sea-level, falls to zero at about 600 m. At this relatively low altitude therefore, snow will occur on many occasions, and may cover the ground when nothing but rain is to be seen at sea-level. Hence the rate of increase in the frequency of snow and of snow-cover with altitude is unusually rapid.

Approximations can be made: inland, in the central lowlands, it appears that 2–3% only of the total precipitation falls as snow (including sleet, that is melting as it falls) whereas at 600 m, the proportion probably approaches 15% and in eastern Scotland may approach 20%; but again, a considerable proportion of this is likely to melt as it falls. Snow, including sleet, may be observed to be falling on an average of about 20 days yearly in the northern English lowlands; this is likely to rise to 60–70 days at 600 m. Inland in the lowlands, it will be seen to lie on perhaps 10 mornings a year; this figure will rise to about 70 at 600 m in northern England, and to 100 or more in northeastern Scotland (see Tables XVI, XVII).

In summer, altitude ensures that the mean July temperature falls to about 10.5°C at 700 m and, on the prevailingly windy uplands, this probably represents the limit to which trees might grow. There is evidence for this statement in the number of dead trees to be found towards that altitude; and the upland plantings of the Forestry Commission for the most part are not carried above 500 m.

But more significant from the human standpoint is the rapid rate at which the length, as well as the accumulated warmth, of the growing season decreases. If we adopt the familiar 6°C as the threshhold above which growth of the most characteristic vegetation will be maintained, the "growing season" in the north Midlands and northern England or southern Scotland can be said to last from about mid-March to early November, almost eight months. In the extreme southwest, beside the sea, growth is rarely entirely checked; at most, for a few days only in a cold spell. Even in the inland parts of Devonshire the length of the effective growing season exceeds 9 months (Fig.3).

But at 600 m in the Pennines and southern Scotland, it diminishes to scarcely 5 months (mid-May–mid-October). The rate of decrease is about 14 days for each 100 m, and this is more rapid than in any other temperate land for which we have data. Moreover, if we take out the "accumulated temperature" during this growing season as a rough, but useful means of comparison, at 600 m it is only about 4/10 of that prevailing at inland locations near sea-level.

Hence the highest altitude at which any land reclaimed for cultivation can be found in Britain lies almost exactly at 600 m in northern England (in extreme upper Teesdale). In

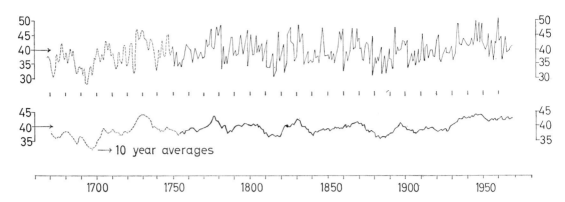

Fig.3. Central and northern England: to illustrate the incidence of favourable seasons. Variations of the accumulated temperature during the growing season for the years 1671–1968. Units are in month-degrees above 5.5°C at the prevailing arable limit (180–220 m). The "normal growing season" is late March to the end of October.

the driest parts of eastern Scotland, the limit is about 500 m. That these limits are found a little below the treeline corresponds with the characteristic features of upland settlement in the Alps and other European mountains; for example, in the driest parts of central Norway the respective limits can be put at about 900 and 1,100 m.

BELASCO (1952) has discussed the characteristics of air-masses over the British Isles. In January, mean temperature is lowest in polar continental air deriving from northern Russia with anticyclonic curvature. In July and August, mean temperature is highest in tropical continental air coming directly from North Africa or, sometimes, with anticyclonic curvature from southeast Europe. It is notable that there is a considerable proportion of days for which the characteristics are indeterminate or transitional; these are mainly anticyclonic days and occur more frequently in southern England than in northern Scotland, as might be expected.

Some comment is necessary on the results of the arrival of the less frequent air-masses. It is notable that for the period mid-November–mid-February the highest temperatures that have been recorded have been brought by maritime-tropical air which from time to time reaches England from far down the Atlantic, south of Madeira, occasionally 20°–25°N. It is, therefore, readily understood that during this winter period the highest maximum temperatures that have been recorded come from the areas subject to föhn effects when such air streams arrive, especially along the North Wales coast, and the south shore of the Moray Firth; occasionally at Dublin, or east of the Pennines. This is a very characteristic climatic effect; it is best developed when a warm southerly wind is crossing a mountain range lying athwart the current, with high pressure persistent to the eastward.

The most remarkable instance was that of December 2, 1948; when the temperature reached 65° (18.3°C) at Achnashellach in the far northwestern Scottish Highlands, and on the same day 63° (17.2°C) at the very exposed lighthouse on Cape Wrath, which itself stands 110 m above the sea. These are the highest December maxima officially recorded in any part of Britain. In January maxima of 63° (17.2°C) have been recorded on the north coast of Wales. In southeastern England no maxima exceeding 15.5°C have been officially reported in either December or January.

TABLE II

MAXIMUM RECORDED TEMPERATURES FOR EACH MONTH[1]

| Month | Place recorded | °C | °F |
|-------|----------------|-----|-----|
| January | Rhyl, Aber (North Wales) | 17.2 | 63 |
| | Durham, Dublin | 16.7 | 62 |
| February | Cambridge | 19.4 | 67 |
| | London | 18.9 | 66 |
| March | Wakefield and Whitby (Yorks), Cromer (Norfolk) | 25.0 | 77 |
| | Wealdstone (also London) | 23.9 | 75 |
| April | London | 29.4 | 85 |
| | Cambridge | 28.3 | 83 |
| May | Tunbridge Wells | 32.8 | 91 |
| | London | 32.2 | 90 |
| June | London | 35.6 | 96 |
| | Oxford | 35.0 | 95 |
| July | Tonbridge (Kent) (1868) | 38.0 | 100.5 |
| | nr. London | 36.1 | 97 |
| | Cambridge | 35.6 | 96 |
| August | Isleworth near London | 37.2 | 99 |
| | North Essex | 36.1 | 97 |
| September | Northamptonshire | 34.4 | 94 |
| October | London | 28.9 | 84 |
| | Cambridge | 28.3 | 83 |
| November | Cambridge (also London) | 21.1 | 70 |
| December | Achnashellach (northwestern Scotland) | 18.3 | 65 |

[1] *Source: Monthly Weather Report* and scattered earlier records.

From mid-February onward to early November, almost all the highest temperatures on record have been observed at stations in the southeast, in anticyclonic weather, with dry warm air of North African or southeast European derivation, that can be described as continental tropical. Under such conditions, the air crossing by the shortest sea passage is but little modified, and inland, London with the southeastern and eastern midland counties, enjoys cloudless skies. Extremes on record, observed under standard conditions, with the place of observation, are given in Table II. It will be noted that, rather rarely, the dry continental air can move northward with little modification. Hence on two occasions, in 1876 and 1948, maxima of 90° (32.2°C) have been officially reported from southern Scotland. It is evident that the possibility exists that over a long period of time such a maximum might be attained even in the north of Scotland, as 89° (31.8°C) has been observed beyond Inverness. In Ireland, an inland maximum of 90° (32.2°C) has been recorded once in the period 1901–1940.

The lowest temperatures occur in open valley-bottoms inland, on clear calm nights when the ground is well covered with fresh snow, following the arrival of continental arctic air between December and February. During the remaining months, an outburst of maritime-arctic air appears more likely to provide the needful airmass, which occasionally will build up into a cold anticyclone over Scotland. Extremes on record are quoted in Tables II and III; some, however, must be stated as approximate, as they were observed

TABLE III

MINIMUM RECORDED TEMPERATURES FOR EACH MONTH[*1]

| Month | Place recorded | °C | °F |
|-------|----------------|-----|-----|
| January | southeastern Scotland (inner Berwickshire) | −29 | −20[*2] |
| February | eastern Highlands (Braemar) | −27.2 | −17 |
| March | eastern Highlands (Logie Coldstone) | −22.8 | − 9 |
| April | northwestern England (near Penrith) | −15.0 | + 5 |
| May | eastern England (Lynford) (sandy soil) | − 9.4 | +15 |
| June | eastern England (Santon Downham) (sandy soil)[*3] | − 5.6 | +22 |
| July | southeastern Scotland[*4] (West Linton) | − 2.2 | +28 |
| August | eastern Highlands (Balmoral) | − 2.8 | +27 |
| September | central Highlands (Dalwhinnie) | − 6.7 | +20 |
| October | central Highlands (Dalwhinnie) | −11.7 | +11 |
| November | eastern Highlands (Braemar) | −23.3 | −10 |
| December | southeastern Scotland (inner Berwickshire) | −29 | −20[*2] |

[*1] *Source: Monthly Weather Report* and scattered earlier records.
[*2] Approximate.
[*3] Also central Highlands.
[*4] Also eastern Highlands.

under non-standard conditions. For Ireland, as might be expected, extremes have not fallen so low, but at one or two inland stations extreme minima have been reported at −4°F (−20°C). Fahrenheit readings are quoted as they were formerly standard.

One of the most notable features of the British climate, however, lies in the extent to which the incidence of severe frost is mitigated in favourable locations adjacent to the sea. The well known principles that have long been observed by continental climatologists on the slopes above Lake Garda, operate very effectively on the steeply sloping coastal margins of Cornwall, Pembrokeshire or the Isle of Man, as well as on the smaller offshore islands. As a result early daffodils can be gathered and marketed before the end of January. Along the southern Irish coast, and, more remarkably, in corners sheltered from the wind along the steeply sloping shores beside the coastal lochs of northwestern Scotland, killing frosts are rare; there is a famous garden in which a variety of subtropical evergreens flourish at Inverewe, in nearly 58°N beside Loch Broom on the west coast.

Further inland, the extent to which frost damage can be avoided by choice of a more elevated site is noteworthy. The observations kept at Malvern, a wellknown resort overlooking the Severn valley, on an east-facing slope at 120 m, indicate that the frost-free season is likely to average about six weeks longer than near Worcester in the broad valley itself, about 15 km distant.

Indeed, the extraordinary effectiveness of small changes of site, of aspect, of soil, and of provision of shelter or drainage in regard to the incidence of frost, makes it unusually difficult to provide statistics that will be representative of all possible cases. In many districts almost every field has its own microclimate, and must be treated on its individual merits. For this reason Table IV has been compiled in order to present data that are characteristic of a wide area inland.

This table has been derived by averaging the data available from a number of inland stations within the Midland counties. It should not be regarded as precisely applicable to any single location, at which the climate is likely to be affected by local relief, soil and vegetation, the proximity of buildings or water surfaces, exposure to wind and other minor factors. The data given will, however, be found to be closely representative of the characteristics of open, gently undulating and cultivated farmland over a wide area between 52°30′–53°N, 1°–3°W, at an altitude averaging 50 m above sea level. For the details at individual stations, of which about 400 send in monthly summaries, the *Monthly and Annual Weather Reports*, published by the Meteorological Office since 1884 should be consulted. Since 1941, the Irish Meteorological Service publishes data for the Republic.

## Comments based on the statistics—the average distribution of the meteorological elements

### Solar radiation

Intermittent measurements of the intensity of solar radiation have frequently been made, and for long the old blackbulb thermometers were to be seen. Regular daily measurements with modern instruments have only been kept at a few stations in recent years. Records of the duration of bright sunshine are, however, widely kept, and for a few stations data are available since 1881, or a little earlier. They are discussed later in this section (p.129). The Campbell–Stokes sunshine recorder does not register sunshine at very small intensities of radiation; the card stops burning when the direct radiation is less than about 20 mW/cm². Otherwise, the total receipt of direct radiation has been shown to be almost exactly proportional to the recorded duration of sunshine.

Daily instrumental measurements of the normal-incidence solar radiation have been maintained at Kew Observatory since 1932. Measurements of total and diffuse radiation on a horizontal surface were added in 1947, and since then several other stations have been so equipped. Available official material has been listed by JACOBS (1961). More recently, MONTEITH (1966) has discussed these records and has drawn attention to the considerably higher values recorded at Aberporth, on the Welsh coast south of Aberystwyth. Comparisons between the several available records, point to interesting differences consistent with differences in air pollution. Part of this pollution is the normal inland "natural aerosol" and part is due to smoke. In recent years there has been a noteworthy improvement in the ratio between the "bright sunshine" recorded in the centre of London and that in the surrounding country. A corresponding improvement can be observed elsewhere, notably in Lancashire. Radiation receipt in central London is measured and is still about 5% below that at Kew, an outer southwestern suburb; on a brief record Kew appears about 4% below Bracknell, 25 km west-southwest of Kew.

The radiation income in the cloudy climate, but unusually clean air in northern Scotland is demonstrably more favourable than was earlier thought.

Tables V and VI give the average monthly and annual values of daily radiation, both "total" and "diffuse", received on a horizontal surface, and of bright sunshine over the same period (1958–1968). For Aberporth, the period is 1959–1968. There is some support for the high values recorded at Aberporth from those kept at Valentia on the coast of southwestern Ireland. It also appears that the high values are characteristic of a very narrow coastal strip.

TABLE IV

THE GENERAL CHARACTERISTICS OF THE INLAND CLIMATE OF ENGLAND[1]

| | Jan. | Feb. | Mar. | Apr. | May | June | July | Aug. | Sept. | Oct. | Nov. | Dec. | Year |
|---|---|---|---|---|---|---|---|---|---|---|---|---|---|
| Mean pressure 1931–1960, reduced to sea level[2] (mbar) | 1,015 | 1,014 | 1,012 | 1,013 | 1,015 | 1,016 | 1,015 | 1,014 | 1,017 | 1,013 | 1,013 | 1,011 | 1,014 |
| Mean temperature (°C), 1931–1960 | 3.4 | 3.9 | 5.9 | 8.4 | 11.4 | 14.6 | 16.2 | 16.0 | 13.7 | 10.1 | 6.7 | 4.7 | 9.6 |
| Average daily range (max.-min.) (°C) | 5.8 | 6.2 | 7.8 | 8.8 | 9.8 | 9.8 | 8.8 | 9.0 | 8.4 | 7.4 | 6.0 | 5.6 | |
| Warmest month on record (°C) (1723–1965) | 7.5 (1916) | 7.9 (1779) | 9.2 (1957) | 10.6 (1865) | 15.1 (1833) | 18.2 (1846) | 18.8 (1783) | 18.6 (1947) | 16.6 (1729) | 12.8 (1921) | 9.5 (1818) | 8.1 (1934) | 10.6 (1949) |
| Coldest month on record (°C) (1723–1965) | −3.1 (1795) | −1.9 (1947) | 1.2 (1785) | 4.7 (1837) | 8.6 (1740) | 11.9 (1916) | 13.4 (1816) | 12.9 (1912) | 10.5 (1807) | 5.3 (1740) | 2.3 (1782) | −0.8 (1890) | 6.8 (1740) |
| Average extremes of temperature to be expected (°C) | 11 | 12 | 15 | 19 | 23 | 25 | 27 | 26 | 23 | 18 | 14 | 12 | 28 |
| | −7 | −6 | −5 | −3 | 0 | 4 | 6 | 6 | 3 | −1 | −3 | −6 | −9 |
| Probable absolute extremes to be expected over 100 years (°C) | 15 | 17 | 22 | 26 | 29 | 33 | 34 | 34 | 32 | 27 | 20 | 15 | 34 |
| | −20 | −19 | −15 | −9 | −3 | 1 | 3 | 2 | −2 | −6 | −12 | −20 | −20 |
| Frequency of days with air frost (min. 0 °C or below) | 15 | 13 | 10 | 5 | 1 | <0.1 | – | – | <0.1 | 2 | 7 | 12 | 65 |
| Rainfall: average monthly fall in mm., characteristic of undulating lowland[3] (1916–1950) | 65 | 48 | 44 | 49 | 56 | 48 | 68 | 67 | 58 | 70 | 67 | 60 | 700 |
| Average number of days with 0.2 mm or more falling | 15 | 14 | 15 | 13 | 14 | 12 | 13 | 15 | 13 | 17 | 16 | 17 | 174 |
| Relative humidity (average for 24 h) | 89 | 89 | 82 | 78 | 74 | 77 | 79 | 81 | 82 | 84 | 86 | 86 | 85 |
| Probable average evaporation-loss from a standard open tank (mm) (Meteorological Office, 6 ft. square) | 3 | 8 | 21 | 42 | 66 | 82 | 78 | 64 | 42 | 21 | 8 | 4 | 440 |

TABLE IV (*continued*)

| | Jan. | Feb. | Mar. | Apr. | May | June | July | Aug. | Sept. | Oct. | Nov. | Dec. | Year |
|---|---|---|---|---|---|---|---|---|---|---|---|---|---|
| Probable average number of days with "fog" (vis. <1 km) or "dense fog" (vis. <200 m) in open rural locations | 3 | 2 | 2 | 0.8 | 0.3 | 0.3 | 0.3 | 0.3 | 1 | 2 | 4 | 4 | 20 |
| at the early morning observation | 1 | 0.5 | 0.2 | rare | rare | rare | rare | rare | 0.1 | 0.5 | >1 | >1 | 5 |
| Bright sunshine (h)[4] | 50 | 70 | 105 | 150 | 180 | 190 | 170 | 165 | 130 | 90 | 57 | 43 | 1,400 |
| Days with snow observed to fall[5] | 5 | 4 | 4 | 2 | <0.5 | – | – | – | – | <0.5 | 1 | 3 | 20 |
| Days with snow lying at 09h00[6] | 3.5 | 3.5 | 2.5 | <1 | – | – | – | – | – | – | <1 | 2 | 12 |
| Days with thunder heard | 0.1 | 0.1 | 0.3 | 0.8 | 2 | 2 | 3 | 2 | 1 | 0.4 | 0.2 | <0.1 | 12 |
| Approx. average duration of measurable rainfall in h | 60 | 45 | 50 | 37 | 38 | 38 | 40 | 45 | 40 | 50 | 62 | 60 | 565 |

Probable average frequencies of wind directions, in open country, over the year in percentages:

| north | northeast | east | southeast | south | southwest | west | northwest | calm |
|---|---|---|---|---|---|---|---|---|
| 6 | 11 | 8 | 7 | 13 | 22 | 20 | 10 | 3 |

Average yearly frequency of "ice days" (max. 0 °C or below), about 2.
Approximate range of variation of the annual rainfall, from 60% to 150% of average.
Greatest number of days in succession with measurable rain, approx, 60.
Greatest number of days in succession without measurable rain, approx. 40.
Days with gale, average 3 per annum (inland).
Greatest depth of penetration of frost, about 75 cm.

[1] Figures for mean temperatures are based on the "Central England" tables (MANLEY, 1953, 1959); other data are derived by averaging from several Midland stations, away from large towns, cf. also Table X. They are best related to undulating lowland between Birmingham, Nottingham and Manchester, 52°30′–53°N, 2°W.
[2] The diurnal range approximates to 1 mbar in every month.
[3] Average number of days with 1 mm or more falling, approx., 125.
[4] Range of variation of the annual total from 70% to 130% of the average.
[5] Range of variation of the annual total, in round figures, from 2–60.
[6] Range of variation of the annual total, in round figures, from 1–60. Longest continuous duration of snow cover likely to be observed, between 50–60 days.

TABLE V

SOLAR RADIATION (mW/h cm²) ON A HORIZONTAL SURFACE (1958–1965)*

| Place | Annual mean value of daily total radiation | Daily diffuse radiation | Mean duration (h) of "bright sunshine" 1958–1965 | 1931–1960 |
|---|---|---|---|---|
| Lerwick (island) 60°08′N 1°11′W | 212 | 131 | 2.85 | (2.96) |
| Eskdalemuir 55°19′N 3°12′W | 224 | 135 | 3.26 | (3.31) |
| Cambridge 52°12′N 0°8′E | 262 | 155 | 3.97 | (4.08) |
| Kew (outer London suburb) 51°28′N 0°19′W | 248 | 142 | 4.15 | (4.16) |
| ¹Aberporth (Welsh coast) 52°08′N 4°34′W | 295 | 163 | 4.23 | (4.40) |

* *Source:* Meteorological Office data, *Monthly Weather Report.*
¹ 1959–1968.
² Shorter records indicate that these totals average about 5% lower than Kew in central London, and about 4% higher at Bracknell, 25 km west-southwest of Kew.

TABLE VI

MONTHLY MEAN VALUES OF DAILY RADIATION ON A HORIZONTAL SURFACE (mW/h cm²; 1958–1968)

| Station | * | Jan. | Feb. | Mar. | Apr. | May | June | July | Aug. | Sept. | Oct. | Nov. | Dec. |
|---|---|---|---|---|---|---|---|---|---|---|---|---|---|
| Lerwick | (a) | 27 | 80 | 168 | 310 | 398 | 462 | 415 | 321 | 207 | 100 | 37 | 16 |
| | (b) | 21 | 55 | 105 | 184 | 247 | 279 | 270 | 210 | 129 | 68 | 27 | 14 |
| | (c) | 0.9 | 1.7 | 2.7 | 4.4 | 4.6 | 5.2 | 4.4 | 3.5 | 3.1 | 1.9 | 1.1 | 0.5 |
| Eskdalemuir | (a) | 63 | 111 | 189 | 308 | 405 | 444 | 366 | 315 | 236 | 134 | 65 | 39 |
| | (b) | 35 | 73 | 120 | 188 | 243 | 268 | 256 | 211 | 146 | 84 | 41 | 26 |
| | (c) | 1.6 | 2.2 | 2.9 | 4.1 | 5.1 | 5.4 | 4.3 | 4.1 | 3.5 | 2.7 | 1.8 | 1.5 |
| Cambridge | (a) | 69 | 118 | 227 | 333 | 450 | 517 | 453 | 378 | 285 | 167 | 81 | 61 |
| | (b) | 45 | 78 | 131 | 205 | 263 | 284 | 278 | 223 | 160 | 98 | 53 | 39 |
| | (c) | 1.8 | 2.1 | 3.8 | 4.4 | 6.1 | 7.0 | 5.5 | 5.3 | 4.6 | 3.3 | 1.9 | 1.8 |
| Kew | (a) | 59 | 104 | 217 | 311 | 432 | 481 | 436 | 371 | 283 | 164 | 76 | 43 |
| | (b) | 41 | 69 | 122 | 183 | 239 | 257 | 260 | 211 | 153 | 93 | 50 | 32 |
| | (c) | 1.7 | 2.1 | 3.9 | 4.6 | 6.4 | 7.0 | 6.2 | 5.7 | 5.0 | 3.5 | 2.0 | 1.5 |
| Aberporth | (a) | 73 | 140 | 366 | 392 | 509 | 573 | 516 | 441 | 315 | 174 | 86 | 58 |
| (1959–1968) | (b) | 50 | 89 | 151 | 217 | 269 | 288 | 293 | 239 | 169 | 98 | 58 | 39 |
| | (c) | 1.7 | 2.7 | 4.0 | 5.1 | 6.5 | 7.4 | 5.8 | 5.7 | 4.8 | 3.2 | 2.0 | 1.8 |

¹ *Source:* Meteorological Office data, *Monthly Weather Report*, 1958–1968.
* a = total; b = diffuse; c = mean daily bright sunshine.

**Pressure**

The average distribution of pressure for each month has been mapped in the *Climatological Atlas* (METEOROLOGICAL OFFICE, 1952), and in other publications of the Meteorological Office; earlier sources include for example Buchan's maps in *Bartholomew's Atlas of Meteorology* (1899). In detail there are small differences, but the principles remain the same. Throughout the year, pressure falls off, more often than not, from south to north; isobars run from southwest or west to east or northeast; in most months the mean isobars display a very slight northward curvature over the land areas, southward over the seas. The mean gradient of pressure from Sussex to Shetland is greatest (12 mbar) in January, after which it diminishes to barely 2 mbar in May; thereafter it rises to 4 mbar in July, 6 mbar in August, a little less in September, then steadily to nearly 10 mbar in December. Over the greater part of England and Ireland, pressure is highest in June. Up to 1940, pressure in the winter months in the Shetlands has been a little lower during this century than last, i.e., there has been a slight increase in the westerly gradient. (See Table VII.) These facts reflect the tendency for stronger winds in winter; and that it is in the spring months that the chances are greatest that the pressure gradient will be reversed, so that, with higher pressure to the northward, easterly or northeasterly winds will prevail for some time. It is in April, that at the majority of stations the frequency of winds from the northeasterly quadrant almost equals that from the southwest. In June the chance that the Azores anticyclone will spread over western and southern Britain reaches its maximum; accompanying this are the highest sunshine durations, decreased amount of cloud, and the lowest frequency of rain. July and August bring, in general, rather more unsettled and cloudy weather, especially in August towards the north; but September's slightly slacker pressure gradient reflects the frequent tendency for a spell of relatively quiet, dry weather during that month with slightly less cloud. At the majority of stations June is on the average the least cloudy, December the most cloudy of the months.

The slight tendency for anticyclonic curvature of the isobars over the land implies that the prevailing wind direction veers a little towards the east coasts of Ireland and of Great Britain. This is linked with a tendency to greater stability, divergence and subsidence east of the uplands. But these tendencies are very slight; the variations from year to year are great. The mean isobars indeed do little more than indicate that winds from some southwesterly or westerly point prevail on slightly more than half the days.

Variability of mean pressure is greater to the north, especially in winter. In the cold month of January, 1963, mean pressure (1,030 mbar) was nearly 25 mbar above normal at Stornoway, and 24 mbar above at Lerwick, but 6 mbar above normal at Scilly. In August 1947, 13 mbar above normal at Lerwick and 11 at Stornoway, 3 mbar above at Scilly. Such distributions commonly arise from the intermittent development and persistence of blocking anticyclones in unusual northern locations, for example, off the Hebrides or over the Norwegian Sea.

In each issue of the *Monthly Weather Report* the mean distribution of pressure for the month is shown, against the normal distribution. Some months in which the distribution has been particularly abnormal include January, 1963 and 1959; February, 1932, 1947 and 1955; March 1938; April 1938; May 1946; June 1925; July 1955; August 1947 and 1955; September 1959; October 1921; November 1938, and December 1890 (see Fig.2).

TABLE VII

CLIMATOLOGICAL DATA: ATMOSPHERIC PRESSURE AND TEMPERATURE (1931–1960)*

| Station | Jan. | Feb. | Mar. | Apr. | May | June | July | Aug. | Sept. | Oct. | Nov. | Dec. | Year |
|---|---|---|---|---|---|---|---|---|---|---|---|---|---|
| *Atmospheric pressure (mbar):* | | | | | | | | | | | | | |
| Lerwick | 06.4 | 08.7 | 12.0 | 11.0 | 15.6 | 13.1 | 10.7 | 10.9 | 10.2 | 09.0 | 07.3 | 05.5 | 10.0 |
| Stornoway | 07.0 | 10.0 | 11.6 | 11.8 | 15.2 | 13.5 | 11.0 | 11.1 | 10.6 | 09.4 | 07.4 | 05.9 | 10.4 |
| Aberdeen (Dyce) | 09.4 | 11.4 | 13.4 | 12.8 | 16.1 | 14.2 | 11.6 | 12.1 | 12.1 | 11.1 | 09.2 | 08.2 | 11.8 |
| Eskdalemuir | 11.0 | 12.9 | 14.0 | 13.9 | 15.8 | 14.7 | 12.3 | 12.8 | 13.3 | 12.6 | 10.6 | 10.2 | 12.8 |
| Manchester Airport | 12.8 | 14.3 | 14.7 | 15.1 | 16.0 | 15.7 | 13.9 | 13.9 | 14.9 | 14.1 | 12.2 | 12.3 | 14.2 |
| Waddington | 13.1 | 14.3 | 15.0 | 15.0 | 16.1 | 15.9 | 14.0 | 14.1 | 15.1 | 14.4 | 12.4 | 12.6 | 14.3 |
| Yarmouth (Gorleston) | 13.4 | 14.6 | 15.2 | 15.0 | 16.0 | 16.1 | 14.4 | 14.2 | 15.3 | 14.5 | 12.9 | 13.2 | 14.6 |
| London (Kew Obs.) | 14.5 | 15.4 | 15.5 | 15.7 | 16.2 | 16.7 | 15.2 | 15.0 | 16.1 | 15.4 | 13.6 | 14.1 | 15.3 |
| Plymouth (Mt. Batten) | 14.7 | 16.0 | 14.7 | 16.2 | 16.0 | 17.1 | 16.0 | 15.5 | 16.5 | 15.5 | 13.5 | 14.2 | 15.5 |
| Aldergrove (N. Ireland) | 10.4 | 12.9 | 13.0 | 14.3 | 15.5 | 15.0 | 13.2 | 13.1 | 13.6 | 12.7 | 10.4 | 09.7 | 12.8 |
| Dublin (Ireland, 1941–1960) | 12.2 | 13.9 | 14.2 | 15.7 | 15.6 | 15.7 | 14.4 | 12.4 | 13.9 | 14.4 | 12.2 | 10.4 | 13.8 |
| Valentia (Ireland) | 12.3 | 14.7 | 12.5 | 15.5 | 15.3 | 16.4 | 15.4 | 14.7 | 14.8 | 14.1 | 11.7 | 11.8 | 14.1 |
| | | | | | | | | | | | | | |
| *Mean temperature (°C):* | | | | | | | | | | | | | |
| Lerwick | 3.1 | 2.9 | 3.9 | 5.4 | 7.8 | 10.0 | 12.0 | 12.1 | 10.6 | 8.2 | 5.9 | 4.4 | 7.2 |
| Stornoway | 4.3 | 4.4 | 5.7 | 7.0 | 9.3 | 11.6 | 13.3 | 13.3 | 11.8 | 9.3 | 6.9 | 5.5 | 8.6 |
| Aberdeen (Dyce) | 2.4 | 2.8 | 4.5 | 6.6 | 9.0 | 12.0 | 14.0 | 13.6 | 11.7 | 8.8 | 5.6 | 3.7 | 7.9 |
| Eskdalemuir (upland, 242 m) | 1.4 | 1.8 | 3.6 | 5.8 | 8.9 | 11.7 | 13.3 | 12.9 | 10.8 | 7.6 | 4.6 | 2.8 | 7.1 |
| Manchester Airport | 3.3 | 3.4 | 5.9 | 8.2 | 11.7 | 14.2 | 15.8 | 15.6 | 13.6 | 10.3 | 6.8 | 5.1 | 9.5 |
| Waddington | 3.0 | 3.6 | 5.6 | 8.3 | 11.2 | 14.3 | 16.3 | 16.1 | 13.8 | 10.1 | 6.4 | 4.2 | 9.4 |
| Yarmouth (Gorleston) | 4.1 | 4.2 | 5.6 | 8.1 | 10.8 | 14.1 | 16.3 | 16.5 | 14.8 | 11.3 | 7.7 | 5.3 | 9.9 |
| London (Kew Obs.) | 4.2 | 4.4 | 6.6 | 9.3 | 12.4 | 15.8 | 17.6 | 17.2 | 14.8 | 10.8 | 7.2 | 5.2 | 10.5 |
| Plymouth (Mt. Batten) | 6.2 | 5.8 | 7.3 | 9.2 | 11.7 | 14.5 | 15.9 | 16.2 | 14.7 | 11.9 | 8.9 | 7.2 | 10.8 |
| Aldergrove (N. Ireland) | 3.7 | 4.2 | 5.9 | 7.8 | 10.5 | 13.3 | 14.7 | 14.6 | 12.7 | 9.7 | 6.6 | 4.9 | 9.1 |
| Dublin (Ireland)[1] | 4.4 | 4.7 | 6.4 | 8.1 | 10.5 | 13.5 | 15.1 | 14.9 | 12.9 | 10.2 | 7.1 | 5.5 | 9.4 |
| Valentia (Ireland) | 6.9 | 6.8 | 8.3 | 9.4 | 11.4 | 13.8 | 15.0 | 15.4 | 14.0 | 11.6 | 9.1 | 7.8 | 10.8 |
| | | | | | | | | | | | | | |
| Additional stations: | | | | | | | | | | | | | |
| Braemar (upland, 340 m) | 0.6 | 1.0 | 2.8 | 5.2 | 8.3 | 11.4 | 13.1 | 12.6 | 10.2 | 7.0 | 3.8 | 2.0 | 6.5 |
| Renfrew (Glasgow Airport) | 3.1 | 3.9 | 5.6 | 7.8 | 10.7 | 13.6 | 15.1 | 14.8 | 12.6 | 9.5 | 6.1 | 4.4 | 8.9 |
| Durham | 2.8 | 3.3 | 5.1 | 7.4 | 10.1 | 13.2 | 15.3 | 14.9 | 12.7 | 9.4 | 6.1 | 4.1 | 8.7 |
| Cambridge | 3.4 | 3.9 | 6.1 | 8.8 | 11.9 | 15.1 | 17.0 | 16.7 | 14.3 | 10.4 | 6.7 | 4.5 | 9.9 |
| Oxford | 3.7 | 4.2 | 6.4 | 9.1 | 12.0 | 15.3 | 17.1 | 16.9 | 14.4 | 10.5 | 7.0 | 4.9 | 10.1 |
| Aberystwyth | 4.5 | 4.3 | 6.2 | 8.0 | 10.9 | 13.5 | 14.9 | 15.2 | 13.5 | 10.6 | 7.6 | 5.8 | 9.6 |

* *Source:* Meteorol. Office, ANONYMOUS (1962) and W.M.O. Climatological Normals for Ireland.
[1] Dublin 1931–1960, approximate reduction from Meteorological Office data combined with Irish Meteorological Service data over 30 years.

Extremes on record for pressure at sea-level, range from 1,055 mbar at Aberdeen in January 1902 and 925 mbar in Perthshire in January 1884.

The mean pressure distribution indicated by the maps makes it very evident that the centres of a majority of the active depressions move from west to east, in latitudes north of Scotland, or from southwest to northeast past the Hebrides. Such depressions are occompanied by troughs of varying intensity crossing the British Isles. The frequency with which secondaries to the above systems cross Britain, and the frequency with which Atlantic depressions move towards Europe on the southern flank of Britain, from the

Bay of Biscay across France is notable; such systems give rise to the not uncommon easterly winds with rain and, in winter, to some of the heaviest snowfalls throughout southern England.

## Wind conditions in general

Wind roses for a number of stations are given in the *Climatological Atlas*, and detailed statistics of frequency and strength at coastal stations can be found in the series of *Coastal Pilots* published by the Hydrographic Department of the Admiralty (London), and the handbooks for seamen, *Weather in Home Fleet Waters*.

While surface winds are often influenced by the local relief and the neighbourhood of the sea, or of extensive estuaries and lakes, over much of Britain the frequencies can be generalised in a statement; the probability is that the prevailing wind on 6 days out of 10 will be from between south and west; on rather more than 2 from between north and east; considerably less than one, calm; the remaining few occasions will be of south to east wind, or north to west, from both of which directions winds are, in the open, relatively infrequent.

Strength of wind is not easy to generalise; in open country free from obstacles, such as a large airfield, the average speed of the wind throughout the year is considerably greater than it would be in a built-up area, or in more wooded "residential" country. At representative airfields by day it is commonly of the order of force 3 (10 knots, 5.4 m/sec). At sea-coast stations the strongest surface winds will tend to blow from seaward, in whichever direction that lies; winds off the land will be more gusty, but much less strong overall.

The range of the wind speed between gusts and lulls is likewise very much greater at inland stations than over the sea; it becomes worst at those anemograph stations that lie in country much broken up by buildings, woods and other obstacles (Table VIII). Winds of gale force for any given day are reported when the average speed of the wind over a short period of ten minutes or so exceeds force 8, that is 33 knots (18 m/sec). At a few inland stations in the more sheltered districts towards the southeast, gale force by this definition has not yet been recorded, for example at Kew Observatory. But at the majority of stations well inland, both in Ireland and southern Britain, gale force will be reached on two, perhaps three days each year; and rather more in Scotland, where the trend of the hill ranges tends to canalise the flow. Maximum gusts are another matter; even well inland, occasional gusts may almost attain the strength of the wind at sea and cause widespread structural damage. From the map, the frequency of days with gale will be evident; the frequency is everywhere higher on the west coast, averaging more than 20 days a year at almost all stations, rising to over 30 towards the northwest and to upwards of 40 locally on the northwestern and northern Scottish islands. This is to be expected, as the strongest winds come from points between south and west.

The average frequency of days with gale for each month is shown in Table IX.

Statistics of gusts and duration of gales are given in some detail in the *Monthly and Annual Weather Reports* of the Meteorological Office. Extreme gusts in excess of 100 miles/h (85 knots) have been registered at many coastal stations, and at a few inland, chiefly in Scotland. In February 1968 a record of 136 miles/h (117 knots) was measured at Kirkwall in the Orkneys.

TABLE VIII

AVERAGE WIND FORCE, OR SPEED, BASED ON OBSERVATIONS TWICE DAILY, GENERALLY AT 07H00 AND 13H00, OVER PERIODS OF 10 YEARS*

| Station | Jan. | Feb. | Mar. | Apr. | May | June | July | Aug. | Sept. | Oct. | Nov. | Dec. | Year |
|---|---|---|---|---|---|---|---|---|---|---|---|---|---|
| *Wind speed in knots:* | | | | | | | | | | | | | |
| Lerwick | 19 | 18 | 14 | 14 | 13 | 13 | 10 | 10 | 13 | 15 | 17 | 18 | 14 |
| Holyhead | 16 | 15 | 13 | 13 | 11 | 11 | 11 | 11 | 13 | 16 | 13 | 15 | 13 |
| Aberdeen (Observ.) | 9 | 8 | 8 | 7 | 7 | 6 | 6 | 5 | 6 | 7 | 8 | 9 | 7 |
| Renfrew | 9 | 9 | 9 | 7 | 7 | 7 | 7 | 7 | 7 | 9 | 6 | 7 | 7 |
| Manchester Airport | 11 | 11 | 10 | 10 | 9 | 9 | 9 | 9 | 10 | 10 | 10 | 11 | 10 |
| (15 years) | | | | | | | | | | | | | |
| London (Kew Observ.) | 6 | 6 | 6 | 6 | 6 | 6 | 5 | 5 | 4 | 5 | 6 | 7 | 6 |
| | | | | | | | | | | | | | |
| *Wind speed in beaufort:* | | | | | | | | | | | | | |
| Stornoway | 3.5 | 3.1 | 2.7 | 3.1 | 2.8 | 2.9 | 2.5 | 2.6 | 2.9 | 3.3 | 3.1 | 3.3 | 3.0 |
| Tynemouth | 2.6 | 2.6 | 2.3 | 2.4 | 2.1 | 2.1 | 2.0 | 2.0 | 2.1 | 2.4 | 2.6 | 2.7 | 2.3 |
| Yarmouth (Gorleston) | 3.1 | 3.3 | 2.9 | 3.1 | 3.0 | 2.8 | 2.5 | 2.7 | 2.8 | 3.1 | 3.4 | 3.5 | 3.0 |
| Plymouth (Mt. Batten) | 3.7 | 3.7 | 3.5 | 3.4 | 3.3 | 3.1 | 3.1 | 3.1 | 3.1 | 3.7 | 3.7 | 3.7 | 3.5 |
| Scilly | 4.8 | 4.9 | 4.3 | 4.0 | 3.6 | 3.7 | 3.8 | 3.2 | 3.8 | 4.2 | 4.2 | 4.8 | 4.2 |
| Cardiff | 2.4 | 2.5 | 2.6 | 2.7 | 2.5 | 2.4 | 2.3 | 2.1 | 2.2 | 2.4 | 2.4 | 2.4 | 2.4 |
| Birkenhead (Bidston) | 3.7 | 3.4 | 3.1 | 3.1 | 2.7 | 3.0 | 2.7 | 2.9 | 2.9 | 3.0 | 3.2 | 3.5 | 3.1 |
| Douglas (Isle of Man) | 3.5 | 3.1 | 3.0 | 3.0 | 2.8 | 2.9 | 2.7 | 2.7 | 2.8 | 2.9 | 3.0 | 3.3 | 3.0 |
| | | | | | | | | | | | | | |
| Ireland (30 years): | | | | | | | | | | | | | |
| Dublin (Phoenix Park) | 3.0 | 2.8 | 2.5 | 2.6 | 2.3 | 2.3 | 2.4 | 2.3 | 2.5 | 2.5 | 2.6 | 2.9 | 2.6 |
| Valentia | 4.1 | 3.9 | 3.9 | 3.7 | 3.6 | 3.5 | 3.5 | 3.5 | 3.5 | 3.8 | 3.8 | 4.0 | 3.7 |

\* *Sources:* Meteorol. Office, tables from *Weather in Home Fleet Waters* and Admiralty Pilots for British and Irish coasts, and CROWE (1962).

Southeasterly and easterly gales on the North Sea coast, and along the English Channel, can be severe, but have not been known to attain the strength of the southwest and westerly gales on the west facing coasts. Northerly gales blow more rarely, but attained exceptional strength in the great storm of 31 January, 1953. At Lerwick in the Shetlands a wind speed of 70 knots (81 miles/h) was reported, with gusts up to 94 knots (111 miles/h). It is therefore not possible to state with certainty that the strongest winds will invariably come from the southwestern quadrant.

*Localised occurrences of unusually strong winds*

Apart from the occasional violent gusts that may occur inland in association with the passage of a well-marked cold front, knowledge of the behaviour of air crossing mountain barriers under conditions that ensure that stability prevails at some upper levels, makes it very evident that from time to time, extremely severe gales can blow down the lee-side of the larger hill ranges and cause much damage.

In the above-named instance (January 31, 1953), the extraordinary damage to the Scottish forests caused by the severe northerly gale was the more remarkable as it was very much more marked on the south-facing slopes. Likewise, the Sheffield gale of February 18, 1962 (AANENSEN, 1965) caused an extraordinary amount of structural damage. This was a very violent westerly gale whose effects were largely confined to the

TABLE IX

AVERAGE NUMBER OF "DAYS WITH GALE" (WIND OF BEAUFORT 8 OR MORE AT ANY TIME OF THE DAY)[1]

| Station | Years of obser-vation | Jan. | Feb. | Mar. | Apr. | May | June | July | Aug. | Sept. | Oct. | Nov. | Dec. | Year |
|---|---|---|---|---|---|---|---|---|---|---|---|---|---|---|
| Lerwick | 15 | 9 | 6 | 2 | 2 | 0.9 | 0.5 | 0.2 | 0.5 | 1 | 5 | 4 | 7 | 38 |
| Stornoway | 8 | 10 | 6 | 4 | 3 | 1 | 1 | 0.5 | 0.7 | 3 | 7 | 5 | 7 | 48 |
| Wick | 29 | 5 | 3 | 2 | 1 | 0.7 | 0.3 | 0.1 | 0.3 | 0.8 | 2 | 3 | 5 | 23 |
| Inverness | 20 | 2 | 1 | 1 | 0 | 0 | 0 | 0 | 0 | 1 | 1 | 1 | 1 | 8 |
| Renfrew | 20 | 2 | 1 | 0.7 | 0.3 | 0.1 | 0.1 | 0.1 | 0.1 | 0.3 | 0.9 | 0.5 | 1 | 7 |
| Inchkeith | 13 | 2 | 1 | 0.3 | 0.5 | 0.2 | 0.2 | 0 | 0.1 | 0.2 | 1 | 0.5 | 1 | 8 |
| Margate | 10 | 1 | 0.8 | 0.3 | 0.3 | 0.2 | 0.1 | 0.1 | 0.2 | 0.4 | 0.7 | 0.9 | 0.9 | 6 |
| London (Greenwich) | 10 | 0.4 | 0.2 | 0.1 | 0.1 | 0.1 | 0 | 0 | 0 | 0.1 | 0.1 | 0.4 | 0 | 2 |
| London (Kew) | 10 | none observed | | | | | | | | | | | | |
| Plymouth (Mt. Batten) | 17 | 3 | 1 | 0.5 | 0.3 | <0.1 | 0.1 | 0.3 | 0.2 | 0.6 | 1 | 2 | 3 | 12 |
| Scilly | 27 | 5 | 4 | 2 | 2 | 0.3 | 0.4 | 0.4 | 0.5 | 0.7 | 2 | 3 | 5 | 25 |
| Cardiff | 20 | 0.7 | 0.3 | 0.1 | 0.3 | 0 | 0.1 | 0 | 0.1 | 0.1 | 0.1 | 0.3 | 0.5 | 3 |
| Holyhead | 25 | 3 | 2 | 1 | 1 | 0.5 | 0.1 | 0 | 0.3 | 0.7 | 2 | 3 | 4 | 18 |
| Birkenhead (Bidston) | 20 | 2 | 1 | 0.6 | 0.4 | 0.1 | 0.3 | 0.1 | 0.2 | 0.5 | 1 | 0.8 | 2 | 9 |
| Douglas (Isle of Man) | 20 | 4 | 2 | 1 | 0.9 | 0.3 | 0.5 | 0.1 | 0.4 | 0.6 | 2 | 3 | 3 | 18 |
| Ireland | | | | | | | | | | | | | | |
| Dublin (Phoenix Pk.) | 30 | 2 | 1 | 0.5 | 0.7 | 0.2 | 0.2 | 0.4 | 0.4 | 0.7 | 1 | 1 | 2 | 10 |
| Valentia | 20 | 3 | 2 | 0.7 | 0.5 | 0.2 | 0.1 | 0.2 | 0.4 | 0.7 | 2 | 2 | 4 | 16 |

[1] Data for other stations use "hourly wind exceeding 33 knots"; see METEOROLOGICAL OFFICE, 1964, 1965.

eastern flanks of the Pennines. The explanation in both cases lay in the rapid spread of an anticyclone that followed the passage of a depression, and the development of a subsidence inversion at a high altitude, giving the situation that favours the production of "standing waves" and severe downslope winds on the lee side of the hill-ranges; the Grampians in Aberdeenshire, the Pennines for Sheffield.

The best localised example of such phenomena is that provided by the so-called "helm wind" that has for centuries been recognised as a characteristic feature of the northern Pennines at times when the pressure distribution provides fresh northeast to east winds across the north of England (MANLEY, 1945b). From the standpoint of a surface observer, the phenomena bear resemblance to those operative when a stream of water flows over a submerged weir. Given that the stream exceeds a certain velocity, thereafter standing-waves are formed at, and downstream from the obstacle. For similar phenomena to develop when an air-stream crosses a range of hills, there should be an inversion layer at a suitable altitude above the crest of the range, which itself should be smooth, and not too steep-sided, or broken up by valleys or subsidiary ranges.

In addition to the passage of squalls and the results produced by transverse hill barriers that give rise from time to time to unusual acceleration of the surface wind, a variety of

Fig.4. Thunder: annual average number of days with thunder heard, 1931–1960. (Based on Meteorological Office data.)

other phenomena can cause local intensification of the air flow and sometimes damage. With a steep pressure gradient and isobars running parallel to a coast backed by ranges of hills, convergence in the surface air stream may give rise to local gales. For example, with a strong southerly wind convergence of the streamlines off the northwest coast of Anglesey, can provide a local gale; likewise, a strong easterly wind blowing in the Irish Sea can become a northeasterly gale along the coast of the Isle of Man. In bad weather, local gales and squalls are notorious among the Scottish islands.

It is not possible to give statistics with regard to such local occurrence. The development of a lee-wave disturbance such as that of February, 1962, which gave wind speeds in parts of Sheffield that were estimated to be far in excess of 100 miles/h (85 knots) appears to require conditions that might only prevail once in several decades. Suffice it to point out that the fact that winds of this intensity have not yet been recorded at a number of inland stations, does not preclude the possibility that they can occur. The distribution of anemometers, especially in the north and northwest of Scotland, is scanty and we are far from knowing all the local possibilities of wind behaviour.

Localised winds of another type arise in occasional small-scale tornadoes. As far as is known these do not in any way attain the magnitude or frequency of those in the United States. Damage in the form of uprooted trees, fallen chimneys and blown-out windows is usually confined to a very narrow strip or strips running across country, of the order of 10–30 m in width. They are essentially instability phenomena, whose incidence appears to be closely related to the regions of maximum frequency of thunder heard (Fig.4). Scattered reports of such phenomena in various places come in, by estimation on about one day in three years; such reports tend to come most often from the east Midlands in the warmer months, but a good many have been reported from elsewhere. On December 8, 1953, a minor tornado crossed the western suburbs of London and caused considerable structural damage; in June, 1937, damage amounting to about £20,000, mostly broken windows, was reported from Birmingham (LAMB, 1957).

*Coastal and mountain breezes: föhn*

The British Isles experience all these on a scale sufficient to modify the climate appreciably in particular districts.

Sea-breezes are a normal concomitant of the warmer months of the year when there is sufficient difference of temperature between the sea and the adjacent land, and convection is not suppressed by the presence of unusually warm subsiding air above, in which event the daytime sea breeze will not develop. In general the sea-breeze is unlikely to develop if the gradient is such that the wind comes directly off the land and exceeds, roughly, force 3 (8–12 knots, or 10–15 miles/h).

The distribution of the mean surface temperature over the British seas shows that in summer the tongue of relatively cool water extends southward down the east coast of Britain; hence, on this coast the day time sea-breezes are well developed. In June, July and August the water temperature off the North Sea shores is a little lower than that off the shore of the Irish Sea. At the same time, westerly winds off the land, by reason of friction, are likely to be less strong on the east coasts; whereas on the west coast, the sea-breeze merely intensifies the strength of the prevailing wind. With light westerly winds, skies inland are a little less cloudy on the eastern side of the country, day-time temperatures inland rise a little higher and the temperature gradient across the coast is sharper. It is quite common to find that on summer days when the maximum a few kilometres inland rises to between 27° and 30°C, it will not exceed 20° at most on the east coast.

The distance to which the afternoon sea breeze will penetrate, depends on the extent to which it is favoured or opposed by the overall isobaric gradient. On the east coast, if the westerly land wind does not exceed force 3, summer sea breezes will undercut it for about 4 hours in the afternoon, and may penetrate for between 5–8 km from the coast.

On the south coast, the prevailing westerly wind of summer acquires a slight thermal component, so that by day it tends to back to the southwest, and to veer northwest and fall to light during later night hours. In fine weather, therefore, the sky adjacent to the coast tends to be a little clearer by day, with cloud building up a few kilometers inland.

By night, such cloud development as there is will be more over the sea. As a result the coasts everywhere show a greater sunshine duration than is recorded a few kilometers inland. The difference is of the order of 5–10% on the year.

Occasionally when the isobaric gradient reinforces the thermal effect, sea breezes may become perceptible on warm afternoons many kilometres inland. There is evidence that in hot summer weather with light east winds, the sea breeze from the southern North Sea can be perceived in London in a slight freshening of the air about five hours after noon (WALLINGTON, 1964).

The low stratus cloud that can accompany a sea breeze from time to time is discussed elsewhere, under cloudiness.

Land-breezes are in general very slight, except occasionally after very cold winter nights when the ground is snow-covered.

Mountain breezes are to be observed as small-scale evening katabatic flow in quiet weather, and the normal valley breeze by day is occasionally observed; but, in general, in these northern latitudes the weather by day, especially in Scotland, is rarely calm enough for such effects to prevail. The winds resulting from the greater circulation obscure or suppress such local developments.

Föhn effects, however, are noteworthy. They are most common when an anticyclone has become centred to the eastward, so that the prevailing wind lies between south and south-southeast. In such circumstances, subsiding air above leads to the existence of an inversion-layer at anything from 5,000 to 10,000 ft. (1,500–3,000 m). Frequently there is a second such layer at a lower level, marked by more or less extensive North Sea stratus cloud with a base at 500–700 m, summit 1,000–1,500 m.

In these circumstances standing-wave phenomena develop over and in the lee of the mountains. In the regions where the dry warm air is descending, the temperature rises unexpectedly high, accompanied by the characteristic gusty wind and very low humidities.

The most frequent developments of this kind affect the coastal resorts and airfields along the south shore of the Moray Firth in northeastern Scotland; they become most noticeable in the cooler months, from October to April. Temperatures exceeding 15°C in December and January have been known; a maximum of 22.3°C was observed at Kinloss and Forres on 9 March, 1948. Occasionally northwest Scotland can be affected. In a strong southeast airstream on December 2, 1948 a maximum of 18.3° was observed at Achnashellach in Ross-shire, with 17.2° beside the lighthouse at Cape Wrath, 115 m. above the sea. Under similar conditions a maximum of 22.3°C was observed at Cape Wrath on October 22, 1965.

Further south, the highest midwinter maxima under similar circumstances are recorded along the North Wales coast—17.8°C in December and 17.2°C in January have been observed. High winter temperatures in descending air on the margin of an anticyclone have also been known at Durham and at Dublin in eastern Ireland. Statistics are not available, but in an average year there are likely to be 5–10 days on which föhn developments will be both noticeable and widespread.

**Temperature conditions in general**

The distribution of mean temperature in the British Isles is governed by (*1*) latitude;

(2) distance from the continent, from which the drier air may spread giving unusual warmth in summer, cold in winter; (3) proximity to, or distance from the sea-coast; (4) altitude.

Local modifications, thereafter, can be ascribed to aspect and the effects of minor relief features, soil, proximity to water bodies, and the works of man in the form of drainage, shelter-belts, building and, most markedly, the extension of the built-up areas of towns and cities together with their production of smoke.

The effect of latitude does not solely derive from the declining altitude of the sun or from increased proximity to the arctic ice; it becomes the more effective as we find that the further north, the greater are the chances that any place will be affected by the stronger winds, the greater amount of cloud, the shorter duration of bright sunshine, and the more frequent precipitation associated not only with the passage of active fronts, but also the greater frequency with which instability of the surface air prevails. As the principal tracks of active Atlantic depressions lie between the Hebrides and Iceland, fronts are more active and frontal passages more frequent throughout the northwest and north, from Mayo to Sutherland and the Shetlands, while winds are on the whole stronger and more changeable. Eastern Scotland and northeast England owe much to the existence of the uplands and mountains that provide, within the prevailing westerly stream, a certain amount of orographic shelter, while Wales provides the like shelter for much of the west and northwest Midlands. The small relatively dry area round Dublin should also be noted; this is the only part of Ireland with an annual average rainfall very slightly below 750 mm.

The operation of the several factors can be observed in the trend of the mean isotherms for each month. In the midwinter months, the isotherms show, in common with those of western Europe, a marked trend from south to north, expressive of distance from the cold continental air supply; in detail, the coasts are everywhere a little warmer than the interior. The coldest winters at sea level lie within the east Midlands, but altitude is sufficient to ensure that central Scotland at higher levels is considerably colder.

In spring the trend of the isotherms becomes rather east-southeast to west-northwest, but with a sharp fall of mean temperature on the coasts adjacent to the chilly waters of the North Sea. The most forward spring growth is found in the sheltered inland valleys of the southwest; south Devonshire in this respect is about a month ahead of Edinburgh. In summer the effects of inland location become marked; compare the relatively cool Isle of Man, in July, with the warmth of Yorkshire or that of southeast Ireland.

The isotherms retain a general west–east trend in the earlier autumn, when the effect of the coastal seas ensures that the coastal stations remain warmer than inland, especially towards the southwest.

The warmest summers are found in the Thames Basin and inland Kent and Sussex. Hence London, with its additional "urban effect" which raises the mean summer temperature over the inner suburbs and the city by about 0.5°–0.7°C, has almost always the highest mean and extreme temperatures in Britain (CHANDLER, 1965). But the warmest autumn means are found beside the southwest coasts.

From northwest to southeast, summer averages rise while those of winter fall; the annual range of monthly mean temperature accordingly increases from about 9°C in Donegal and the smaller Outer Hebrides to 14°C or nearly so inland in the southeast of England, 12°C towards eastern Scotland and 11°C in southeast Ireland. The effect of the Irish Sea

TABLE X

COMPARISON OF THE STATISTICS OF MEAN TEMPERATURE (°C) OVER 3 LENGTHY PERIODS
OF TIME FOR CENTRAL ENGLAND (cf. Table IV)

| Period | Jan. | Feb. | Mar. | Apr. | May | June | July | Aug. | Sept. | Oct. | Nov. | Dec. |
|---|---|---|---|---|---|---|---|---|---|---|---|---|
| 1721–1800[1] | 2.9 | 3.8 | 5.0 | 7.9 | 11.3 | 14.5 | 16.1 | 15.8 | 13.6 | 9.4 | 5.8 | 3.8 |
| 1801–1880[2] | 3.0 | 4.1 | 5.3 | 8.0 | 11.2 | 14.3 | 15.8 | 15.4 | 13.1 | 9.7 | 5.7 | 4.0 |
| 1881–1960[3] | 3.7 | 4.0 | 5.5 | 8.0 | 11.3 | 14.2 | 16.0 | 15.5 | 13.3 | 9.7 | 6.3 | 4.4 |

*Standard deviations for individual months from the above means:*

| Period | Jan. | Feb. | Mar. | Apr. | May | June | July | Aug. | Sept. | Oct. | Nov. | Dec. |
|---|---|---|---|---|---|---|---|---|---|---|---|---|
| 1721–1800 (°C) | 2.0 | 1.7 | 1.6 | 1.3 | 1.2 | 1.1 | 1.1 | 1.0 | 1.1 | 1.2 | 1.3 | 1.6 |
| 1801–1880 | 1.9 | 1.8 | 1.3 | 1.2 | 1.3 | 1.2 | 1.2 | 1.1 | 1.1 | 1.2 | 1.4 | 2.0 |
| 1881–1960 | 1.8 | 1.9 | 1.4 | 1.2 | 1.1 | 1.0 | 1.2 | 1.2 | 1.1 | 1.3 | 1.4 | 1.7 |

[1] Period of scattered records of variable quality, imperfect instruments, exposures and hours of reading.
[2] Period with reliable records carefully reduced and standardised by intercomparison.
[3] Period of standardised exposures with observations accepted by the Meteorological Office.

in ensuring a somewhat milder winter over much of Ireland becomes evident. The mean daily range of temperature is likewise greater inland and towards the southeast. On the whole there is less wind, a greater chance of quiet anticyclonic weather, and clearer skies with more sunshine; this applies throughout the year.

The average extremes for each month follow a broadly similar pattern when the isotherms are plotted on a small-scale map. But the diversity of small-scale relief, and even more that of the soils, gives rise to local variations in the minima, on clear nights, of remarkable significance. To show this is quite beyond the capacity of normal isothermal maps. It can however be demonstrated, without any special instrumentation, by comparing the observations from existing normal climatological stations (cf. MANLEY, 1944).

In Fig.5 the average extreme minima for each month over a period of years have been generalised from a number of inland stations distributed over the English Midlands.

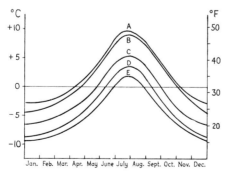

Fig.5. Course of the average extreme minimum air temperature by months in differing exposures inland·
A = inner parts of large cities; B = favoured slopes; C = normal open unsheltered lowland; D = frost-hollows; E = exceptional lowland sites on sandy soils. Applicable for England.

TABLE XI

AVERAGE NUMBER OF DAYS WITH "AIR FROST" (SCREEN MINIMUM 0°C OR BELOW)
RECORDED OVER 10 YEARS, 1956–1965*

| Place | Number of days |
|---|---|
| Eastern England (to illustrate variations over a broad agricultural lowland with no marked features): | |
| Cambridge (Botanic Gardens) | 65 |
| Cambridge (Agric. Botany Station) | 60 |
| Boxworth, 15 km west | 63 |
| Cardington, 40 km west | 68 |
| Wyton, 20 km northwest | 58 |
| Santon Downham (sandy, 45 km northeast) | 102 |
| Mildenhall, 35 km northeast | 57 |
| *Other stations:* | |
| Lerwick (Shetland Is.) | 49 |
| Aberdeen-Dyce (agricultural N.E. Scotland) | 79 |
| Eskdale muir (south Scotland, upland) | 95 |
| London (Wisley, rural outskirts to southwest) | 58 |
| Renfrew (lowland, by Glasgow) | 60 |
| Balmoral (eastern Scottish Highlands, valley site) | 119 |
| Moor House (Pennines, 550 m near limit for tree growth) | 131 |
| Oxford (south Midlands, somewhat sheltered) | 56 |
| Plymouth (coastal, S.W. Devonshire) | 29 |
| Waddington (Lincolnshire, agricultural) | 57 |
| Manchester Airport (Cheshire, lowland) | 51 |
| Stornoway (Isle of Lewis, Hebrides) | 47 |
| Aldergrove (N.E. Ireland, inland) | 54 |
| Cape Wrath (northern coast, Scotland) | 27 |
| Scilly (Islands) off southwest Cornwall | 4 |

* The very great differences between an exposed coastal locality, such as Cape Wrath and an inland valley among the mountains (Balmoral) will be evident. For southern Ireland inland, 35 to 50 is probable. For characteristic monthly frequencies inland in England see Table IV.

It follows that the statistical chances of frost damage in any given location are very hard to evaluate. There is evidently a long period in an average year during which radiation nights will allow the temperature to fall to a point which may be just above, quite nearby, or just below 0°C, for a varying interest depending on the local microclimate.

For the decade 1956–1965 statistics are available and the average number of days with a screen minimum below 0°C can be given, in order to show the very large differences that can be found within a few kilometres (Table XI), for example, around Cambridge.

Coastal stations likewise show remarkably varied annual totals of days with minima below 0°C. A station at the mouth of a wide flat valley will show a much higher frequency of frost than one situated on a favourable slope or at the foot of a narrow well-wooded coombe (MANLEY, 1944).

These facts go to show that it is quite difficult to generalise with regard to the occurrence of killing frost, or the mean dates of the last frost in spring or the first in autumn. Over much of the English Midlands, and the central Scottish lowland, the characteristic average can be put at 60–70 nights each winter between mid-October and early

May. But if choice is made of individual locations over short sequences of years this number might be halved or doubled.

*Average and absolute extremes of temperature*

The highest temperatures that have been recorded in the British Isles approximate to those on the adjacent continent; from a trustworthy instrument at Tonbridge, in Kent, 38° (100.5°F) was reported in July 1868. There have been a number of occasions with observations between 35° and 37° at inland stations in southeastern England and the east Midlands, notably in August 1911 and August 1932. In Scotland and Ireland 32°C has barely been exceeded in 100 years (90°F has been registered at officially recognised stations on two occasions in Scotland, once in Ireland).

The lowest temperatures on record, taken under standard conditions, come from valley stations in the eastern Scottish Highlands; −27.4°C at Braemar in February 1895. Earlier unofficial observations give reason to believe that −30°C may have been touched in some upland Scottish valleys in the past. For England, −24°C has been observed at Buxton, also in February 1895, with −23°C in central Wales in January 1940. From records kept before the days of Stevenson screens, in combination with what we know regarding the location of unusually severe frost-hollows, the local occurrence of −25°C in southern England is not at all beyond possibility.

For Ireland the lowest yet recorded is −20°C; rarely has the minimum temperature at any of the rather limited number of inland observing stations fallen below −15°C.

The average annual maximum within the built-up area of London reaches 32°C; 30°C over most of the southeast and east Midlands away from the coast; 28°C in northern England, 26°C at Edinburgh and Glasgow, and in the lowlands of eastern Ireland. The average annual minimum varies much more widely. At a characteristically exposed lowland station in the English Midlands −9°C or −10°C can be expected, but on rising ground −7°C will probably not be passed, with −5°C at many coast resorts, or in the towns. Along the coasts of the relatively narrow Cornish peninsula −2°C is only attained every other year; and even in the Hebrides and Shetlands the average annual minimum may remain above −5°C. By contrast, the probable average extreme minimum, representative of high-lying valleys in the Scottish Highlands, can be put at −15°C (Table XII). On clear cold radiation nights the severity of frost at many stations depends greatly on whether there has been a deep and freshly fallen snow cover. Hence there is a very considerable element of chance in the incidence of really low temperatures. The effect of a deep snow cover may be estimated by comparing the extent to which valley minima fall below those on the adjacent slopes; if this difference is 2°C on calm nights with the ground free from snow, it will be of the order of 5°C if the ground is well covered.

*Overall totals of "accumulated temperature" during the growing season*

The accumulated temperature for any day is the amount, in degrees, by which the mean for the day exceeds some adopted value; in Britain 42°F (5.6°C) is used, it being assumed that this is the temperature at which the growth of characteristic plants can be said to begin and to be maintained. It is not by any means a perfect measure, but it makes a useful yardstick with which to compare the effectiveness of various places from the

TABLE XII

AVERAGE DAILY MAXIMA AND MINIMA (°C), 1931–1960[1]

| Station | Jan. | Feb. | Mar. | Apr. | May | June | July | Aug. | Sept. | Oct. | Nov. | Dec. | Year |
|---|---|---|---|---|---|---|---|---|---|---|---|---|---|
| Lerwick | 5.0 | 5.0 | 6.0 | 8.1 | 10.5 | 12.6 | 14.4 | 14.4 | 12.9 | 10.1 | 7.8 | 6.2 | 9.4 |
|  | 1.2 | 0.9 | 1.8 | 2.8 | 5.1 | 7.4 | 9.7 | 9.7 | 8.4 | 6.2 | 4.1 | 2.7 | 5.0 |
| Stornoway | 6.3 | 6.6 | 8.2 | 9.8 | 12.4 | 14.6 | 15.9 | 16.0 | 14.4 | 11.7 | 9.1 | 7.3 | 11.0 |
|  | 2.4 | 2.2 | 3.1 | 4.2 | 6.3 | 8.6 | 10.6 | 10.6 | 9.1 | 7.0 | 4.8 | 3.6 | 6.0 |
| Aberdeen (Dyce) | 4.9 | 5.7 | 7.7 | 10.5 | 12.9 | 16.0 | 17.9 | 17.5 | 15.5 | 12.0 | 8.3 | 6.0 | 11.2 |
|  | −0.2 | 0.0 | 1.2 | 2.8 | 5.1 | 8.0 | 10.1 | 9.6 | 7.9 | 5.6 | 2.9 | 1.4 | 4.5 |
| Eskdalemuir (upland, | 4.0 | 4.8 | 7.4 | 10.4 | 14.1 | 16.6 | 17.7 | 17.3 | 14.9 | 11.2 | 7.6 | 5.3 | 10.9 |
| 242 m) | −1.3 | −1.3 | −0.2 | 1.3 | 3.8 | 6.8 | 8.9 | 8.6 | 6.7 | 4.1 | 1.6 | 0.2 | 3.3 |
| Manchester Airport | 5.5 | 6.0 | 9.6 | 12.5 | 16.4 | 18.7 | 19.9 | 19.7 | 17.4 | 13.6 | 9.3 | 7.3 | 13.0 |
|  | 1.1 | 0.8 | 2.2 | 3.9 | 7.0 | 9.7 | 11.7 | 11.5 | 9.8 | 7.0 | 4.3 | 2.9 | 6.0 |
| Waddington | 5.1 | 6.4 | 9.4 | 12.6 | 16.1 | 19.1 | 20.9 | 20.8 | 17.9 | 13.5 | 8.7 | 6.1 | 13.1 |
|  | 0.9 | 0.8 | 1.8 | 3.9 | 6.4 | 9.6 | 11.7 | 11.4 | 9.7 | 6.6 | 4.0 | 2.3 | 5.8 |
| Yarmouth (Gorleston) | 6.0 | 6.4 | 8.1 | 10.9 | 13.5 | 17.1 | 19.5 | 19.6 | 17.9 | 14.0 | 9.9 | 7.2 | 12.5 |
|  | 2.1 | 2.0 | 3.0 | 5.3 | 8.1 | 11.1 | 13.2 | 13.4 | 11.8 | 8.7 | 5.6 | 3.4 | 7.3 |
| London (Kew Obs.) | 6.4 | 7.1 | 10.1 | 13.2 | 16.9 | 20.3 | 21.8 | 21.4 | 18.5 | 14.1 | 9.9 | 7.4 | 13.9 |
|  | 1.9 | 1.8 | 3.1 | 5.3 | 8.0 | 11.4 | 13.4 | 13.1 | 11.0 | 7.5 | 4.8 | 3.0 | 7.0 |
| Plymouth (Mt. Batten) | 8.2 | 8.1 | 10.1 | 12.3 | 15.1 | 17.7 | 19.0 | 19.3 | 17.7 | 14.6 | 11.3 | 9.2 | 13.6 |
|  | 4.1 | 3.5 | 4.6 | 6.1 | 8.4 | 11.3 | 13.0 | 13.0 | 11.7 | 9.2 | 6.6 | 5.1 | 8.0 |
| Aldergrove (N. Ireland) | 6.0 | 6.8 | 9.2 | 11.8 | 14.9 | 17.5 | 18.4 | 18.3 | 16.1 | 12.6 | 9.1 | 6.9 | 12.3 |
|  | 1.5 | 1.5 | 2.7 | 3.9 | 6.1 | 9.2 | 11.0 | 10.7 | 9.2 | 6.8 | 4.1 | 2.9 | 5.8 |
| Dublin (Ireland) | 7.6 | 8.2 | 10.4 | 12.8 | 15.1 | 18.3 | 19.6 | 19.4 | 17.5 | 14.2 | 10.5 | 8.6 | 13.5 |
|  | 1.2 | 1.2 | 2.4 | 3.5 | 5.9 | 8.7 | 10.6 | 10.4 | 8.5 | 6.2 | 3.7 | 2.4 | 5.4 |
| Valentia (Ireland) | 8.9 | 9.0 | 10.7 | 12.2 | 14.3 | 16.6 | 17.6 | 17.9 | 16.6 | 13.9 | 11.3 | 9.8 | 13.2 |
|  | 4.9 | 4.6 | 5.9 | 6.6 | 8.5 | 11.0 | 12.4 | 12.9 | 11.4 | 9.3 | 6.0 | 5.8 | 8.3 |
| *Additional stations* | | | | | | | | | | | | | |
| Braemar (upland 340 m) | 3.7 | 4.2 | 6.6 | 9.4 | 13.3 | 16.4 | 17.6 | 17.1 | 14.4 | 10.5 | 6.8 | 4.8 | 10.4 |
|  | −2.5 | −2.3 | −1.0 | 1.1 | 3.4 | 6.4 | 8.6 | 8.1 | 6.1 | 3.6 | 0.8 | −0.8 | 2.6 |
| Renfrew | 5.4 | 6.6 | 9.0 | 11.9 | 15.4 | 18.0 | 19.0 | 18.8 | 16.4 | 12.6 | 8.8 | 6.6 | 12.4 |
|  | 0.8 | 1.2 | 2.3 | 3.8 | 6.1 | 9.1 | 11.1 | 10.8 | 8.8 | 6.3 | 3.4 | 2.2 | 5.5 |
| Durham | 5.6 | 6.2 | 8.8 | 11.7 | 14.7 | 18.0 | 19.8 | 19.3 | 17.0 | 13.1 | 9.0 | 6.7 | 12.5 |
|  | 0.1 | 0.3 | 1.4 | 3.2 | 5.5 | 8.5 | 10.7 | 10.4 | 8.5 | 5.7 | 3.3 | 1.5 | 4.9 |
| Cambridge | 6.3 | 7.2 | 10.5 | 13.7 | 17.2 | 20.5 | 22.2 | 22.0 | 19.2 | 14.6 | 9.9 | 7.3 | 14.2 |
|  | 0.6 | 0.5 | 1.6 | 3.8 | 6.5 | 9.7 | 11.8 | 11.5 | 9.5 | 6.1 | 3.5 | 1.7 | 5.6 |
| Oxford | 6.5 | 7.3 | 10.5 | 13.6 | 16.9 | 20.2 | 21.7 | 21.6 | 18.8 | 14.4 | 10.0 | 7.5 | 14.1 |
|  | 1.0 | 1.1 | 2.3 | 4.6 | 7.2 | 10.4 | 12.4 | 12.2 | 10.1 | 6.7 | 4.0 | 2.2 | 6.2 |
| Aberystwyth | 6.9 | 6.8 | 9.3 | 11.3 | 14.5 | 16.8 | 17.8 | 18.1 | 16.4 | 13.4 | 10.0 | 8.1 | 12.5 |
|  | 2.1 | 1.8 | 3.1 | 4.8 | 7.3 | 10.2 | 12.0 | 12.2 | 10.6 | 7.9 | 5.2 | 3.5 | 6.7 |

[1] *Basis:* Meteorological Office published data, and reductions by the author in order to provide 30-year approximations for Irish stations with changes of site.

standpoint of general farming. In Celsius degrees the total reaches 1,600 in favourable areas; diminishing to 1,200 in marginal arable areas on the flanks of the uplands, and to less than 800 around the highest inhabited parts of Britain. Fig.3 demonstrates the range of variation of the annual total (1,200 day-degrees = 40 month-degrees, the unit used). This total would be characteristic of the upland margin of land normally ploughed; and would lie at about 250 m for the southern Midlands, 180 m on the Pennines.

TABLE XIII

CLIMATOLOGICAL DATA: HUMIDITY AND PRECIPITATION AT REPRESENTATIVE STATIONS, 1931–1960*

| Station | Jan. | Feb. | Mar. | Apr. | May | June | July | Aug. | Sept. | Oct. | Nov. | Dec. | Year |
|---|---|---|---|---|---|---|---|---|---|---|---|---|---|
| *Relative humidity (%) averaged through 24 h:* | | | | | | | | | | | | | |
| Lerwick | 93 | 93 | 91 | 88 | 88 | 88 | 92 | 92 | 93 | 92 | 93 | 94 | 93 |
| Stornoway | 90 | 89 | 88 | 87 | 86 | 83 | 87 | 89 | 88 | 89 | 88 | 89 | 89 |
| Aberdeen (Dyce) | 88 | 84 | 84 | 81 | 82 | 81 | 83 | 84 | 84 | 86 | 88 | 89 | 87 |
| Eskdalemuir | 90 | 87 | 85 | 83 | 80 | 83 | 85 | 87 | 88 | 89 | 90 | 91 | 89 |
| Manchester (Airport) | 89 | 89 | 82 | 78 | 74 | 77 | 79 | 81 | 82 | 84 | 86 | 86 | 85 |
| Waddington | 91 | 86 | 83 | 80 | 79 | 79 | 79 | 80 | 80 | 87 | 91 | 91 | 87 |
| Yarmouth (Gorleston) | 86 | 83 | 81 | 79 | 80 | 78 | 78 | 80 | 80 | 81 | 86 | 88 | 84 |
| London (Kew Obs.) | 88 | 84 | 79 | 73 | 72 | 70 | 72 | 75 | 80 | 85 | 88 | 89 | 81 |
| Plymouth (Mt. Batten) | 89 | 87 | 85 | 81 | 81 | 82 | 85 | 85 | 85 | 86 | 87 | 89 | 87 |
| Aldergrove (N. Ireland) | 91 | 88 | 85 | 83 | 81 | 82 | 84 | 85 | 90 | 89 | 90 | 92 | 89 |
| Dublin (09h00) | 87 | 85 | 85 | 78 | 77 | 76 | 80 | 83 | 85 | 86 | 87 | 87 | 83 |
| (Ireland, 1941–1960) | | | | | | | | | | | | | |
| Valentia (09h00) | 84 | 84 | 82 | 78 | 76 | 80 | 84 | 84 | 84 | 85 | 85 | 84 | 83 |
| (Ireland, 1941–1960) | | | | | | | | | | | | | |
| | | | | | | | | | | | | | |
| *Amount of precipitation (mm):* | | | | | | | | | | | | | |
| Lerwick | 122 | 96 | 78 | 76 | 58 | 63 | 79 | 80 | 99 | 118 | 127 | 133 | 1129 |
| Stornoway | 107 | 75 | 63 | 65 | 52 | 68 | 87 | 88 | 97 | 118 | 111 | 110 | 1041 |
| Aberdeen (Dyce) | 77 | 54 | 52 | 50 | 62 | 53 | 92 | 73 | 65 | 90 | 91 | 78 | 837 |
| Eskdalemuir | 175 | 112 | 98 | 97 | 87 | 108 | 131 | 120 | 136 | 148 | 153 | 162 | 1527 |
| Manchester Airport | 77 | 53 | 45 | 49 | 57 | 61 | 79 | 81 | 67 | 78 | 80 | 72 | 799 |
| Waddington | 54 | 42 | 37 | 37 | 46 | 51 | 62 | 58 | 47 | 53 | 62 | 48 | 597 |
| Yarmouth (Gorleston) | 57 | 42 | 37 | 36 | 38 | 44 | 57 | 56 | 53 | 65 | 68 | 52 | 605 |
| London (Kew Obs.) | 53 | 40 | 37 | 38 | 46 | 46 | 56 | 59 | 50 | 57 | 64 | 48 | 594 |
| Plymouth (Mt. Batten) | 105 | 77 | 73 | 55 | 65 | 58 | 71 | 80 | 82 | 94 | 115 | 115 | 990 |
| Aldergrove (N. Ireland) | 80 | 52 | 50 | 48 | 52 | 68 | 94 | 77 | 80 | 83 | 72 | 90 | 846 |
| Dublin | 70 | 50 | 51 | 41 | 64 | 57 | 65 | 84 | 78 | 64 | 65 | 79 | 768 |
| (Ireland, 1941–1960) | | | | | | | | | | | | | |
| Valentia (Ireland) | 164 | 107 | 103 | 74 | 86 | 81 | 107 | 95 | 122 | 140 | 151 | 168 | 1398 |
| | | | | | | | | | | | | | |
| Additional stations (1916–1950): | | | | | | | | | | | | | |
| Braemar | 102 | 68 | 57 | 57 | 66 | 45 | 76 | 79 | 76 | 103 | 97 | 96 | 924 |
| Renfrew | 118 | 81 | 64 | 59 | 67 | 61 | 79 | 85 | 91 | 119 | 105 | 106 | 1035 |
| Durham | 59 | 45 | 42 | 43 | 51 | 46 | 71 | 69 | 60 | 61 | 63 | 58 | 668 |
| Cambridge | 49 | 35 | 33 | 45 | 46 | 37 | 61 | 48 | 52 | 50 | 53 | 42 | 551 |
| Oxford | 60 | 45 | 42 | 48 | 52 | 43 | 61 | 57 | 56 | 64 | 66 | 58 | 652 |
| Aberystwyth (Frongoch, 136 m) | 94 | 74 | 81 | 59 | 60 | 76 | 88 | 108 | 81 | 119 | 106 | 116 | 1062 |
| Snowdon (Grib Goch)** | 494 | 339 | 266 | 245 | 222 | 284 | 360 | 407 | 392 | 472 | 434 | 445 | 4360 |

* *Source:* Meteorological Office, ANONYMOUS (1962) and W.M.O. Climatological Normals for Ireland.
** Altitude 713 m, wettest of the official reporting stations in north Wales mountains.

## Humidity

Throughout the British Isles there is as a rule sufficient moisture in the atmosphere to ensure that the relative humidity is comparatively high, and the evaporation rate by day is rarely excessive. Indeed, from early November to the beginning of March, evaporation of moisture from the ground surface is so slow that in many years the soil and vegetation remain perceptibly damp almost throughout. In Tables IV and XIII, averages for relative humidity refer to the whole period of 24 h each day, except at Dublin and Valentia.

Statistics have been compiled by the Meteorological Office to show the daytime distribution of average relative humidity and vapour pressure, based on the observations at 13h00. This is a useful figure, inasmuch as it is reasonably representative of daytime experience by those who are out-of-doors. In mid-winter the range is from just over 85% on most of the coasts to just under 80% locally inland. In early summer (June) the averages are lowest; they still attain 80% locally on the coasts, but decrease to just under 60% inland. It is interesting to note that inland the relative humidity in April is generally lower than that for September, and everywhere considerably lower than in October; likewise, May shows lower values than July. These values for 13h00 are mapped in the *Climatological Atlas*.

Throughout the year the average daytime values inland, even in the drier and warmer southeastern counties of Ireland, are in excess of those in the eastern Highlands of Scotland, and till more so by comparison with the English Midlands. This is significant from the standpoint of perception by those who reside in Ireland, associated as it is with slightly less sunshine, more cloud, diminished evaporation, and rather more rain falling on more days than over the greater part of agricultural England.

## Precipitation

In common with much of the lowlands of northwestern Europe, over the greater part of the British Isles precipitation is relatively frequent, but not heavy. About three-quarters of the English lowlands receive less than 750 mm as an annual average; with this can be included a narrow strip forming the lowland fringe of eastern Scotland, and a very small coastal area around Dublin in eastern Ireland (Table XIII).

For the "standard period", 1916–1950, the mean annual rainfall taken over the whole area of England has been calculated by the British Rainfall Organisation (Meteorological Office) to be 32.7 inches or 831 mm, for Wales 53.4 inches or 1,356 mm, and for Scotland 55.9 inches or 1,421 mm. For Ireland as a whole, by extrapolation from earlier data, about 47 inches or 1,200 mm can be estimated for the same period. It is thus evident that the uplands and mountains receive very much more rain than the lowlands and for example, in the western Highlands of Scotland the summit of Ben Nevis receives upwards of 150 inches (3,800 mm); to the westward it is thought that the average fall may exceed 200 inches (5,000 mm) locally, among the mountains at the head of the sea lochs; it may also attain that figure around the summit of Snowdon in north Wales. Falls in excess of 170 inches (4,200 mm) are known in a small area in the central Lake District. The excessive rainfall here and elsewhere is attributable to the local effects of convergence of the moist air from the southwest into valleys radiating seaward from a central group of summits.

Other small areas where rainfall exceeds 2,000 mm are found among the uplands of southwest Scotland, southwest and western Ireland, south and central Wales, the northern Pennines and Dartmoor in Devonshire. The resultant windswept, sour and uncultivable moorland with widespread and deep hill-peat or "blanket-bog" on all the flatter areas is characteristic. The evidence of the volume and extent of the glaciers in the Ice Age indicates that a correspondingly heavy snowfall of Atlantic origin in all these mountain areas from the latitude of south Wales northward, provided the source of the greater part of the British and Irish lowland ice-sheets.

Diminution of precipitation to leeward is, however, quite rapid. In the lee of north Wales, a small area towards Liverpool receives less than 700 mm. Near Inverness, parts of the pleasantly cultivated shores of the inner Moray Firth have less than 600 mm, and it has already been suggested that this is about 30% less than would be likely to fall if there were no mountains to the south and west.

Statistics of the occurrence, distribution and frequency of heavy and prolonged orographic rains each year are given in the annual volumes of *British Rainfall*. In the wettest districts, several instances have occurred of falls exceeding 50 inches (1,270 mm) in a single month, that is about 30% of the year's average total.

As a general rule, it can be said that the wettest month of the year averages about twice the rainfall of the driest month. On an average, from 5 to 6% of the annual total comes in the driest month, frequently June, at some places March or April; and 11 to 12% in the wettest, frequently December but in some places October, November, or January. In the east Midlands, however, where orographic effects are largely absent, the frequency of thunder-rains in July and August is sufficient to bring up the averages in those months; at Cambridge for example July is the wettest month, based on the averages for 1916–1950.

Long records indicate that the variability of the annual rainfall is likely to range from about 60 to 150% of the average. In eastern Kent in 1921 less than 240 mm fell near Margate, with 318 mm at London.

In the lowlands the number of rain-days, or days with "measurable rain" (0.2 mm or more), both in Great Britain and in Ireland, rises slowly northwestward. The annual average, about 160 by the lower Thames estuary, is 167 days at London, 189 at Edinburgh, 206 near Glasgow, 263 at Stornoway. This is in considerable measure a consequence of the greater frequency of showery weather in the more unstable maritime-polar air that comes over the northern seas. In Ireland there is a similar increase from about 180 in the extreme southeast, to upwards of 260 on the northwest coast.

In most places December is the month with the highest average number of days with measurable precipitation; it has also the lowest evaporation.

"Measurable rain" of 0.2 mm represents little more than 10 minutes of normal light rainfall. The British Rainfall Organisation has also tabulated "wet days" on which more than 1 mm falls, which are more likely to be noticed by the public. The proportion of "wet days" to "rain days" is about 65% in the southeast and Midlands, but it rises to 90% in the mountainous districts of northwest Ireland and western Scotland (Table XIV), that is from about 110 to 240 days. With regard to "very wet" days with rainfall exceeding 10 mm, the annual total, less than 15 near the Thames estuary and in the Fenland, rises to 110 or more locally in the wet west Highlands.

TABLE XIV

AVERAGE NUMBER OF DAYS WITH RAIN (1 MM OR MORE) FOR EACH MONTH (PERIOD GENERALLY 20 YEARS)[1]

| Station | Jan. | Feb. | Mar. | Apr. | May | June | July | Aug. | Sept. | Oct. | Nov. | Dec. | Year |
|---|---|---|---|---|---|---|---|---|---|---|---|---|---|
| Lerwick | 26 | 21 | 21 | 19 | 16 | 15 | 16 | 17 | 20 | 24 | 24 | 25 | 244 |
| Stornoway | 23 | 16 | 16 | 17 | 15 | 14 | 14 | 17 | 19 | 22 | 20 | 21 | 214 |
| (Aberdeen (Observatory) | 12 | 11 | 11 | 13 | 11 | 10 | 11 | 11 | 11 | 14 | 13 | 14 | 140 |
| Renfrew (17 yr) | 17 | 12 | 10 | 10 | 10 | 10 | 13 | 14 | 13 | 16 | 14 | 16 | 155 |
| Tynemouth | 11 | 9 | 10 | 10 | 10 | 8 | 11 | 12 | 9 | 12 | 12 | 13 | 129 |
| Yarmouth (Gorleston) | 12 | 10 | 9 | 12 | 9 | 8 | 9 | 9 | 10 | 11 | 12 | 13 | 123 |
| Birkenhead (Bidston) | 13 | 9 | 9 | 11 | 11 | 9 | 11 | 14 | 12 | 15 | 12 | 13 | 139 |
| London (22 yr) (Kew Obs.) | 11 | 8 | 8 | 11 | 9 | 8 | 9 | 10 | 9 | 10 | 9 | 11 | 113 |
| Cardiff | 15 | 11 | 11 | 12 | 12 | 9 | 11 | 13 | 11 | 15 | 14 | 15 | 149 |
| Plymouth | 17 | 12 | 12 | 11 | 10 | 7 | 11 | 10 | 12 | 14 | 14 | 17 | 147 |
| Dublin (35 yr) | 13 | 11 | 10 | 11 | 11 | 11 | 13 | 13 | 12 | 12 | 12 | 13 | 142 |
| Valentia | 21 | 15 | 15 | 14 | 14 | 12 | 15 | 16 | 16 | 18 | 19 | 20 | 195 |

[1] *Source:* METEOROLOGICAL OFFICE (1967) and *Monthly Weather Report.*

TABLE XV

AVERAGE MONTHLY DURATION OF MEASURABLE RAINFALL, IN HOURS (PERIOD 1951–1960)[1]

| Station | Jan. | Feb. | Mar. | Apr. | May | June | July | Aug. | Sept. | Oct. | Nov. | Dec. | Year |
|---|---|---|---|---|---|---|---|---|---|---|---|---|---|
| Lerwick | 98 | 76 | 57 | 56 | 50 | 51 | 59 | 55 | 60 | 78 | 92 | 106 | 838 |
| Stornoway | 105 | 62 | 58 | 60 | 52 | 55 | 59 | 57 | 56 | 85 | 83 | 101 | 833 |
| Aberdeen (Dyce) | 58 | 57 | 50 | 43 | 40 | 39 | 61 | 51 | 38 | 54 | 61 | 69 | 621 |
| Eskdalemuir (south Scotland, 242 m) | 120 | 89 | 79 | 74 | 67 | 76 | 91 | 90 | 86 | 93 | 116 | 149 | 1130 |
| Manchester (Airport) | 64 | 48 | 48 | 41 | 35 | 47 | 52 | 55 | 50 | 54 | 58 | 71 | 623 |
| London (Kew) | 45 | 42 | 36 | 27 | 30 | 28 | 27 | 31 | 33 | 40 | 43 | 45 | 427 |
| Plymouth (Mt. Batten) | 75 | 53 | 58 | 37 | 41 | 33 | 36 | 42 | 47 | 52 | 69 | 73 | 616 |
| Aldergrove (north Ireland) | 80 | 59 | 51 | 45 | 43 | 55 | 59 | 62 | 53 | 54 | 66 | 95 | 722 |
| Loch Sloy (wet mountain station, in west Scottish Highlands) | 144 | 107 | 105 | 95 | 77 | 88 | 104 | 108 | 105 | 139 | 143 | 207 | 1422 |

[1] After METEOROLOGICAL OFFICE, 1967.

## Duration of rainfall

Records are available from about 300 stations of the average number of hours during which measurable rain falls. A short selection is given in Table XV.

The significance of these changes, from relatively dry southeast to wetter northwest, in the public mind is much greater than might appear. An expression is given by the ratio between the number of hours when the sun is shining during the daytime, and the number when measurable rain is falling. On the favoured Sussex coast sunny hours are nearly eight times as many as the rainy hours; but in the wet mountain regions of western Scotland the sunny and rainy hours are only about equal in number. The fact that this ratio is nearly eight times as great on the Sussex coast must make an enormous difference to the impression that is made on the average visitor, especially if he comes from one of the wetter northern districts. This ratio, on the Sussex coast, approaches twice that which is recorded in Edinburgh, and one and a half times that on most of the coast of Cornwall. Inland, the ratio exceeds 6 at Cambridge and London; between 4 and 5 over much of the northern lowlands of England and around Edinburgh and Dublin; about 3 around Glasgow (see Table XV; cf. Table XX).

*Intensity of rainfall; long-continued heavy rains*

Statistics of the incidence of rainfall are abundant, thanks to the remarkably good network of rain gauges that stems from the work of the British Rainfall Organisation founded in 1860. The prevailing intensity with which rain falls in the lowlands is not great; overall, about 1.3 mm/h. This is about one-third of the intensity that is characteristic of the Mediterranean, or of eastern North America.

Heavy falls in short periods occur, chiefly in summer thunderstorms; local intensities have frequently been observed in excess of 50 mm/h. Several stations have recorded falls within 1 hour of 75–100 mm. Almost every county in Great Britain can show one or more records of a fall of 100 mm in a day. The greatest recorded fall within one day occurred in south Dorset in July, 1955; 280 mm were recorded in a prolonged series of thunderstorms over a small area close to the coast. A series of violent summer thunderstorms converging over high ground can give rise to severe flooding in steeply-falling streams and narrow valleys with consequent devastation. Notable events of this kind befell on Exmoor in August 1952; upwards of 230 mm fell over a period of a few hours in the afternoon, causing the Lynmouth floods. Similar catastrophic upland thunderstorms have been known elsewhere, in which local falls of rain of the order of 150–200 mm have been recorded; a number have occurred in the Pennines, doing much damage.

Falls exceeding 100 mm in a day occur among the mountains as long-continued orographic rainstorms, generally in association with strong winds and marked convergence ahead of slow-moving warm fronts in deep depressions. Most of these have taken place among the mountains towards the west coast, where several stations have reported falls of between 180 and 220 mm in the past. It is, however, noteworthy that the most prolonged orographic falls of this type have not quite attained in one day the amount recorded in occasional thunderstorms; and it is noteworthy that hitherto the heaviest falls of the latter type have all been recorded towards the southwest (Dorset, Somerset and Devon).

Very heavy orographic rains intensified by the effects of surface convergence have been recorded from the east side of Britain. Over 150 mm fell in one day on the uplands of Berwickshire in August 1948. Probably the greatest series of floods on record in the northeast of Scotland were the "Moray floods" of August 1829, associated with continued convergence of the air over a period of more than one day into the Moray Firth.

*Origin of the rainfall*

Recent studies at Keele University by SHAW (1962) on the incidence of rain in the North Midlands, have shown—for the 5 years 1956–1961—the proportion that could reasonably be ascribed to different synoptic situations. Rather more than half the rain can be described as frontal; divided fairly equally between warm front, warm sector and cold front. About one quarter is attributable to rainfall in unstable maritime-polar air; and only about 3% can be ascribed to "thunderstorm rains".

The significance of this appears to be, that if the "instability" contribution were much increased in a period of zonal flow, associated perhaps with a cooling phase of climate, the precipitation would increase much more markedly on the northwestern Highlands, and probably on the Pennines and Wales. This view received some support from the events of the generally unsettled year, with a very cool summer, in 1954; rainfall was 50% above normal in the western Pennines, but only 10% above normal in the southeast.

While in the uplands it is not easy to make a clear attribution of rainfall to orographic, cyclonic, convectional or instability causes, as each category must be partly influenced by the others, it can on the whole be considered that in the wettest mountain areas, the orographic factor is responsible for upwards of three-quarters of the whole precipitation. Moreover, a part of the decrease towards eastern England and Scotland arises from "orographic shelter".

This is demonstrable if we consider the probable rainfall that would be expected if Britain were an entirely flat plain. Choosing stations as free as possible from orographic effects, the isohyets would probably run southwest–northeast, or more nearly south-southwest–north-northeast, and would range from under 600 mm in the extreme southeast to over 1,100 mm in the far northwest. This would suggest that the lowlands of northeastern England and eastern Scotland now receive, on the average, about 10% less rain than if the country were flat; part of Herefordshire, the north Wales coast with northwestern Cheshire and Liverpool 15–20% less, and small areas along the south shores of the Moray Firth perhaps 30% less than if the mountains were not present.

*Frequency of droughts and dry spells*

Absolute drought is defined as a spell of 15 days or more without measurable rain on any day. On the average, at least one such spell can be expected each year in the southeast, but they are much more rare in the far northwest, where as many as 20 dry days in succession have not hitherto been recorded. South of the Scottish Highlands, most of Britain has experienced a spell of more than 30 days without rain, rising to 40 locally. In a small strip along the coast of Kent and Sussex, a spell of over 60 days without rain was recorded in the spring of 1893. Maps are given in the *Climatological Atlas of the British Isles* (METEOROLOGICAL OFFICE, 1952).

Drought in Ireland is less likely, although a total of 30 days without rain has been slightly exceeded over parts of the east and southeast.

**Snowfall and snow cover**

Snow or sleet is likely to fall whenever a sufficient depth of cold air is present; this means that the wet-bulb temperature of the surface layer should not exceed 3°C. Surface winds from some point between southeast through northeast to north are most likely; occasionally northwest. Westerly winds, with a longer Atlantic travel, are rarely cold enough for snow, except at high altitudes in Scotland.

Throughout the inhabited parts of the British Isles, the mean temperature of the coldest winter month lies above the freezing point, but there are at least six months during which it lies below 10°C. The chance that some of the precipitation on any one day will comprise snow or sleet rather than rain increases from about 0.3% in a month averaging 10°C to upwards of 70% in a month averaging 0°C. The frequency increases rapidly below 4°C. As large parts of the lowlands have a January mean of the order of 4°C, and as the range of variation of the mean temperature lies from 4°C above that normal to about 6°C below, it will be evident that from October to May exceptionally wide variations occur in the frequency with which snow falls, both from month to month and year to year. At quite low altitudes in southern England, snowflakes have been observed in every month from September to June, and from the north Midlands there are credible reports that melting snowflakes, rather than hail, were observed on one occasion in July 1888.

The extreme dates of occurrence on record for the London area since 1811 are 25 September, 1885, and 27 May, 1821.

But so frequently are the temperatures "marginal" that there are many occasions when an alert observer will record sleet, that is melting snowflakes, that others nearby on the same day will fail to note. It is therefore necessary to treat the observations with careful attention to the quality of the station. At a first-class airfield keeping a night watch, the average number of days with snow or sleet observed will be nearly double that which may be noted at a minor climatological station nearby, at which instruments are only visited once daily. Compare, near London, Kew Observatory (5 m) with an average of 16.2 days with snow or sleet observed (1931–1960); Croydon Airfield (67 m), 19.0; Wisley (32 m) 9.0; Bromley (64 m), 10.3; these latter stations are of minor rank. There is however a sufficiently constant relationship between stations to enable satisfactory maps to be drawn.

Snow frequency increases from south to north, from west to east, and with altitude. It also increases on all those uplands that lie exposed either to the unstable maritime Arctic airstreams, or to the continental outflow that, if cold, acquires instability as it crosses the unfrozen North Sea. Instability snow and sleet showers are therefore more frequent adjacent to the North Sea coast, and sometimes in north Wales, north Norfolk and east Kent. Eastern and northern Ireland are similarly affected, but less often as the additional crossing of the Irish Sea which remains slightly warmer than the North Sea, means that at the surface such showers may be of cold rain or hail, rather than sleet or snow.

*Variations in the frequency, or annual number of days, of snowfall*

Since 1668 the occurrence of snowfall, including sleet, has been recorded by a long succession of observers in the London area, whose records can be standardised.

Fluctuations in the total of days for each winter can be stated in terms of percentage of the average. That of 1694–1695 gave over three times the normal total; those of 1784–1785, 1878–1879, 1916–1917, 1962–1963 almost three times. By contrast there have been winters during which some observers did not record snow at all; careful comparison,

TABLE XVI

FREQUENCY OF DAYS WITH SNOW OR SLEET OBSERVED TO FALL, AT REPRESENTATIVE INLAND STATIONS WITH A CONSISTENT RECORD: 1931–1960*

| Station | Jan. | Feb. | Mar. | Apr. | May | June | July | Aug. | Sept. | Oct. | Nov. | Dec. | Year |
|---|---|---|---|---|---|---|---|---|---|---|---|---|---|
| London (Kew Obs., 5 m) | 4.9 | 5.2 | 2.7 | 0.9 | 0.1 | – | – | – | – | <0.1 | 0.3 | 2.1 | 16.2 |
| London (Hampstead, 135 m) | 6.2 | 6.7 | 4.1 | 1.3 | 0.3 | – | – | – | – | 0.1 | 0.5 | 3.1 | 22.3 |
| Cambridge, 13 m | 4.5 | 5.1 | 2.5 | 0.6 | 0.2 | – | – | – | – | 0.1 | 0.3 | 1.7 | 15.0 |
| Huddersfield Oakes, 232 m (Pennine slopes E.) | 7.5 | 8.5 | 5.2 | 2.8 | 1.0 | <0.1 | – | – | – | 0.3 | 1.3 | 4.9 | 31.5 |
| Stonyhurst, 115 m (Pennine slopes W.) | 6.2 | 6.0 | 3.8 | 1.5 | 0.3 | – | – | – | – | 0.1 | 0.7 | 3.1 | 21.7 |
| Eskdalemuir (upland 70 km S. of Edinburgh, 242 m) | 11.7 | 11.1 | 8.0 | 4.3 | 1.5 | <0.1 | – | – | <0.1 | 1.1 | 2.7 | 7.5 | 47.9 |
| Braemar, 340 m (E. Highlands, near Balmoral, 283 m) | 10.8 | 10.4 | 7.7 | 5.7 | 2.3 | <0.1 | – | – | 0.1 | 1.3 | 2.6 | 6.5 | 47.4 |
| Rhayader (Central Wales) 231 m, 1918–1949 | 5.2 | 4.0 | 3.6 | 1.8 | 0.4 | – | – | – | – | 0.3 | 1.1 | 3.2 | 19.6 |
| Exeter (Devon) S.W. England, 32 m 1946–1965 | 4.5 | 4.5 | 1.9 | 0.4 | – | – | – | – | – | – | 0.3 | 1.9 | 13.5 |
| Dublin (E. Ireland) 48 m, 1921–1940 | 2.8 | 2.4 | 2.4 | 1.5 | 0.3 | – | – | – | – | 0.1 | 1.0 | 1.5 | 12.0 |

* *Source: Monthly Weather Report*; calculations and estimates for gaps by the author.

however, has not revealed a winter completely free, if account is taken of the London region; but there have been occasional winters with not more than three days, such as 1862–1863; and 1685–1686, 1733–1734, 1942–1943, 1956–1957 were also nearly free (Fig.6 and Table XVI).

The interest of the long series of London observations lies in the fact that the fluctuations are broadly representative of those over the country generally.

Ireland lies at a greater distance from the source-regions, continental or arctic, from which the colder air-masses of winter arrive. Hence there will be many occasions when snow falls in Scotland or England, but in the same air mass merely sleet or cold rain will be observed at Irish stations. In general the Irish Sea is a degree or so warmer than the North Sea throughout the winter months.

Fig.6. Snow cover: annual average number of mornings with snow cover, 1931–1960. Isopleths at 5, 10, 15, 20, 30, 50 and 100 days. (Mountains cannot be shown in detail, especially in Scotland.) Basis: *Meteorological Office Monthly Weather Reports* with additional data.

*Snow cover*

For the same reason, namely that the coldest month has a mean temperature a little above the freezing-point, snow cover in the British Isles is in general intermittent, and of brief duration; but exceedingly variable from season to season (MANLEY, 1939).

A day with snow lying is recorded when at 09h00 more than half the ground in sight at the level of the observer is covered. Hence in the milder parts of the country this will not always imply a whole day's duration of snow cover. Much depends on how far the ground surface may already have been chilled. But in the uplands, the total of days with snow cover at 09h00 is probably very close to the number of days duration. Along the south coast the average decreases from about four mornings a year in Sussex to less than two in Cornwall. Around London at lower altitudes, though not within the closely built-up area, this rises to about seven mornings a year; it has however varied from 1 to 56 within the past three decades.

Over much of the Midlands, and as far north as York and Lancaster, between eight and twelve mornings a year is characteristic; and the Scottish lowlands lying by the Forth and Clyde are very similar. The effect of the sea is felt everywhere; even in the Shetlands, snow lies on an average of twelve mornings or so, although it can be expected to fall on upwards of 40 days, twice as many as in the English Midlands. In Ireland the distribution is similar, but the frequency lower; much of central Ireland has an average of about five mornings a year, and on the southwest coasts snow cover becomes quite rare and brief.

The number of days with snow-cover rises rapidly with altitude, by reason of lower temperatures; but also, because of the greater quantity that falls, especially on those hills

TABLE XVII

FREQUENCY OF DAYS WITH "SNOW-LYING" (GROUND MORE THAN HALF COVERED AT 09H00), AT REPRESENTATIVE INLAND STATIONS, 1931–1960

| Station | Jan. | Feb. | Mar. | Apr. | May | June | July | Aug. | Sept. | Oct. | Nov. | Dec. | Year | Range of variation |
|---|---|---|---|---|---|---|---|---|---|---|---|---|---|---|
| London (Hampstead) | 4.4 | 4.3 | 2.0 | 0.1 | – | – | – | – | – | – | 0.1 | 1.6 | 12.5 | 0–61 |
| Cambridge | 4.3 | 3.9 | 1.4 | <0.1 | – | – | – | – | – | – | 0.1 | 1.2 | 10.5 | 0–51 |
| Huddersfield Oakes (Pennine slopes E.) | 8.4 | 8.5 | 4.2 | 0.4 | <0.1 | – | – | – | – | <0.1 | 0.5 | 3.6 | 25.6 | 3–69 |
| Stonyhurst (Pennine slopes W.) 1921–1949 | 5.6 | 5.5 | 2.1 | 0.1 | <0.1 | – | – | – | – | <0.1 | 0.2 | 1.9 | 15.4 | 3–56 |
| West Linton (250 m, 25 km from Edinburgh) 1912–1949 | 10.9 | 9.0 | 7.0 | 1.3 | 0.1 | – | – | – | – | 0.6 | 2.8 | 6.1 | 37.8 | 13–88 |
| Eskdalemuir (70 km S. of Edinburgh) | 9.4 | 7.9 | 4.6 | 0.6 | 0.1 | – | – | – | – | 0.1 | 0.9 | 4.7 | 28.3 | 5–68 |
| Braemar (E. Highlands) 1913–1949 | 16.7 | 15.8 | 11.3 | 3.1 | 0.5 | – | – | – | <0.1 | 0.8 | 2.7 | 8.3 | 59.2 | 28–122 |
| Rhayader (Central Wales) 1918–1949 | 4.6 | 4.0 | 3.0 | 0.9 | <0.1 | – | – | – | – | – | 0.5 | 2.3 | 15.4 | 2–60 |
| Exeter (Devon) 1946–1965 | 2.7 | 1.7 | 0.2 | – | – | – | – | – | – | – | – | 0.9 | 5.5 | 0–36 |
| Dublin (E. Ireland) 1921–1940 | 1.5 | 0.9 | 0.9 | – | – | – | – | – | – | – | – | 0.7 | 4.0 | 0–20 |

*Source: Monthly Weather Report:* calculations and adjustments by the author.

and mountains that lie exposed to the east or northeast wind. "Orographic snowfall" can be extremely heavy and prolonged on occasion on these northern uplands; two or three such falls in succession can give a very great accumulation, capable of lasting for some time. Moreover, on the open windswept uplands such falls are accompanied by severe drifting. At the beginning of March, 1947, 60 inches (150 cm) was the prevailing depth in upper Teesdale, and also on the uplands of northeast Wales. In a single fall lasting 56 h, 42 inches (110 cm) fell at Durham in February, 1941.

Extremely heavy orographic snowfall can accumulate on the southwestern uplands. With a depression centred over western France and a strong cold east wind blowing down-Channel, Dartmoor presents a barrier across which severe blizzards can occur. Even on the upland of west Cornwall heavy drifting snowfalls, though rare, are not unknown, as in February, 1955; for similar reasons, occasional heavy snow may fall along the south coast of Ireland.

Apart however from the uplands and mountains, upon which orographic uplift adds to the precipitation, the maximum depth of snow that can accumulate is considerably less: from a variety of sources, it appears that 60 cm of level snow is probably about the largest fall that can ever be expected on low ground, and this perhaps once in 50–100 years. Much of lowland Lancashire received a snowfall approaching this depth at the end of January, 1940. There are reports of falls of the same order in London in February, 1579, over East Anglia in December, 1836, over parts of Oxfordshire in January 1881.

In round figures the rate of increase with altitude in the average annual number of days with snow-cover can be put at about 8 per 100 m of rise. It is rather more rapid (about 11) in northeast Scotland, less rapid in south Wales and Devonshire (about 5). There are considerable local variations, as some hill slopes and summits are likely to carry less snow on account of drifting, so that the local duration may be less. Hence generalised figures such as the above must be approximate only (Fig.6).

It is probable that the most serious snowfalls from the standpoint of the disturbance they cause, are those which from time to time affect the whole of the southern counties of England, bordering the Channel. Famous "blizzards" of this type befell in 1814, 1836, 1881, 1891 and 1963. In each case, a vigorous depression moving eastward into central France was accompanied by a very cold continental outflow on its northern flank.

These facts are mentioned because, although in many years snowfalls are of quite negligible importance over most of the British Isles, no part of the islands is wholly free from the possibility of very heavy falls; and they can occur at unusual seasons, such as those produced in a "polar-air low" developing in an arctic-maritime outflow in springtime. Exceedingly heavy snowfall was reported in northwest Ireland in April 1917, in east Suffolk in April 1919; in Lancashire and west Yorkshire in May 1935, and in north Scotland in June 1749.

The range of variation in the number of days with snow cover can be expressed; if $N$ is the normal, the maximum at low levels in the south and midlands appears likely to be $N + 50$; at 250 m this probably rises to $N + 60$; at 500 m in northern England, to $N + 70$, and above 1,000 m, to $N + 80$ or more.

On rare occasions, therefore, continuous cover will last for two months, as in 1963, over much of the lowland country of England and perhaps one month in central Ireland; but the probability of such an event appears to be of the order of once, possibly twice, in a century.

On the mountains, the probability of an average of 100 days with snow cover is attained at 600 m in the eastern Scottish Highlands. Snow cover at this altitude is, therefore, fairly reliable from early January to the end of March and winter skiing is now a regular development at several resorts. It is however still quite common for an influx of warm maritime tropical air with heavy rain to remove the whole of the snow up to 1,000 m or so. Even on the summit of Ben Nevis (1,343 m) the temperature has been known to rise to 9°C in January, although the mean for the month is —4.5°C.

Small drift-accumulations in the steep, narrow and much shaded north-facing gullies may last throughout the year, both on Ben Nevis, and on Braeriach in the Cairngorms, where they persist at an altitude of 1,150 m. Up to 1933, these had not been known completely to disappear. In the September of that year, however, they did, and since then there have been at least seven occasions when the last of them has melted completely in the late summer, sometimes in September. This response is in accordance with the prevailing climatic amelioration in northwest Europe and the glacier retreat in Norway. It is considered that the theoretical climatic snowline for Ben Nevis would now be found at about 1,620 m.

Knowledge of the overall fluctuations of temperature since 1670 leads to the belief that no true glaciers are likely to have developed in Scotland in historic times.

## Hail

Soft hail is a frequent occurrence during windy and showery westerly weather in the cooler months. Rain and hail showers are very characteristic of maritime polar air behind a depression, when the surface air is not quite cold enough for sleet or snow. They occur most often further north, along the west coasts and on the uplands. Statistics of the frequency are not very reliable, as small hail mixed in rain, or sometimes with snow, is not always either observed or recorded and its effects are insignificant.

True hail accompanies the development of towering cumulo-nimbus clouds with thunderstorms or thundery showers, and is similarly distributed in space and time. While as a rule such phenomena do not attain the violence that is observed in more southern lands, occasionally hailstorms of considerable severity, causing for example, widespread damage to greenhouses, have been known. In such storms hailstones have been reported up to the size of small plums at various localities in the east and southeast, and also from the flanks of the southern Pennines. All these areas, as the maps in the *Climatological Atlas* show, display a relatively high frequency of thunder, averaging up to 20 days yearly in parts of the east Midlands, along the Trent Valley.

The greatest chance of severe hail is probably in the early summer, but the number of really serious falls of a kind likely to cause damage is fortunately small and generally very localised. Hailstones are said to have accumulated to a depth approaching 2 ft. in a small area in the centre of Tunbridge Wells after a violent storm in August 1954.

## Thunderstorms

The frequency of "days with thunder heard" is shown in the accompanying map (Fig.4). From this and the table of frequencies (Table IV), it will be evident that by far the greater number occur inland in the summer months. They are associated with daytime convection and are frequent if cool air at higher levels is over-running warm and humid air below. As a whole, therefore, they are considerably more common inland.

Towards the exposed west coasts however, winter thunder is by no means rare; it is associated with unusual instability in maritime-polar air during the cooler months, with as a rule strong westerly winds. It is especially characteristic of the mountainous fringe of northwest Scotland and western Ireland, where indeed the winter frequency generally exceeds the summer frequency.

Averaged over the year, the number of days just exceeds 20 along the Trent Valley in the east Midlands; but over most of southern Britain inland, 10–15 is more general, the great majority in the months May to August. In Scotland and Ireland, small areas only exceed 10 days.

Thunderstorms are occasionally very prolonged, sometimes as the result of the convergence of several storms. Occasional storms, prolonged far into the night, have occurred in southern England, and are attributable to the arrival of warm humid air following a day of very high temperatures in France; such air, moving north, becomes unstable at high levels over England. Occasional catastrophic rains on the uplands, with severe flooding, have followed the convergence of several slow-moving storms under such circumstances.

**Evaporation**

Evaporation has been measured over many years at a small number of stations, using open tanks; these tend to give rather higher figures than normal ground surfaces. Lysimeters have been operated in a few locations, and elsewhere run-off studies have been made. Evaporation from different types of vegetation cover has been studied. More recently PENMAN (1948) has attacked the problem of estimating the evaporation from the land-surface generally by deduction from the prevailing temperature, humidity, wind speed and sunshine duration. More measurements are needed.

It is evident that under British conditions, the loss by evaporation during the six warmer months generally exceeds the rainfall, wherever the total rainfall does not exceed 700 mm or thereabouts. In a normal year, nearly seven-eighths of the measured evaporation takes place from early April to mid-October. This means that from an average ground surface in the south Midlands, the loss during that period is about 350 mm while the precipitation will average about 330 mm.

In a cool and cloudy summer evaporation may decrease to 270 mm while precipitation increases to, say, 450 mm. On the other hand, a very hot dry summer will see these figures reversed; evaporation 450 mm, precipitation possibly as low as 200 mm. Evaporation from differing surfaces, under varying wind speeds, varies considerably, and while the loss from such surfaces is generally rather less than that from the standard tank (6 ft, or 1.8 m square), there is no constant ratio. Available observations since 1950 have recently been summarised and reviewed by F. J. Holland in the volume *British Rainfall, 1961* (METEOROLOGICAL OFFICE, 1967). Hydrological studies are being extended.

**Cloud, fog and visibility**

The proportion of the sky covered by low cloud (base <2000 m) and the total covered by cloud have been observed for many years, following international practice. British skies as a whole are cloudy, and the proportion of the sky covered by low cloud increases from southeast to northwest. Among the mountains it increases still more markedly. On the whole, skies are clearest in late spring and early summer, and cloud amounts tend to reach a maximum in November and December (Table XVIII).

There are local variations; in general by day there is less cloud on the coasts than inland, and in the spring and early summer low stratus cloud is a little more common on the northeast coast of Britain than elsewhere, on account of the frequency of light easterly winds at that season which become cooled over the North Sea after leaving the continent. Fog in the meteorological sense, that is horizontal visibility of less than 1 km, occurs with much the same frequency as on the adjacent continent. It may take the form of radiation fog, and this tends to be most frequent inland in the late autumn and early winter months, when, given quiet weather, fog formed overnight may last through the succeeding day, especially in broad, damp, river basins such as that of the River Thames. Otherwise, advection fogs are most common on the coast, especially in the late spring and early summer, when warm air from the land may cross water that is considerably cooler. In general the water of the North Sea off the English and Scottish coasts is cooler than that off the Dutch and German coasts; hence fog, lifting during the day with a slight increase in the wind speed to form very low stratus, is a frequent accompaniment, when at this season, pressure becomes high over Germany and the isobaric gradient across the North Sea is slack. At the same season, for similar reasons, they occur on the south coast of Cornwall and Devon when further east the skies remain clear.

Low stratus cloud over high ground, or "hill fog", most commonly occurs in autumn and winter when maritime-tropical air, or returning maritime-polar air with a long fetch over the Atlantic, approaches from the warm seas to the southwest and moves on to the cooler land, so that cloud base is lowered.

TABLE XVIII

AVERAGE AMOUNT OF CLOUD (OKTAS) BASED ON OBSERVATIONS AT STATED HOURS
(PERIOD OF OBSERVATIONS, GENERALLY 10 YEARS)

| Station | Time | Jan. | Feb. | Mar. | Apr. | May | June | July | Aug. | Sept. | Oct. | Nov. | Dec. |
|---|---|---|---|---|---|---|---|---|---|---|---|---|---|
| Lerwick | 07h00 13h00 | 6.1 | 6.4 | 6.3 | 6.2 | 6.2 | 6.0 | 6.6 | 6.4 | 6.1 | 6.2 | 6.2 | 6.3 |
| Stornoway | 07h00 13h00 | 6.2 | 6.3 | 5.8 | 6.0 | 5.9 | 6.1 | 6.7 | 6.4 | 6.4 | 6.5 | 6.0 | 6.0 |
| Aberdeen[1] | 07h00 13h00 | 5.0 | 5.3 | 5.4 | 5.8 | 5.7 | 5.4 | 5.7 | 5.6 | 5.4 | 5.3 | 5.2 | 5.2 |
| Yarmouth (Gorleston) | 07h00 13h00 | 5.8 | 6.2 | 5.2 | 5.9 | 5.6 | 5.3 | 5.3 | 5.1 | 5.2 | 5.6 | 6.2 | 6.0 |
| London (Kew) | 07h00 13h00 | 6.0 | 6.2 | 5.7 | 5.8 | 5.4 | 5.2 | 5.0 | 5.2 | 5.4 | 5.6 | 6.2 | 5.7 |
| Plymouth (Mt. Batten) | 07h00 13h00 | 5.9 | 5.9 | 5.5 | 5.6 | 5.6 | 5.3 | 5.5 | 5.5 | 5.4 | 5.9 | 6.0 | 6.1 |
| Dublin[1] | 09h00 | 5.9 | 5.8 | 5.7 | 5.6 | 6.0 | 5.5 | 5.8 | 5.6 | 5.4 | 5.4 | 5.7 | 6.1 |
| Valentia | 07h00 13h00 | 6.2 | 5.8 | 5.7 | 5.7 | 5.5 | 6.0 | 6.4 | 6.2 | 5.8 | 6.2 | 6.2 | 6.0 |
| Renfrew | 07h00 13h00 | 6.2 | 6.2 | 6.2 | 6.3 | 6.1 | 6.2 | 6.4 | 6.6 | 6.2 | 6.3 | 6.2 | 6.3 |
| Birkenhead (Bidston) | 07h00 13h00 | 5.6 | 5.8 | 5.4 | 5.6 | 5.6 | 5.1 | 5.6 | 5.5 | 5.5 | 5.7 | 5.7 | 5.9 |
| Cardiff | 09h00 | 5.5 | 5.9 | 5.4 | 5.7 | 5.4 | 5.0 | 5.1 | 5.4 | 5.0 | 5.5 | 5.9 | 5.8 |

*Source:* METEOROL. OFFICE (1964, 1965), and W.M.O. climatic data.
[1] From older records kept at Aberdeen (Observatory) and Dublin (Phoenix Park).

TABLE XIX

FREQUENCY OF FOG AT INLAND STATIONS AT 07H00 (VISIBILITY < 1 KM)

| Station | Jan. | Feb. | Mar. | Apr. | May | June | July | Aug. | Sept. | Oct. | Nov. | Dec. | Year |
|---|---|---|---|---|---|---|---|---|---|---|---|---|---|
| Renfrew | 5 | 3 | 1 | 1 | 1 | 0.3 | 0.4 | 1 | 3 | 4 | 5 | 4 | 28 |
| Greenwich | 7 | 5 | 5 | 1 | 0.2 | 0.1 | 0 | 0.1 | 2 | 4 | 7 | 8 | 39 |
| Dublin (09h00) (Phoenix Park) | 2 | 2 | 2 | 0.6 | 0.2 | 0.1 | 0.2 | 0.5 | 0.7 | 1 | 2 | 2 | 13 |

The radiation fogs formed over the land, especially in the broad river basins that are occupied by the great cities or "conurbations", become very much thicker and more unpleasant on account of the discharge of smoke and other products of combustion into them. Impurities accumulate beneath the inversion that is commonly developed in the first 200–300 m or so above the ground; the fog itself becomes denser and more difficult to dissipate because less penetrable by radiation from above. Visibility in a serious London, Manchester or Glasgow fog can diminish to less than 10 m, by daylight, and less than 5 m after dark. The impurities—oxides of sulphur, unburnt carbon particles, hydrocarbon from motor exhaust—are irritating to the throat and nose and, apart from their danger to health, the resultant unpleasantness is never to be forgotten by those who have experienced a really bad city fog. Fortunately the evidence supports the view that in London at least, really bad city fogs are less frequent and unpleasant than they were 50–60 years ago. But the 4 day London fog of December 5–8, 1952, during the quiet anticyclonic weather of that month, was shown to have caused about 4,000 more deaths than would have been expected had the weather been more normal, on account of the unusual concentration of irritant impurities, largely compounds of sulphur, in the surface atmosphere. Subsequent legislation has led to much improvement.

Apart from the serious fogs, it has already been indicated that visibility is at times much diminished by the smoke-drift off the great cities especially in quiet, misty winter weather with gentle winds. Smoke-drift from large built-up areas such as London or Glasgow can occasionally affect the visibility down-wind for more than 100 km from the source. It is this effect of the towns and cities that is responsible for the extensive areas in which poor visibility is found, which may extend over the adjacent sea, for example, in the Thames estuary and Liverpool Bay.

In Scotland the effect of Glasgow is especially noticeable; the smoke-drift at times is sufficiently canalised by the trend of the hills to cause visibility to be affected as far away as Montrose on the east coast.

Excellent visibility is most frequently found in those areas that are most distant from sources of atmospheric pollution; notably, in the Scottish Highlands. Further south it is relatively much more frequent in summer than in winter, except perhaps in the far southwest. Ireland, like northern Scotland, is also more remote and, except near Dublin and Belfast, the proportion of days with good visibility is relatively high.

**Dust storms: haze**

Incipient dust-storms are not unknown; on the fine black soils of the Fenland in late April and early May, considerable drifting on the surface soil has occurred, especially when cool air from the northeast is accompanied by clear skies and strong sunshine.

Under such circumstances considerable turbulence develops in the surface layers, with the result that in sunny weather with easterly or northeasterly winds, especially in spring, haze is widespread; it is well-known in Cornwall, being the more noticed as visibility is otherwise often good.

Smoke-haze from London and the large industrial areas is capable of spreading far to leeward, under conditions that limit upward convection and dissipation. On rather rare occasions it may spread for some miles with sufficient density to curtail the duration of bright sunshine; but in general, it is perceived through the presence of a belt of diminished visibility radiating from the city. In summer the sea-breeze from the mouth of the Tees may sometimes carry the smoke of the steel and chemical works inland for 15–20 km, so that drivers northward bound through Yorkshire find that the visibility decreases from 20–30 km to less than 5 km in a very short distance; after driving a further 10 km they emerge again into clearer air. Smoke-haze spreading at higher altitudes is known to curtail from time to time the shorterwave radiation from the sun; it is considered that this has become something of a disadvantage in the Lea Valley, north-northeast of London, and it is certainly a serious disadvantage from the standpoint of agriculture in southeast Lancashire to the east of Bolton and Manchester.

**Bright sunshine**

The duration of bright sunshine, that is sunshine of sufficient intensity to be registered by the Campbell–Stokes instrument, is recorded at a large number of stations. Health resorts vie with each other in claiming exceptional durations. The desire for sunshine appears to be especially developed among city-dwellers in temperate climates. As the proportion of the British population that can be called "urban" is greater than in any other country of its size, and has now been so for upwards of a century, the climate tends to be unduly stigmatised. The duration of bright sunshine in the lowlands compares well with that of Belgium, The Netherlands, Denmark or northern France, and country-dwellers find little to complain of.

At a few stations along the Channel coast from the Isle of Wight eastward, over 1,800 h, just over 40% of possible, are recorded as an annual average. By contrast, in the northern Shetlands less than 1,100 hours, or 24% of possible, are recorded. Everywhere around Britain itself the flatter off-shore islands, together with a narrow coastal strip, at best only a few km wide, experience slightly clearer skies by day and in consequence record about 10% more bright sunshine by comparison with even the most favoured inland stations. The contrast becomes still more marked when the country is mountainous.

*Average annual duration of bright sunshine (1931–1960) from coastal to inland stations*

*South coast:* Worthing 1,821 h; Tunbridge Wells 1,632 h; London (Kew) 1,514 h; (City) average 1,360 h.

*West coast:* Isle of Man 1,584 h; Blackpool 1,519 h; Stonyhurst 1,320 h; Bolton (industrial town) 1,197 h.

*West Scotland:* Tiree 1,450 h; Oban 1,241 h; Fort William (inland valley) 1,002 h.

*East Scotland:* Arbroath 1,505 h; Edinburgh 1,384 h; Perth 1,312 h.

*Ireland:* Dublin 1,440 h*; Shannon 1,370 h*; Armagh 1,265 h; Rosslare (extreme southeast) 1,700 h*. (* means approximate averages for 1931–1960.)

It will thus be evident that the agricultural lowlands of Britain, away from the sea, receive about 1,500 hours in the southern Midlands, between 1,300–1,400 hours in northern England and south-central Ireland, about 1,300 hours in Scotland and north-central Ireland. The eastern counties are generally more favoured, but an exception can be made for the coastal strip to the lee of the north Wales mountains and coastal Lancashire. Herefordshire, Shropshire and west Cheshire are also relatively favoured. By

TABLE XX

AVERAGE DAILY DURATION OF BRIGHT SUNSHINE, 1931–1960*

| Station | Jan. | Feb. | Mar. | Apr. | May | June | July | Aug. | Sept. | Oct. | Nov. | Dec. | Year |
|---|---|---|---|---|---|---|---|---|---|---|---|---|---|
| Lerwick | 0.8 | 1.8 | 2.9 | 4.4 | 5.3 | 5.3 | 4.0 | 3.8 | 3.5 | 2.2 | 1.1 | 0.5 | 3.0 |
| Stornoway | 1.1 | 2.2 | 3.5 | 4.7 | 6.3 | 5.8 | 4.1 | 4.3 | 3.7 | 2.5 | 1.5 | 0.9 | 3.4 |
| Aberdeen | 1.6 | 2.7 | 3.3 | 4.8 | 5.7 | 6.0 | 5.0 | 4.7 | 4.3 | 3.0 | 1.8 | 1.3 | 3.7 |
| Renfrew | 1.1 | 2.1 | 2.9 | 4.7 | 6.0 | 6.1 | 5.1 | 4.4 | 3.7 | 2.3 | 1.4 | 0.8 | 3.4 |
| Eskdalemuir | 1.4 | 2.3 | 3.1 | 4.3 | 5.6 | 5.6 | 4.5 | 4.2 | 3.3 | 2.5 | 1.7 | 1.1 | 3.3 |
| Manchester Airport (1945–1968 only) | 1.2 | 1.9 | 3.2 | 4.3 | 5.6 | 5.9 | 4.9 | 4.7 | 3.7 | 2.8 | 1.5 | 1.2 | 3.4 |
| Cranwell (8 miles south of Waddington) | 1.8 | 2.5 | 3.6 | 5.2 | 6.2 | 6.6 | 6.0 | 5.6 | 4.6 | 3.3 | 2.0 | 1.6 | 4.1 |
| Yarmouth (Gorleston) | 1.7 | 2.5 | 4.0 | 5.5 | 6.7 | 7.2 | 6.8 | 6.1 | 5.1 | 3.6 | 1.9 | 1.5 | 4.4 |
| London (Kew Obs.) | 1.5 | 2.3 | 3.6 | 5.3 | 6.4 | 7.1 | 6.4 | 6.1 | 4.7 | 3.1 | 1.8 | 1.3 | 4.2 |
| Plymouth (Mt. Batten) | 1.9 | 2.9 | 4.3 | 6.1 | 7.1 | 7.4 | 6.4 | 6.4 | 5.1 | 3.7 | 2.2 | 1.7 | 4.6 |
| Aldergrove (N. Ireland) | 1.5 | 2.3 | 3.3 | 5.0 | 6.3 | 6.0 | 4.4 | 4.4 | 3.6 | 2.6 | 1.8 | 1.1 | 3.5 |
| Dublin-Phoenix Park[1] (Ireland) | 1.8 | 2.6 | 3.7 | 5.3 | 5.8 | 6.0 | 5.5 | 4.9 | 4.3 | 3.1 | 2.3 | 1.5 | 3.9 |
| Valentia (Ireland) | 1.4 | 2.3 | 3.8 | 5.4 | 5.9 | 5.8 | 5.1 | 4.8 | 4.3 | 2.9 | 2.2 | 1.3 | 3.8 |
| *Additional representative stations* | | | | | | | | | | | | | |
| Inverness | 1.4 | 2.3 | 3.5 | 4.5 | 5.4 | 5.5 | 4.4 | 4.2 | 3.8 | 2.8 | 1.6 | 1.0 | 3.4 |
| Braemar (1933–1960) | 0.8 | 2.0 | 3.0 | 4.3 | 5.4 | 5.5 | 4.6 | 4.0 | 3.5 | 2.1 | 1.0 | 0.6 | 3.1 |
| Edinburgh | 1.7 | 2.7 | 3.6 | 4.9 | 5.8 | 6.3 | 5.2 | 4.6 | 4.2 | 3.1 | 1.9 | 1.4 | 3.8 |
| Durham | 1.7 | 2.4 | 3.3 | 4.6 | 5.4 | 6.0 | 5.1 | 4.7 | 4.1 | 3.0 | 1.9 | 1.4 | 3.7 |
| York | 1.3 | 2.1 | 3.2 | 4.7 | 6.1 | 6.4 | 5.6 | 5.1 | 4.1 | 2.8 | 1.6 | 1.1 | 3.7 |
| Stonyhurst (Lancs.) | 1.4 | 2.1 | 3.3 | 4.8 | 6.2 | 6.3 | 5.1 | 4.9 | 3.8 | 2.8 | 1.6 | 1.1 | 3.6 |
| Cambridge | 1.7 | 2.5 | 3.8 | 5.1 | 6.2 | 6.7 | 6.0 | 5.7 | 4.6 | 3.4 | 1.9 | 1.3 | 4.1 |
| Oxford | 1.7 | 2.6 | 3.9 | 5.3 | 6.1 | 6.6 | 5.9 | 5.7 | 4.4 | 3.2 | 2.1 | 1.6 | 4.1 |
| Cardiff | 1.8 | 2.7 | 3.9 | 5.7 | 6.3 | 7.0 | 6.1 | 6.0 | 4.7 | 3.4 | 2.0 | 1.5 | 4.3 |
| Aberystwyth | 1.8 | 2.8 | 4.1 | 5.4 | 6.5 | 6.6 | 5.1 | 5.3 | 4.5 | 3.4 | 1.9 | 1.5 | 4.1 |
| Worthing (Sussex) | 2.3 | 2.9 | 4.5 | 6.2 | 7.4 | 8.1 | 7.2 | 7.0 | 5.5 | 4.0 | 2.4 | 2.0 | 5.0 |

* *Sources:* Meteorol. Office, ANONYMOUS (1962), Irish Stations based on Meteorological Office and Irish Meteorological Service data.

[1] Based on older records, and exposure at Valentia imperfect in winter.

comparison with the northwest, the northeast coast tends to lose a little in late spring and early summer, on account of low stratus cloud off the North Sea; but it gains a little in autumn. Loss of sunshine within London and the larger cities was formerly serious; locally annual totals fell 25% below those in adjacent country. Recent legislation has led to marked improvement and within London and Manchester losses are about half those prevailing before 1930.

For average daily duration of sunshine by months, see table XX.

## Ground frost

Ground frost, as it is called, has long been recorded, on those nights when thermometers capable of radiating freely and exposed at the surface of the ground, over short grass, have shown a minimum below freezing-point. Under British conditions such observations are of interest as there are many cool clear nights when the air is sufficiently stirred for the screen minimum not to fall to freezing point, yet on the ground the thermometer, suitably exposed, will do so. On such nights damage to the leaves of young growing plants, e.g., potatoes, may occur. But much depends on the details of the exposure, and how freely the grass thermometer can radiate, so that these "ground frost" records are not easily standardised; hence they must on the whole be regarded chiefly as indicators than otherwise of the characteristics of particular sites.

In general, at inland stations, the air temperature in the screen falls below 0°C (the freezing point) on 50–70 mornings yearly; at the same selection of stations the mean number of "ground frosts" is about 85–90. Some exceptional sites may be mentioned. At Malham Tarn House (1,297 ft., 396 m) on the Yorkshire Pennines, the site of the thermometer is on a south-facing slope above a lake and the annual average frequency, 90 days with minimum air temperatures below 0°C, is relatively low.

But radiation on clear nights from these uplands is very free, and accordingly the number of "grass minima" below freezing-point is very much greater, averaging 150–160. Such totals are comparable with the numbers recorded in the Highland valleys of Scotland.

The frequency of "ground frosts" is probably a fairly close index of the number of occasions on which the air temperature in an unusually well-developed local frost hollow, would itself fall below the freezing-point.

This is borne out by studies made in an exceptionally severe frost hollow, a dry valley at Rickmansworth about 25 km northwest of London (HAWKE, 1944).

## Atmospheric optical phenomena: aurora

These are of little significance from the standpoint of the climatologist and will not be discussed here. Away from the glare of towns, aurora is said to be observed in southern England on about five nights each year. Northward this average rises to over thirty at Aberdeen, and to nearly 100 in the Shetlands.

## Acknowledgements

Acknowledgement is due to the Controller of Her Britannic Majesty's Stationery Office for permission to reproduce officially published data, and to the Meteorological Office for assistance in locating material. Up-to-date reductions from published data where necessary have been calculated by the author.

## References

AANENSEN, C. J., 1965. Gales in Yorkshire, February 1962. *Geophys. Mem., Meteorol. Office*, 108: 44 pp.

ANONYMOUS, 1962. Climatological normals (CLINO) for CLIMAT and CLIMATSHIP stations for the period 1931–1960. *W.M.O. Tech. Publ.*, 117 (T.P.52), 358 pp.

BELASCO, J. E., 1952. Characteristics of air masses in the British Isles. *Geophys. Mem., Meteorol. Office*, 87: 34 pp.

BILHAM, E. G., 1938. *Climate of the British Isles*. MacMillan, London, 347 pp.

BIRKELAND, B. J., 1957. Homogenisierung der Temperaturreihe Greenwich 1763–1840. *Geofys. Publikasjoner, Norske Videnskaps Akad.*, XXI: 1–17.

BRAZELL, J., 1968. *London Weather*. H.M. Stationery Office, London.

BROOKS, C. E. P., 1949. *Climate Through the Ages*. Benn, London, 2nd ed., 395 pp.

BRUNT, D., 1925. Periodicities in European weather. *Phil. Trans. Roy. Soc. London, Ser. A*, 225: 247–302.

BUCHAN, A., 1893. The temperature of London 1763–1892 and the temperature of N.E. Scotland 1764–1892. *J. Scot. Meteorol. Soc., Ser. 3*, 9: 231.

BUCHAN, A., 1898. The mean atmospheric pressure and temperature of the British Isles. *J. Scot. Meteorol. Soc., Ser. 3*, 10: 3–41.

CHANDLER, T. J., 1965. *Climate of London*. Hutchinson, London, 292 pp.

CROWE, P. R., 1962. *Climate of Manchester and its region*. In: C. F. CARTER (Editor), *British Association Survey*. Manchester Univ. Press, Manchester, pp.17–46.

DIXON, F. E., 1953. Weather in old Dublin. *Dublin Historical Record*, 13: 94–107.

GLASSPOOLE, J., 1928. Two centuries of rain. *Meteorol. Mag.*, 63: 1–6.

HAWKE, E. L., 1944. Thermal characteristics of a Hertfordshire frost hollow. *Quart. J. Roy. Meteorol. Soc.*, 70: 23–48.

JACOBS, W., 1961. Radiation recording in the Meteorological Office. *Meteorol. Mag.*, 90: 284–289.

LAMB, H. H., 1957. Tornadoes in England, 1950. *Geophys. Mem., Meteorol. Office*, 99: 1–38.

LAMB, H. H., 1963. On the nature of certain climatic epochs which differed from the modern normal. *Proc. W.M.O.–UNESCO Symp. Clim. Changes, Rome, 1961—UNESCO Arid Zone Res.*, 20: 125–150.

LAMB, H. H., 1964. *The English Climate*. English Univ. Press, London, 212 pp.

LAMB, H. H., 1966. *The Changing Climate: Selected Papers*. Methuen, London, 236 pp.

MANLEY, G., 1939. Snowcover in Britain. *Quart. J. Roy. Meteorol. Soc.*, 64: 2–27.

MANLEY, G., 1940. Snowfall, British Isles. *Meteorol. Mag.*, 75: 41–48.

MANLEY, G., 1944. Topographical features and the climate of Britain. *Geograph. J.*, 103: 241–263.

MANLEY, G., 1945a. The effective rate of altitudinal change in temperate Atlantic climates. *Geograph. Rev.*, 25: 408–417.

MANLEY, G., 1945b. The helm wind of Crossfell. *Quart. J. Roy. Meteorol. Soc.*, 71: 197–219.

MANLEY, G., 1952. *Climate and the British Scene*. Collins, London, 314 pp.

MANLEY, G., 1959. Temperature trends in England, 1698–1957. *Arch. Meteorol., Geophys. Bioclimatol., Ser. B*, 9: 413–433.

MANLEY, G., 1961. Early meteorological observations. *Endeavour*, 21: 43–50.

MANLEY, G., 1964. Evolution of climatic environment. In: J. A. W. WATSON and J. B. SISSONS (Editors), *The British Isles: A Systematic Geography*. Nelson, London, pp.152–176.

MANLEY, G., 1966. Climate in Britain (Evening discourse). *Proc. Intern. Geograph. Congr., 20th, 1964*. Nelson, London, pp.34–45.

METEOROLOGICAL OFFICE, 1952. *Climatological Atlas of the British Isles*. H.M. Stationery Office, London, 139 pp.

METEOROLOGICAL OFFICE, 1964, 1965. *Weather in Home Fleet Waters and the N.E. Atlantic*. H.M. Stationery Office, London, I: 265 and II: 275 pp.

METEOROLOGICAL OFFICE, 1967. *British Rainfall, 1961*. H.M. Stationery Office, London, Part I, pp.1–109, Part II, pp.1–66, Part III pp.1–53 (In one volume).

MILL, H. R., 1928. Climate of Britain. In: A. G. OGILVIE (Editor), *Great Britain Regional Essays*. University Press, Cambridge, pp.1–18.

MILLER, A. A., 1962. *Climatology*, 9th ed. Methuen, London, 328 pp.

MONTEITH, J. L., 1966. Local differences in the attenuation of solar radiation over Britain. *Quart. J. Roy. Meteorol. Soc.*, 92: 254–262.

MOSSMAN, R., 1897. The meteorology of Edinburgh. *Trans. Roy. Soc. Edinburgh*, 38: 681–755.

MOSSMAN, R., 1900. The meteorology of Edinburgh. *Trans. Roy. Soc. Edinburgh*, 39: 63–207.

PENMAN, H. L., 1948. Natural evaporation from open water, bare soil and grass. *Proc. Roy. Soc.(London)*, *Ser. A*, 193: 120–145.

SHAW, E. M., 1962. Precipitation in northern England 1956–1960. *Quart. J. Roy. Meteorol. Soc.*, 88: 539–547.

SHELLARD, H. C., 1959. Averages of accumulated temperature and standard deviation of monthly mean temperature over Britain, 1921–1950. *Meteorol. Office Profess. Notes*, 125: 1–30.

SIMPSON, L. P., 1964. Sea breeze fronts in Hampshire. *Weather*, 19: 208–220.

SMITH, L. P., 1963. Significance of climatic variations in Britain. *Proc. W.M.O.—UNESCO Symp. Clim. Changes, Rome, 1961, UNESCO Arid Zone Res.*, 20: 455–463.

STAGG, J. M., 1950. Solar radiation at Kew Observatory. *Meteorol. Office, Geophys. Mem.*, 86.

SUMNER, J., 1954. A study of blocking in the Atlantic–European sectors. *Quart. J. Roy. Meteorol. Soc.*, 80: 402–416.

WALLINGTON, C. E., 1959. The structure of the sea breeze front. *Weather*, 14: 263–270.

# Chapter 4

# The Climate of France, Belgium, The Netherlands and Luxembourg

R. ARLÉRY

## Historical development

The idea of collecting observations concerning the weather and climate in special publications was developed in western Europe in the second half of the 17th century, following the discovery or introduction of the principal meteorological instruments enabling comparatively objective measurement of atmospheric pressure, air temperature, and precipitation.

About 1642, before the invention of the barometer by E. Torricelli, two scholars from Bergerac suggested the production of a meteorological journal. These were Jean Brun, an apothecary, and Deschamps, a doctor, both of them correspondents of Père Mersenne. However, neither the foundation of the Académie des Sciences in 1666, with the support of Colbert, nor the publication from the same year of the *Journal des Savants* furthered this project. It was taken up again without success around 1725, at the time research was being carried out by Réaumur, and only just failed to materialize fifty years later (1775) when Lavoisier's work led the Académie de Médecine to conceive a plan for simultaneous meteorological observations necessary to explain various medical and agricultural problems. A programme was formulated, and nearly a hundred observers were engaged in the scheme, when the revolution of 1789 held up their activity. The first known observations of temperature in France were made in Paris by the astronomer Ismaël Boulliau between 1658 and 1660. At this time the use of Galileo's thermometer, modified around 1640 by physicists at the court of Ferdinand II, Grand Duke of Tuscany, was spreading throughout Europe and the need was beginning to be felt in The Netherlands, France, and England for the construction of comparable instruments. The first measurements of precipitation are generally attributed to Vauban, who in 1685 undertook simultaneous measurements at the Citadel in Lille and at Versailles. They were then continued from 1688 at the Paris Observatory.

A famous series of observations was carried out between 1768 and 1815 at Montmorency by Père L. Cotte, well known for his *Mémoires sur la Météorologie*, which he published in 1788.

The final stage to be noted is the establishment, in 1848, of a meteorological station at Versailles by the Count of Gasparin, appointed to lecture on meteorology at the newly founded Institut Agronomique. The following year saw the publication of the first volume of an *Annuaire Météorologique* in which Haeghens, Ch. Martin and Bérigny resolved to collect each year observations made in various places in France.

The credit for establishing a climatological network covering the whole of France is due to Le Verrier. The meteorological commissions set up in the departments starting from

1876, enabled relatively homogeneous data to be obtained and collected in volumes 2 and 3 of a collection entitled *Annales du Bureau Central Météorologique*, published regularly from 1877 until 1920. These records served as a basis for the first general climatological studies, credited for the most part to A. Angot and published between 1897 and 1914 in volume 1 of the same *Annales*.

According to DUFOUR (1950), the earliest instrumental observations made in Belgium were carried out by Abbé J. Chevalier. These included in particular the maxima and minima of temperature at Brussels between 1763 and 1773; extracts of these can be found in volume 1 of *Mémoires de l'Académie Impériale et Royale des Sciences et Belles-Lettres de Bruxelles*. Abbé T. A. Mann also made observations at Nieuport, and published in the same volume (1780) of the *Mémoires* some notes on the climate of maritime Flanders. Following these pioneers, Dufour further mentions Fr. du Rondeau and the Baron de Poederlé, J. P. Minkelers, J. G. Crahay and J. Kicks, but it is the name of Adolphe Quetelet which remains associated with the classical series of regular observations started in 1833 at the Brussels Observatory. Quetelet, known throughout the world both as a statistician and as a meteorologist, was undoubtedly the first to appreciate fully the advantages to be gained in climatology from a systematic and intelligent application of probability theory, and his work *Sur le Climat de la Belgique* has long been considered as a model of its kind. More recently (1947), in accordance with the resolutions of the Conference of the Directors of the International Meteorological Organization at Warsaw (1935) L. Poncelet and H. Martin made a notable contribution to our knowledge of the climatography of Belgium on the basis of observations during the period 1901–1930.

Dutch sailors were amongst the first to keep meteorological records and in The Netherlands great care has been taken to collect meteorological data, and to preserve and publish them. It need only be remembered that from 1728 it has been possible to publish in full the readings taken by P. van Musschenbroek at Utrecht observatory, and that the well known series of observations of temperature and precipitation made at Zwanenburg between 1735 and 1861 is one of the longest early series in the world. (The temperature observations taken at Zwanenburg have been reduced to the series for Utrecht–De Bilt and the observations of precipitation have been reduced to a series for Hoofddorp by Labrijn in 1945). The Royal Netherlands Meteorological Institute was founded in 1854 with the ardent support of C.H.D. Buys Ballot. Its yearbooks, published regularly from 1866, have enabled Ch. M. A. Hartman and C. Braak to describe the climate of The Netherlands in great detail.

In the Grand Duchy of Luxembourg, the first meteorological observations were made somewhat later, by N. Bodson, lecturer in physical science and mathematics at the Athénée de Luxembourg, who started taking observations during the severe winter of 1837–1838 and probably continued until 1852. LAHR (1950) also draws attention to the observations made by P. J. J. Kerckhoff, a Dutch chemist who held a professorship at the Athénée de Luxembourg from 1837 to 1848 and was in touch with Quetelet. We also owe an uninterrupted series, from 1853 to 1895 to J. F. J. Reuter-Chomé, who was a faithful correspondent of Buys Ballot, the first Director of the Royal Netherlands Meteorological Institute. All the data available up to 1949 have been used by LAHR (1950) in his study of the climate of Luxembourg.

**Climatic factors**

Several reasons can be listed for the temperate climate of the countries with which we are here concerned:

(*1*) Their geographical situation, between latitudes 41°20′ and 53°30′N, places these territories roughly half way between the equator and the North Pole. The minimum duration of the day in winter ranges from 9 h at Bonifacio (most southerly point of Corsica) to 7½ h in the extreme north of The Netherlands, whilst the longest summer days range from 15 h in the south to 16½ h in the north.

(*2*) Along the western borders of this zone, the North Sea, the English Channel and the Atlantic regulate temperature variations. The Mediterranean to the southeast also has a moderating influence. Having a coastline nearly 3,700 km long and a total area of about 617,000 km², these territories benefit greatly from oceanic influences, which particularly favour the western side of the continents.

(*3*) The relief accentuates these advantages. The mountain barriers of the Alps, Jura, and Vosges frequently weaken the continental influences. The Massif Central causes some anomalies, but the greater part of these territories, which rise from the coast to considerable heights on the eastern and southern borders, does not comprise orographic features on a sufficiently large scale to form a barrier against oceanic influences. The small mountain chains of the United Kingdom are relatively far away and constitute only a weak and sporadic barrier.

(*4*) Finally, although its regularity is essentially random, the general circulation of the atmosphere is of prime importance. It is necessary to analyse this circulation to understand the climate and its fluctuations. We know that in temperate latitudes the dominant feature of large atmospheric movements is a "zonal" circulation from west to east (westerlies), i.e., a displacement on a global scale along the parallels of latitude, in the same direction as the rotation of the earth. This zonal circulation has a variable speed and is subject to waves which give rise to secondary circulations. Movements having a meridional component may however be superimposed. The mechanisms by which these various phenomena are maintained are still the subject of much research, and the examination of these would be beyond the scope of this study. It is assumed that the rotation of the earth, its spherical shape, its movement in the solar system, and inequalities in the heat balance at the surface of the globe are the principal causes of the planetary circulation. In order to interpret the meridional circulations, we are led to suppose that these are arranged in cells in each hemisphere, each cell having ascending and descending currents.

In the latitudes of western Europe, the zone between the equatorial and the polar cells is often affected by cyclonic disturbances, which give rise to a transfer of an excess of heat and humidity from tropical regions towards the pole, or a flow of cold air from the polar cap towards the equator. The rhythm of these discontinuous exchanges is irregular. It follows that, from the climatological point of view, the most important characteristic of the general circulation—but also the one most difficult to represent adequately—is, in fact, its variability. One very imperfect way of describing this variability is by attempting to recognize and identify "weather types" on a worldwide scale. These consist of rather arbitrary combinations of meteorological elements. In practice, we examine the synoptic charts and attempt to group together into a "weather type" those situations which are as similar as possible. In this way, we find whole regions, and not just particular localities, having essentially similar characteristics. Since the best criteria for determining "weather types" have not been generally agreed upon, it is not the intention in these pages, to give even a simple description of the principal weather types which affect western Europe, nor any details of how they succeed each other or interact with each other at various times of the year. We shall therefore limit ourselves to a preliminary description of general aerological conditions, and to an examination of seasonal types of circulation, before going on to consider the usual seasonal and monthly evolution of the weather.

**Upper air circulation in general**

In the temperate latitudes, between the 35th parallel and the Arctic Circle, we find throughout the troposphere—i.e., throughout a layer extending up to about 12 km—a general mean airstream from west to east, controlled by the subtropical anticyclones on the one hand, and by the complex low-pressure area, which covers the subpolar territories at high levels, on the other. Although this global aspect of atmospheric movements clearly depends on the mean values of the wind measurements especially over Europe, examination of the daily meteorological situations shows the presence of appreciable deformations in the flow. The streamlines take the form of waves of large amplitude, having a wavelength which may reach several thousand kilometres. Consequently only a limited number are found around each hemisphere. These waves are not stationary but move in the direction of the rotation of the earth and give rise to meridional components in the general circulation, directed sometimes towards the north and sometimes towards the south. These components may occasionally exceed the zonal component. The topography of the pressure field reflects these features of the large-scale atmospheric currents. Each wave corresponds to a trough-ridge system, causing either a flow of cold air towards southerly latitudes, or an advection of warm air towards the pole.

Although these deformations of the circulation are very variable, there are certain preferred positions for each season, shown by the characteristic appearance of the mean fields of flow and pressure. However, the individual meridional components tend to be smoothed out in the mean values, even those which are very marked, so that these features in the zonal flow appear in a greatly modified way by comparison with their day to day appearance. For this reason the corresponding parameters show large dispersions about their mean values.

This variability is more marked in the very lowest layers of the atmosphere—say from the surface up to 2,000 m—than at higher altitudes, because the thermal and orographic effects are greater near the earth's surface.

Spatial disparities in the properties of the atmosphere appear not only in the direction of the upper air flow, but also in its speed. In particular, in the upper troposphere and the lower stratosphere, it is possible to identify cores of strong winds known as jet-streams which have a length of several thousand kilometres, a width of about a thousand kilometres and a depth of only a few kilometres. Such jet-streams, in which the wind often exceeds 60 m/sec along the central axis, are often observed over western Europe. They are generally more frequent and active in the winter, and sometimes they can easily be linked with systems of disturbances at the surface. In spite of their dominant zonal component from west to east, the axes of these jet-streams may undergo marked deformations, and locally may have a completely meridional orientation or may even be reversed. The latter case, which is fairly rare over Europe, tends to occur in autumn or winter.

At higher levels, at a height of about 20 km, the atmosphere is relatively calm, being characterized by light winds showing little organization in direction, especially in the summer.

Above this transition layer there is a regular seasonal swing, known as the "stratospheric monsoon", between the westerly winds of winter, which are strong at times, and a rather moderate but very constant summer flow having an easterly component. At the latitudes of France, Belgium, and The Netherlands, the reversals in direction take place about the middle of May and towards the end of September. In winter, and especially from the first few days of January onwards, there are temporary easterly circulations in the temperate latitudes, associated with warming in the middle and upper stratosphere over subpolar regions.

TABLE I

MEAN TEMPERATURES (°C)

| Station | Pressure level (mbar) | Jan. | Feb. | Mar. | Apr. | May | June | July | Aug. | Sept. | Oct. | Nov. | Dec. | Annual |
|---|---|---|---|---|---|---|---|---|---|---|---|---|---|---|
| De Bilt | 850 | − 3.9 | − 2.6 | − 1.6 | 0.6 | 2.8 | 6.8 | 7.5 | 7.9 | 7.0 | 4.6 | 0.9 | − 1.5 | 2.4 |
| | 700 | −11.6 | −10.1 | − 9.8 | − 7.9 | − 5.5 | − 1.4 | − 0.6 | − 0.6 | − 0.5 | − 2.7 | − 6.6 | − 8.9 | − 5.5 |
| | 500 | −28.0 | −26.8 | −26.6 | −24.2 | −21.9 | −17.3 | −16.2 | −15.6 | −16.2 | −18.8 | −23.4 | −25.5 | −21.7 |
| | 300 | −52.5 | −51.6 | −51.6 | −50.2 | −47.8 | −44.5 | −43.2 | −42.2 | −43.1 | −45.3 | −49.0 | −51.0 | −47.7 |
| Uccle | 850 | − 1.9 | − 2.6 | 0.2 | 0.3 | 4.5 | 7.3 | 9.1 | 8.4 | 8.0 | 5.3 | 1.6 | − 1.0 | 3.3 |
| | 700 | − 9.7 | −10.2 | − 7.8 | − 7.9 | − 4.0 | − 0.8 | 0.7 | 0.0 | 0.4 | − 2.6 | − 5.7 | − 7.8 | − 4.6 |
| | 500 | −26.2 | −26.4 | −24.7 | −24.2 | −19.9 | −16.7 | −15.0 | −15.6 | −15.5 | −18.5 | −22.4 | −24.2 | −20.8 |
| | 300 | −50.6 | −51.0 | −50.0 | −49.6 | −46.5 | −43.5 | −41.4 | −42.0 | −42.0 | −45.0 | −47.8 | −49.7 | −46.6 |
| Brest | 850 | − 0.5 | − 1.8 | 1.2 | 1.3 | 3.8 | 6.5 | 8.2 | 8.3 | 7.5 | 5.2 | 2.1 | 1.4 | 3.6 |
| | 700 | − 8.0 | − 9.5 | − 6.7 | − 6.4 | − 6.0 | − 1.1 | 1.1 | 0.8 | 0.3 | − 1.9 | − 4.9 | − 6.3 | − 4.0 |
| | 500 | −23.8 | −25.7 | −23.3 | −22.7 | −20.1 | −16.6 | −14.1 | −14.3 | −14.9 | −17.7 | −21.0 | −22.6 | −19.7 |
| | 300 | −49.4 | −50.6 | −49.3 | −48.7 | −46.1 | −43.2 | −40.3 | −40.7 | −41.0 | −43.6 | −46.7 | −48.0 | −45.6 |
| Trappes | 850 | − 1.5 | − 2.5 | 0.4 | 1.1 | 4.7 | 7.4 | 9.2 | 9.0 | 7.9 | 4.8 | 1.6 | 0.5 | 3.5 |
| | 700 | − 8.9 | −10.4 | − 7.7 | − 7.1 | − 3.9 | 0.8 | 1.1 | 0.7 | 0.3 | − 2.5 | − 6.0 | − 6.9 | − 4.3 |
| | 500 | −25.0 | −26.5 | −24.0 | −23.3 | −19.4 | −16.0 | −13.9 | −14.4 | −15.0 | −18.1 | −21.5 | −23.1 | −20.0 |
| | 300 | −50.2 | −50.7 | −49.6 | −48.8 | −45.9 | −42.1 | −39.7 | −40.3 | −40.8 | −44.1 | −47.2 | −48.8 | −45.7 |
| Bordeaux | 850 | 0.2 | − 0.9 | 2.7 | 2.8 | 6.3 | 8.8 | 11.1 | 11.0 | 9.9 | 6.6 | 3.6 | 2.1 | 5.3 |
| | 700 | − 7.5 | − 8.9 | − 6.2 | − 5.7 | − 2.5 | 0.4 | 3.3 | 2.9 | 1.7 | − 1.4 | − 4.1 | − 5.7 | − 2.8 |
| | 500 | −23.8 | −25.3 | −22.8 | −22.0 | −18.4 | −15.1 | −12.2 | −12.6 | −13.5 | −17.0 | −20.2 | −22.0 | −18.7 |
| | 300 | −49.5 | −50.4 | −49.1 | −48.5 | −45.5 | −41.8 | −38.9 | −38.9 | −40.0 | −43.1 | −46.4 | −48.2 | −45.0 |
| Nîmes | 850 | − 0.0 | − 0.6 | 2.1 | 3.2 | 7.4 | 10.5 | 13.3 | 12.9 | 10.8 | 6.7 | 3.5 | 1.9 | 6.0 |
| | 700 | − 7.3 | − 8.5 | − 6.3 | − 5.4 | − 1.7 | 1.3 | 4.2 | 3.0 | 2.2 | − 1.4 | − 4.0 | − 5.6 | − 2.5 |
| | 500 | −24.2 | −25.2 | −23.2 | −22.3 | −18.4 | −14.8 | −12.1 | −12.3 | −13.5 | −17.3 | −20.4 | −22.4 | −18.8 |
| | 300 | −49.8 | −50.2 | −49.5 | −49.0 | −45.5 | −42.1 | −39.2 | −38.8 | −40.1 | −43.8 | −46.7 | −48.4 | −45.3 |

TABLE II

MEAN WINDS AT THE 500 MBAR LEVEL[1]

| Station | | Jan. | Feb. | Mar. | Apr. | May | June | July | Aug. | Sept. | Oct. | Nov. | Dec. |
|---------|-----|------|------|------|------|-----|------|------|------|-------|------|------|------|
| Uccle | dd | 306 | 306 | 304 | 287 | 271 | 262 | 276 | 278 | 268 | 264 | 276 | 287 |
| | FF | 7 | 6 | 4 | 6 | 8 | 8 | 9 | 8 | 11 | 7 | 9 | 11 |
| | Vm | 17 | 16 | 17 | 15 | 15 | 15 | 15 | 13 | 16 | 15 | 17 | 17 |
| | S | 43 | 39 | 23 | 41 | 52 | 55 | 63 | 56 | 69 | 47 | 49 | 65 |
| Brest | dd | 305 | 315 | 265 | 320 | 280 | 280 | 285 | 275 | 270 | 290 | 275 | 280 |
| | FF | 8 | 10 | 4 | 5 | 8 | 7 | 10 | 11 | 12 | 8 | 5 | 10 |
| | Vm | 17 | 20 | 14 | 14 | 16 | 14 | 16 | 15 | 17 | 15 | 15 | 17 |
| | S | 46 | 49 | 31 | 39 | 51 | 51 | 63 | 73 | 71 | 50 | 36 | 59 |
| Trappes | dd | 310 | 305 | 285 | 320 | 285 | 265 | 265 | 275 | 270 | 285 | 285 | 285 |
| | FF | 7 | 8 | 4 | 6 | 8 | 7 | 10 | 11 | 10 | 7 | 5 | 9 |
| | Vm | 18 | 18 | 14 | 14 | 15 | 14 | 15 | 15 | 16 | 15 | 15 | 17 |
| | S | 40 | 44 | 31 | 41 | 51 | 52 | 67 | 73 | 63 | 49 | 31 | 55 |
| Bordeaux | dd | 271 | 310 | 285 | 325 | 285 | 280 | 275 | 280 | 275 | 305 | 315 | 300 |
| | FF | 6 | 10 | 4 | 5 | 9 | 8 | 12 | 12 | 10 | 6 | 4 | 7 |
| | Vm | 19 | 20 | 14 | 13 | 14 | 14 | 16 | 16 | 17 | 15 | 15 | 16 |
| | S | 32 | 50 | 28 | 35 | 60 | 59 | 75 | 75 | 59 | 40 | 24 | 44 |
| Nîmes | dd | 315 | 295 | 295 | 290 | 295 | 270 | 275 | 275 | 275 | 295 | 295 | 305 |
| | FF | 7 | 10 | 5 | 6 | 8 | 9 | 13 | 13 | 9 | 5 | 3 | 7 |
| | Vm | 16 | 18 | 14 | 13 | 14 | 15 | 16 | 17 | 15 | 13 | 12 | 15 |
| | S | 43 | 55 | 34 | 45 | 56 | 58 | 81 | 76 | 62 | 36 | 24 | 45 |

[1]dd = direction of the resultant wind, in degrees;
FF = resultant wind speed, in m/sec;
Vm = mean wind speed, in m/sec;
S = constancy, 100 FF/Vm.

To determine the mean flow during a given time interval, a vector mean wind, i.e., the vector mean of the individual winds measured during this interval, is evaluated for each aerological station and for each level. By relating the vector means to the mean pressures calculated for the same time interval at the same standard levels, we obtain a fairly representative picture of the mean flow. It also is convenient to add a parameter of dispersion to the values of the vector mean wind; the most commonly used parameter is the "constancy". This is defined as the ratio, multiplied by 100, of the modulus of the vector mean wind to the mean speed of the wind. The latter quantity is the arithmetical mean of wind speeds regardless of direction.

This parameter would be close to 100 for a completely "constant" flow, and would tend to zero for perfectly random circulation. Thus, near the ground (at an altitude of about 500 m), over the Paris Basin, the constancy in March (period 1956–1960) is only 11, indicating a great variability of the winds both in speed and direction. On the other hand, in July at an altitude of 25 km, the constancy rises to 95. At Uccle (period 1950–1956) the constancy at 850 mbars (about 1,500 m) does not exceed 14 in April but rises to 68 in September.

Side by side with the variability of the winds, mention should be made of the large range of temperatures which may obtain at different levels of the atmosphere (Table I), although the amplitudes of these fluctuations are usually less pronounced than those observed at the surface, particularly in the continental zone. The transfer of the energy of solar radiation to the atmosphere takes place through the medium of the ground, since air is largely transparent to the short wave-lengths which constitute the greater part of solar radiation. As an infrared radiator, the ground is important as a warm or cold source causing transfer of energy by convection in the troposphere.

Other important factors which enter into the heat balance of the atmosphere must also be considered, in particular dynamic phenomena such as adiabatic compressions or expansions associated with vertical movements of the air, as also the energy due to the latent heat of water.

In the 1,500–12,000 m layer of the troposphere the difference between the mean temperatures of the coldest month and the warmest month is of the order of $11°–13°C$. The absolute differences are however much greater. In the Paris region, they reached over a period of eight years (1948–1956):

$43°C$ at 700 mbar (about 3,000 m), i.e., from $+14$ to $-29°C$.
$37°C$ at 500 mbar (about 5,500 m), i.e., from $-4$ to $-41°C$.
$38°C$ at 300 mbar (about 9,000 m), i.e., from $-23$ to $-61°C$.

## TABLE III

RELATIVE FREQUENCIES (%) OF THE DOMINANT COMPONENTS OF WINDS AT 500 MBAR

| Seasons | Wind comp.[1] | Brest | | Trappes | | Bordeaux | | Nimes | |
|---------|---------------|-------|----|---------|----|----------|----|-------|----|
| Spring | N | 25 | ⎫44 | 28 | ⎫43 | 27 | ⎫39 | 29 | ⎫43 |
|        | S | 19 | ⎭ | 15 | ⎭ | 12 | ⎭ | 14 | ⎭ |
|        | W | 36 | ⎫50 | 37 | ⎫49 | 42 | ⎫51 | 39 | ⎫49 |
|        | E | 14 | ⎭ | 12 | ⎭ | 9 | ⎭ | 10 | ⎭ |
|        | calm | 6 | | 8 | | 10 | | 8 | |
| Summer | N | 22 | ⎫34 | 20 | ⎫35 | 20 | ⎫27 | 19 | ⎫28 |
|        | S | 12 | ⎭ | 15 | ⎭ | 7 | ⎭ | 9 | ⎭ |
|        | W | 53 | ⎫60 | 54 | ⎫59 | 62 | ⎫65 | 61 | ⎫62 |
|        | E | 7 | ⎭ | 5 | ⎭ | 3 | ⎭ | 1 | ⎭ |
|        | calm | 6 | | 6 | | 8 | | 10 | |
| Autumn | N | 22 | ⎫40 | 26 | ⎫45 | 30 | ⎫46 | 24 | ⎫40 |
|        | S | 18 | ⎭ | 19 | ⎭ | 16 | ⎭ | 16 | ⎭ |
|        | W | 45 | ⎫52 | 41 | ⎫48 | 37 | ⎫46 | 37 | ⎫48 |
|        | E | 7 | ⎭ | 7 | ⎭ | 9 | ⎭ | 11 | ⎭ |
|        | calm | 8 | | 7 | | 8 | | 12 | |
| Winter | N | 32 | ⎫45 | 30 | ⎫45 | 31 | ⎫41 | 31 | ⎫43 |
|        | S | 13 | ⎭ | 15 | ⎭ | 10 | ⎭ | 12 | ⎭ |
|        | W | 44 | ⎫53 | 36 | ⎫49 | 36 | ⎫54 | 39 | ⎫51 |
|        | E | 9 | ⎭ | 13 | ⎭ | 18 | ⎭ | 12 | ⎭ |
|        | calm | 2 | | 6 | | 5 | | 6 | |

[1] Calm: $\leq$ 5 m/sec.

For a given month and during the same time interval this absolute amplitude is still large; for example, at 500 mbar it reaches 30°C in February and 24°C in August.

The 500 mbar level (about 5,500 m) has mainly been used for characterizing and differentiating the seasonal types of circulation in the upper air (Tables II, III), being far enough from the ground for thermal and orographic effects to be slight, and on the other hand being situated in the middle troposphere, where most of the significant meteorological phenomena (clouds, hydrometeors, etc.) are observed.

**Seasonal types of upper air circulation**

**Winter**

One of the three major waves which appear on a hemispherical scale on the charts of mean flow at the 500-mbar level is situated in the neighbourhood of Europe. The associated ridge extends from the Azores to Iceland and Norway, whilst a trough passes from Mauritania, through Italy and central Europe, to northern Russia. This results in a moderate mean flow from the northwest over western Europe.

The individual situations often depart from this overall pattern. The position of the upper ridge which covers the near Atlantic directly controls the direction and speed of upper windcurrents in Europe. When high pressure spreads over the European continent, strong southwesterlies are observed over France, Belgium, and The Netherlands. This flow is generally maintained for several days. On the other hand, when this ridge intensifies over the Atlantic, and is then associated with a deep trough extending from northern Russia and Scandinavia to the central Mediterranean, the result is strong northwest to north winds, with a large meridional component, in the Europe–Atlantic region. The greatest frequency of winds having a dominant northerly component occurs in the winter. Secondary waves of varying intensity, but having a short wavelength, with a rapid alternation of northwesterly and southwesterly winds, occur over periods of several consecutive days.

Another situation which is fairly characteristic of the winter is when a slow moving or stationary anticyclone becomes established either over northern Europe or over the North Sea and the Norwegian Sea, this anticyclone being associated with a complex low-pressure area situated between the sea area off Portugal and the eastern basin of the Mediterranean. These "blocking" high-level, warm anticyclones sometimes persist for ten to fifteen days. It is indeed in winter that we find the greatest proportion of winds having a dominant easterly component.

Comparison of the monthly mean wind speeds shows that at upper levels the strongest winds are found in winter. In the upper troposphere, in the longitudes of western Europe, the zone between the 45th and 55th parallels is very often affected by a branch of the polar jet-stream.

**Summer**

During this season, the upper winds show, in contrast, a much more marked constancy. This is clearly shown by the constancy figures which exceed 60 in Belgium, The Netherlands, and northern France, and reach 80 in certain months in southern France. Thus although the wind speeds are on average lower than those observed in winter, the modulus of the vector mean wind is the highest of the four seasons.

On the mean upper air charts, and especially at the 500-mbar level, the wave over the eastern Atlantic becomes distinctly weaker, and the flow is practically zonal between the high pressures extending from the Gulf of Mexico to Eritrea and the vast polar cap which narrows towards the higher latitudes.

Individual situations show analogous characteristics, and in particular the decrease in the amplitude of the waves in the streamlines. Moreover, the meridional components of the winds decrease in comparison with the zonal components: it is in the summer that the lowest proportion of winds with a dominant northerly component is found, while winds with a large zonal component show a clear maximum, mainly due to westerly winds because easterlies are very rare during the summer. The daily types of circulation which occur most frequently are either a west–east flow over western Europe or a southwesterly flow when the axis of the anticyclone is displaced towards the European continent. However, in certain cases (but more rarely than during the other seasons) an axis of high pressure over the ocean, associated with a continental and Mediterranean trough, leads to a strong north to northwest airstream, with a disturbed situation at the surface. Sometimes, but rather rarely, an anticyclonic cell is situated over central and western Europe, with light winds showing little ordered arrangement in the middle trophosphere. Finally, a depression occasionally becomes centered over western Europe, off the coast of Portugal or in the Bay of Biscay, giving a south to southwest flow, which is often thundery, from the Mediterranean coast to The Netherlands.

**Autumn and spring**

During these intermediate seasons the upper air flow shows a great variability which is reflected in low constancy values, the minimum usually occurring in the spring. In the spring the upper ridge over the near Atlantic weakens, whilst high pressure increases and extends through western Asia to Novaya Zemlya. The streamline pattern changes rapidly during this season, and strong winds having a dominant meridional component (sometimes southerly but mainly northerly) are almost as frequent as in winter, at the expense of the west–east flow. Thus, a marked ridge is often found over the near Atlantic, associated with a deep trough whose axis runs from eastern Germany or continental Europe to the central basin of the Mediterranean.

Not infrequently, high pressure becomes established in the Scandinavia–Baltic region while an area of low pressure covers the Mediterranean and may even extend to the Bay of Biscay through southern France. This results in an easterly flow over western Europe, which becomes oriented towards the southeast when the trough in the Bay of Biscay extends to the neighbourhood of Iceland. Finally, as in winter, it is not unusual for fast-moving but well marked waves to cross Europe from west to east causing a rapid alternation of northwesterly and southwesterly winds often with predominating meriodional components.

The same upper air patterns are observed in the troposphere during the autumn. However, northwest to north winds are less frequent than in spring, and south to southwest winds are usually the most regular, owing to the Atlantic ridge moving more frequently to the continent. As in the spring and in winter, fairly long spells occur during which the wind direction is characterized by great irregularity in the meridional components.

## The characteristics of the surface circulation and weather by seasons and months

### Winter

In winter the centre of the Azores high pressure zone, or a ridge from this anticyclone, often extends over Spain and sometimes over the southwestern half of France, while an area of low pressure is centred to the southeast of Greenland or in the neighbourhood of Iceland. Disturbances from the Atlantic Ocean move from west to east, the most frequent paths crossing the British Isles before reaching the continent. Sometimes, however, the airstream of the polar front is deviated to the southeast owing to fairly deep depressions covering the northwestern part of the Mediterranean, especially the Gulf of Genoa.

A great variety of situations can arise if an anticyclone persists with its centre over Scandinavia, northern Russia, or central Europe, deflecting the paths of the Atlantic depressions towards the north as they approach the continent.

*December*

In spite of fairly active zonal circulation there is a relatively low frequency of depressions from the west, and north to northwest winds prevail. The latter are usually responsible for the first chilly spells of winter, corresponding to outbreaks of Arctic air whose accumulation results in a more or less sharp drop in temperature. Once established, these weather types are rather persistent. However, in some years an increase in the zonal circulation during the second half of the month may result in the weather becoming milder, with plentiful rain or snow.

*January*

With meridional circulation on a reduced scale, January is usually the most typical of the winter months and the weather is most unsettled. When westerly winds predominate, the weather is rainy and relatively mild. If however an anticyclone is present over the continent, the cold weather of central Europe, Russia or Scandinavia may extend to western Europe as well, though here this cold weather is more liable to be interrupted by sudden milder spells than further to the east. The conflict between oceanic influences from the west and continental influences from the east, gives differing characteristics to the month of January in different years, and is often in evidence in January, resulting in a rapid alternation of the weather conditions during the month.

*February*

Generally, February shows a decrease in the frequency of westerly regimes and may have many spells of settled weather. The most persistent weather types are due either to depressions from the southwest, with relatively mild rainy weather reminiscent of the end of autumn, or to the influence of mild oceanic anticyclones, the sea being warmer than the continent. But if the high pressure of central Europe predominates, the cold weather may persist, and then there is a risk that snow, although less frequent than in January, may, in the Paris Basin for example, become deeper. This is especially the case if depressions from the southern sector interact with the northerly airstreams.

**Spring**

In spring, the Azores anticyclone extends over the ocean frequently. The Icelandic low is still present, with its centre further south than in winter; the mean isobaric gradient is weaker. Disturbances from the west are least frequent and meridional circulations are the most vigorous of the year. In western Europe, the weather is often controlled by a low pressure area covering either the Britisch Isles, France, or central Europe, and by the movement of deep troughs extending from the Icelandic low-pressure centre of action. The low-pressure area of the Mediterranean is often maintained, as is also the ridge extending towards the southwest from the high-pressure area in Scandinavia.

*March*

March is usually an undisturbed month; one half of the weather types are anticyclonic in character, more often continental than maritime. The equalization of temperatures between the ocean and the continent brings a weakening of the westerly flow. The humidity is generally low. A fairly large number of settled and bright days with a large diurnal variation of temperature may lead one to suppose that warm weather has come. But there are also some outbreaks of polar or Arctic air, especially at the beginning of the month, accompanied by showers, and a few disturbances moving from the west, with persistent rain, giving a less uniform character to this month in some years.

*April*

With the warming up of the continent, April is marked by a preponderance of oceanic influences. There is a distinct decrease in European anticyclones, the frequency of which now reaches its annual minimum. The westerly regimes give cold, rainy weather, and weather from the northwest brings Arctic air with gusts, squalls, and heavy showers. Depressions from the southwest sometimes interact with northwesterly airstreams and are accompanied by thundery activity. In spite of the capricious alternation of weather types, April is usually clear on account of the great dryness of the air, which is still more marked than in March; it is in fact usually in April that the relative humidity is lowest.

*May*

Often, May is an uncertain month, in which premature spells of hot weather and cold snaps such as in April, may occur. On average, however, it is one of the months when both weather types and air masses are most constant. Disturbances from the west are the least frequent of the whole year. On the other hand, it is also the time of the highest frequency of southwesterly and southeasterly situations, sometimes interfering with northerly airstreams. This meridional circulation may cause abrupt changes of temperature, the cold polar or Arctic air masses being replaced directly by very warm air from the Mediterranean or even from the Sahara.

**Summer**

In summer, the pressure in the Azores high reaches its maximum value. The centre of the high is further to the north, but its position as regards longitude is rather variable. As the continent becomes warmer than the ocean, the pressures there are lower, and the well developed Atlantic anticyclone may extend a strong ridge over western and central Europe and even over the eastern Mediterranean. The Icelandic low is still present, but is even more displaced towards the north; the activity associated with it is greatly diminished, the pressure gradients are weak, and fewer troughs are found over western Europe. Westerlies predominate, and the flow is less subject to disturbances. Finally Mediterranean depressions are very rare in summer and the southern regions enjoy long spells with clear skies or small amounts of cloud, interrupted only by brief periods of locally violent thundery activity.

*June*

June is frequently a month of transition, when a strong meridional circulation nearly always prevails over the still weak zonal circulation. Disturbances from the northwest are only slightly more rare than in May; the predominance of oceanic anticyclones, associated with the retreat of the westerly depressions, often results in this month being settled and pleasantly warm. However, there is a risk that disturbances from the southwest may cause thunderstorms with hail which present a danger to cultivation of all kinds.

*July*

July usually sees the zonal circulation prevailing over the spring meridional circulation. Depressions from the west again increase in frequency, but they are often rather inactive. In the Paris Basin the "continentalization" of polar oceanic air gives rise to fine days which are distinctly warm but quite pleasant. Except in the south of France, it is however rare for the month to be completely fine. Changes of weather affect the third decade rather more than the second, the latter being on average the least disturbed.

*August*

August is similar to July, but is usually more humid and a little warmer, depending on the region and the year. However, disturbances in the westerly airstream are rather more active and cause long periods of bad weather. In addition to precipitation due to depressions, there is still some precipitation due to thunderstorms, in spite of a slight decrease in the frequency of southwesterly regimes.

There is also the possibility that an anticyclone may become established, so that in some years there is a chance of settled weather. On average, "heat waves" occur during the first fortnight, the second fortnight being more subject to disturbed weather.

**Autumn**

In autumn, the Azores high and the Icelandic low are on average in the same positions as in the spring; however, the westerlies are rather more frequent. The pressure may be high on the continent, being usually centred over southwest Russia. Sometimes, especially towards the end of this season, the western ridge of this European anticyclone is very well developed, extending over the Alps, and joining the Azores high. In other cases, instead of this high pressure barrier which shelters the Mediterranean Basin from disturbances, there may be a polar airstream. Depressions from the west escape the steering effect of the Icelandic low and their paths curve towards the southeast or the south. The Mediterranean depression then dominates the situation and southern France may have long periods of heavy rain.

*September*

September is often a month which retains the summer characteristics, but since the continent now begins to become less warm than the ocean, the Atlantic depressions bring relative warmth which may compensate for the decrease in the length of the days and weakening of radiation received from the sun. The zonal flows are however weaker, and the westerly circulation is still limited to northern latitudes. This allows supply of warm tropical maritime air to the southwesterly and southeasterly airstreams which may make incursions into the southern regions of France and sometimes further north. The absolute humidity of the air decreases. In many regions the cloudiness decreases to its annual minimum, and in the Paris Basin the first two decades of this month are often amongst the most pleasant of the year. On the other hand, a decrease in the zonal circulation during the last decade of September is sometimes accompanied in certain regions by settled periods, but broken by violent "equinoxial storms", often associated with very heavy precipitation on the high ground bordering the Mediterranean (Cévennes, eastern Pyrenees).

*October*

October usually shows an increase in the disturbances from the northern sector, and cold spells mark the approach of winter. However, spells of fine weather occur in certain years, even in the Paris Basin, as pleasant as those of the middle of September. On average, however, this month is mainly characterized by long periods of rain in the coastal regions, with the arrival of disturbances from the west or the southwest. Inland, overcast skies, wet weather, and persistent fog in the first half of the day make October a dull month.

*November*

November is the month when the difference between oceanic and continental factors plays a dominant part; both meridional and zonal circulations increase in intensity. The frequency of disturbances from the west reaches 35–40% at latitude 50°N. Large invasions of cold air maintain secondary meridional circulations. When the weather is not windy or rainy, persistent fog and low stratus often obscure the sun and result in unpleasant drizzle. Even if the precipitation is not very great, this month, the dullest of the year, is dreaded for its humidity, in spite of some exceptions.

## Climatic types

### Mediterranean climate

The climate of Mediterranean regions is the most distinct and for this reason has been placed at the beginning of the present brief review. Mediterranean climate is characteristic of a clearly defined area of France between the shores of the Mediterranean and a mountain belt running from the eastern spurs of the Pyrenees through the Montagne Noire and the Montagnes de l'Espinouse, the precipitous ridge of the Causses du Larzac, Mont-Aigoual, the steep slopes of the Cévennes, and the southern part of the peaks of Vivarais on the right bank of the Rhône, to the southern slopes of the Alps, and from the left bank of the Rhône and Mont Ventoux to the Italian frontier beyond the Principality of Monaco.

The coastal plains of Corsica, most of the inland basins, and the lower slopes of the mountain ranges (below 1,000–1,500 m according to the aspect) enjoy a Mediterranean climate. The frequent summer droughts and the high luminosity of the sky are the best known features which distinguish these regions from the rest of France. Temperatures are relatively high in the summer (22°–24°C on average in July and August) and still moderate in winter (7°–9°C on average in January). Sharp drops in temperature may however occur during this season owing to katabatic winds, often cold and strong, such as the mistral in the Rhône Valley and Provence, or the tramontane in the Roussillon. On average there are 20–30 days with frost in the year on the plains, and still fewer in the coastal strip of the Roussillon and the Côte d'Azur. Snow is very rare, but constitutes a danger—the more to be feared since it is unexpected—and there is a risk that exceptional snowfalls may in a few hours cover the ground locally to a depth of tens of centimetres. Rainfall is relatively high (averaging 700–800 mm/year occurring on about 70–80 days), except on the coastal plains (Crau, Camargue) not directly bordered by hills, which often have less than 600 mm of precipitation in the year. The greatest proportion of rainfall comes from heavy showers at the end of the autumn. In some years considerable rainfall also occurs at the beginning of the winter, with the bad weather sometimes lasting several days (unsettled southeasterly regimes). Oceanic depressions are sometimes the main cause of rainfall in the spring. The hot and marine winds alternate in winter with dry winds (tramontane, cers, mistral) which accentuate the summer droughts.

### Climates under oceanic influences

With the exception of mountainous regions, the greatest part of the area under consideration is subject to oceanic influences (modified by latitude, distance from the sea or relief) which decrease inland where continental tendencies prevail in some years. This is especially so in Alsace, on the plateau of Lorraine, Luxembourg and the Ardennes, Burgundy, and the valley of the Saône.

*Oceanic climate*

(*1*) The purest oceanic climate is found in Brittany and in the lower regions of Normandy. The frequent rainfall (150–180 days per year) is not small in amount during any season, but in October and February reaches maxima associated with disturbances from the Atlantic, with westerly as well as northwesterly and northeasterly regimes. The rain is rarely heavy but falls as drizzle maintaining a high humidity. The mildness of the weather is another characteristic of this climate. The winters can be as mild as on the Mediterranean coast, and the summers are much cooler. The mean temperature in January is of the order of 5° at Rennes and exceeds 6° at Brest; the number of days with frost varies from 15 to 45 according to the year and the distance from the coast. In July the mean temperature is about 17°C.

(*2*) On account of its more southerly position, one would expect an improvement in the oceanic climate of southwestern France, since Atlantic disturbances coming from the west are less frequent at these latitudes. This effect is, however, partially masked by the Pyrenees which cause an intensification of some disturbances (especially from the northwest). These disturbances would be less active if after reaching the continent they continued to move over non-mountainous regions. The climate of Aquitaine is thus less equable than that of Brittany. Rainfall is undoubtedly a little less frequent, but is more abundant when it does fall, mainly in winter and spring. The end of the summer and the beginning of autumn are generally rather dry: the disturbances move parallel to the chain of mountains and large amounts of precipitation occur in the west, in the Basque Country. The barrier of the Pyrenees has a smaller effect on the mean temperatures, which in the summer are 2–4° higher than in Brittany; the winters are less mild, especially inland (Toulouse region). Likewise, except in the immediate approaches to the Pyrenees, cloud amounts are smaller, the sky is brighter and there is more sunshine than in Brittany.

(*3*) Between the estuaries of the Gironde and the Loire, the western central part of France (Saintonge, Charentes, Poitou, Vendée) enjoys a special climate. More favoured than Brittany on account of its more southerly latitude, it is, on the other hand sufficiently far from the Pyrenees to escape the effects of this barrier, and thus has an incontestable advantage over regions situated to the south of the Gironde; there is less precipitation and more sunshine. This advantage decreases as one passes east and southeast towards the Massif Central.

(*4*) The coastal type of oceanic climate, a little less equable in Normandy, shows greater contrasts of temperature in the northern provinces of France (Picardy, Artois), in Flanders and the lowlands of Belgium, and in almost the entire territory of The Netherlands, than in Brittany. Winters become colder, with a lowering of the mean temperature, by about 2–4°, the decrease becoming more marked towards the north and east. The winters are also humid, and the greyness of the skies is added to the lengthening of the nights. The summers, which are hardly as warm as in Brittany, are duller because they are rainy, often more so than in the autumn, whilst the amounts of precipitation are often not very great in the spring. As in Brittany, the weather is sometimes very changeable, the changes being due to storms or heavy squalls which, although less frequent than the mistral, are just as violent and often more disastrous near the coast.

Almost completely devoid of relief, The Netherlands is a very humid country, especially in the autumn it is subjected in turn to gales, rain and fog.

The predominance of maritime influences, which in The Netherlands and in the uplands of Belgium (Brabant, Hainaut) results in uniform temperatures, does not prevent some winters from being fairly severe, especially in the case of persistent snow, and particularly in the northeast of The Netherlands. Neither does it mean that the rainfall regime, with a preponderance of summer precipitation, even fairly near the coast, is independent of that of central Europe.

*Suboceanic climate*

A suboceanic type of climate prevails in approximately one third of French territory, encircling the Massif Central to the west and the north (Quercy, Périgord, Val de Loire, the Paris Basin in the broadest sense, Champagne, Nivernais, Berry), the eastern parts of the Belgian provinces of Namur and Liège, and the former Duchy of Limbourg.

The distance from the sea and the sporadic occurrence of continental influences or modifications due to relief in the neighbourhood of mountain slopes, cause more pronounced annual ranges of temperature, mainly on account of the progressive lowering of winter temperatures. The number of days with frost increases steadily. Except in the neighbourhood of high ground, the amounts of precipitation are smaller. The outstanding characteristics of the rainfall regime are frequent dry springs (especially in Anjou, Maine, and Touraine), a maximum at the end of autumn in the Paris Basin due to depressions, and thundery showers in the summer, which further to the east cause this maximum to be displaced towards the summer.

There is however no essential common feature giving true climatic unity to these various regions. Differences of vegetation, where they exist, are as much due to other geographical factors or to human activity as to differences of climate.

**Subcontinental climate**

Continental influences are felt mainly in Alsace, Lorraine, Luxembourg and in the Ardennes. They may still be noticeable in Argonne and the plateaus of lower Burgundy, and sometimes affect the eastern part of Champagne and regions bordering the valley of the Saône. They are also to be seen in certain depressions in the Massif Central and in the Alps, and in enclosed plains sheltered from the moderating westerly winds where a decrease in the rainfall results in an extreme type of climate. The summer rainfall, which predominates, is often of a thundery character. Inequalities in heating of the ground result in an increase of phenomena due to convection. Winters are dry and comparatively severe, and the absence of snowfall is unusual. Finally, the shortness of the intermediate seasons, autumn and spring, causes the transition to be more abrupt, and the contrasts of temperature more noticeable.

**Mountain climates**

No matter whether we are considering the Vosges, the Jura Mountains, the Alps, the Massif Central, the Pyrenees or mountains in Corsica, the climate at higher levels shows a fairly regular decrease of the mean temperature and a much more irregular increase of rainfall. The proportion of precipitation falling as snow varies with height and according to the type or origin of the disturbance causing the precipitation. The absolute humidity is on average lower than in the plains, and the relative humidity is usually higher in summer than in winter. Differences of latitude, exposure, or orientation, the proximity of regions having a suboceanic or a continental climate, or, as is the case in the southern parts of the Alps, in Corsica and at either end of the Pyrenees, the proximity of the sea, all combine to produce a multitude of local climatic variations. For example, the long, gentle slopes of the Vosges in Lorraine, are exposed to westerly winds and have a plentiful rainfall, whilst on the steeper eastern flank, the climate at the same altitude is colder in winter but warmer and drier in the summer. In general, and especially in depressions, there is a considerable annual and diurnal range of temperature (very cold winters and warm summers). In the mountains themselves, however, the temperatures stabilize to some extent owing to a decrease of the diurnal and annual variations. Local winds (valley and mountain winds) sometimes assume a greater importance than those associated with the general circulation. In the Pyrenees snow may remain throughout the year above 2,800 m, and in the Alps above 2,700 m. The climate of the French Alps has been the subject of outstanding work by BÉNÉVENT (1926). The climate of the Massif Central was studied more recently by ESTIENNE (1956).

**Distribution of climatic elements**

In order to present in a more precise form the features described above, certain parameters whose geographical distributions are worth examining in greater detail, have been chosen; maps of these are given in Fig.1–10.

The scale suitable for the size of this publication is too small to permit an accurate representation of the distribution of the elements shown, and the lines are merely intended to indicate the general trends over the whole of the area. No claim is made to great accuracy, especially in mountainous regions where there are insufficient climatological stations and where the estimation of true values is a delicate matter for those elements for which the laws of variation with height are complex and often unknown. A short description of the maps is given below:

*Temperature.* Fig.1 and 2 (mean temperature in January and July) show clearly that the distribution of temperature in winter is largely controlled by the distance from oceans and seas (the isotherms run roughly parallel to the coasts and temperatures gradually decrease inland), whilst in the summer, the main factor controlling the distribution is the effect of latitude. In this latter season, the isotherms run from west to east and, ignoring some irregularities due to relief, the mean temperature steadily increases from north to south. On the other hand, if we consider the mean daily range of temperature, the greatest effect, due to contrasts between the ocean and the continent, is in the summer. During winter the daily range is largely controlled by latitude and increases from north to south (Fig.3, 4).

*Mean duration of frost-free period.* The map of Fig. 5 has been established by finding the

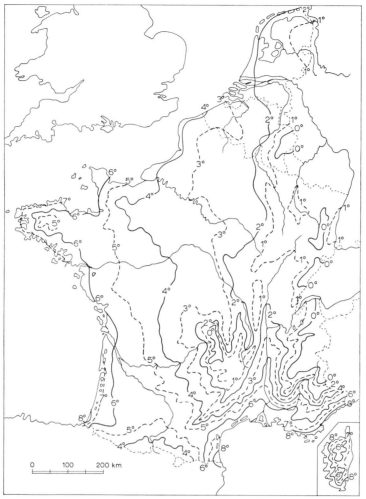

Fig.1. Mean temperature in January (1931–1960).

number of days which elapse from the last spring frost to the first autumn frost (any day on which the minimum shade temperature is equal to or below 0°C, is counted as a day with frost). Maps of the mean and extreme dates of the last spring frost and the first autumn frost, and also of the minimum duration of the frost-free period (1921–1950) have been drawn for France by GARNIER (1954).

*Precipitation.* Fig.6 shows, in a simplified way, the mean annual distribution of precipitation. It can be seen at once that moderate rainfall is characteristic of all the regions. Mean annual totals of more than 2,000 mm only occur in some mountainous regions, and amounts smaller than 550 mm occur in relatively limited areas. Over more than three-quarters of the territory the mean annual amount of precipitation lies between 600 and 800 mm.

The greatest contrasts are found in the Mediterranean region. The coastal plains of the Roussillon, Bas-Languedoc, and the Camargue have the lowest rainfall (at some places the mean annual total is less than 500 mm), but the mountain belt bordering them has more than 800 or 900 mm (more than 1,500 mm at some stations in the Cévennes, the peaks of Vivarais, and the spurs of the Alps Maritimes). Elsewhere it is again over the higher ground that the precipitation is greatest, but the distribution is non-symmetrical.

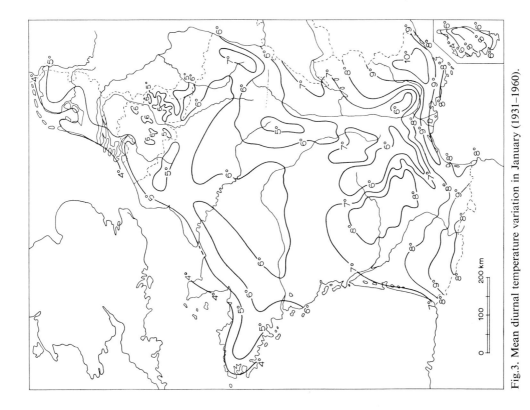

Fig.3. Mean diurnal temperature variation in January (1931–1960).

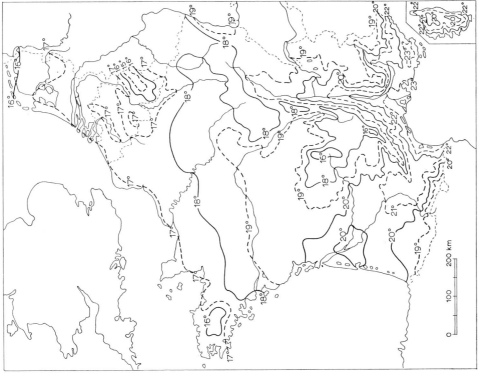

Fig.2. Mean temperature in July (1931–1960).

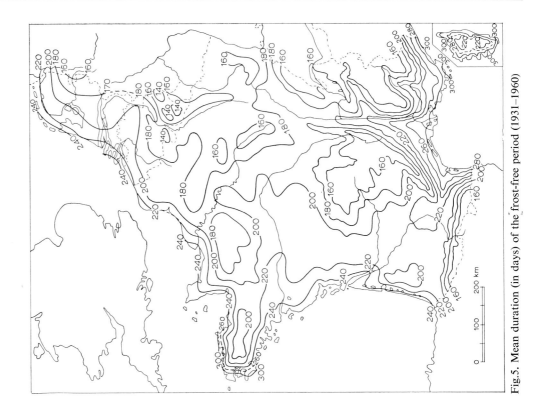

Fig.5. Mean duration (in days) of the frost-free period (1931–1960)

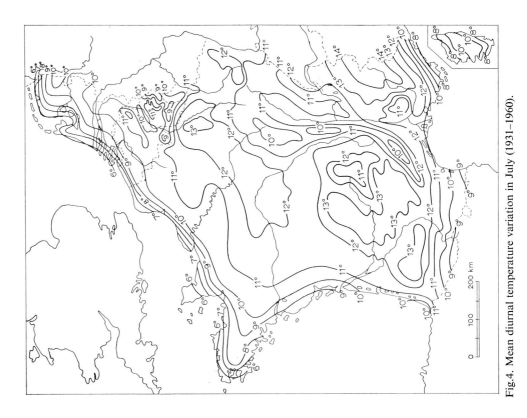

Fig.4. Mean diurnal temperature variation in July (1931–1960).

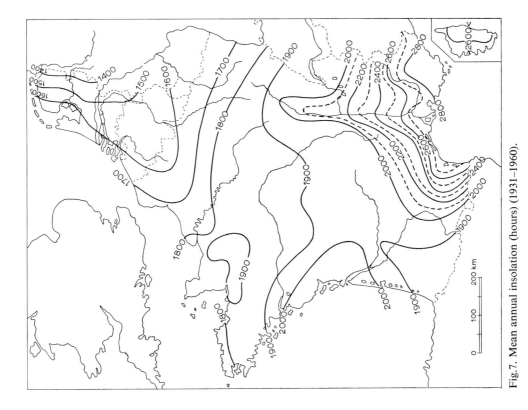

Fig.7. Mean annual insolation (hours) (1931–1960).

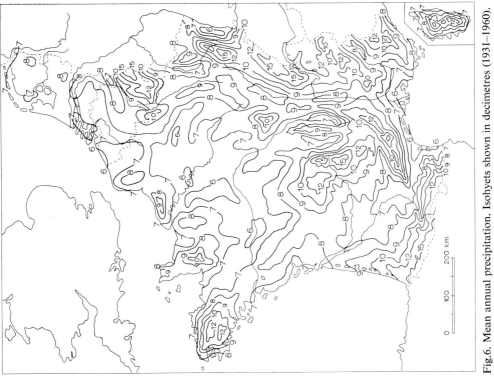

Fig.6. Mean annual precipitation. Isohyets shown in decimetres (1931–1960).

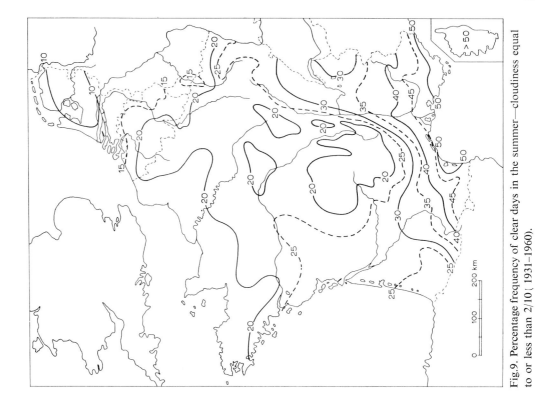

Fig.9. Percentage frequency of clear days in the summer—cloudiness equal to or less than 2/10 (1931–1960).

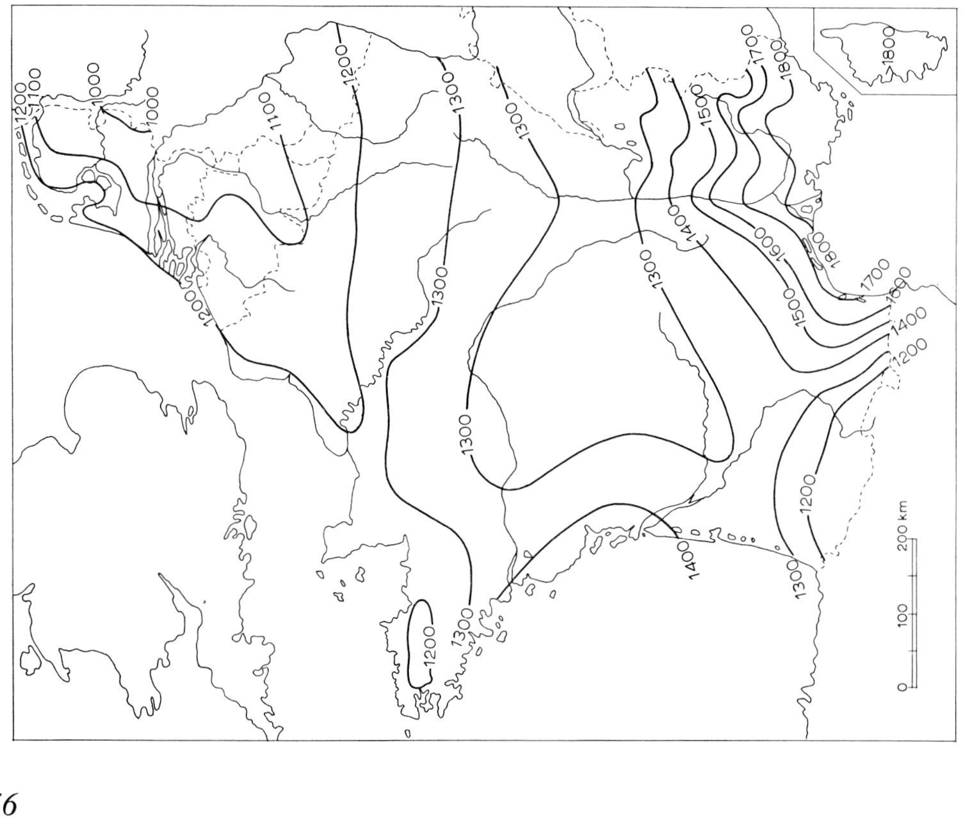

Fig.8. Mean insolation (hours) during the six months April to September (1931–1960).

The Atlantic-facing slopes of the hills of Cotentin, Perche, Poitou, and also the heights of the Ardennes or the western spurs of the Vosges, Morvan, Massif Central, Jura, or the Alps, have a much higher rainfall than the slopes of valleys or depressions which are "sheltered" from the west (the plain in Alsace surrounding Colmar, the sources of the Loire, Limagne, Livradois, Briançonnais, etc.).

In the neighbourhood of the Pyrenees, the annual rainfall may amount to 1,200–1,300 mm even though there are often relatively dry periods from July to September.

The flat parts of the coasts of the North Sea, the English Channel, or Atlantic Ocean immediately bordering the sea, do not have a high rainfall, but where it is at all hilly (the hills of Artois, Caux, Montagne d'Arrée, etc.) the rainfall is higher than in the large inland basins in the region of Paris, the middle Loire, or Aquitaine, where the mean annual totals lie between 600 and 650 mm.

Fig.6 does not show the times of year when the precipitation is usually greatest or smallest. A fairly good idea of the relative frequency of dry and wet months may be gained by referring to the climatological tables of the WORLD METEOROLOGICAL ORGANIZATION (1962). These tables show the limits of the quintiles of the distribution of monthly rainfall during 30 years. For the same month, the six lowest values fall into the first quintile,

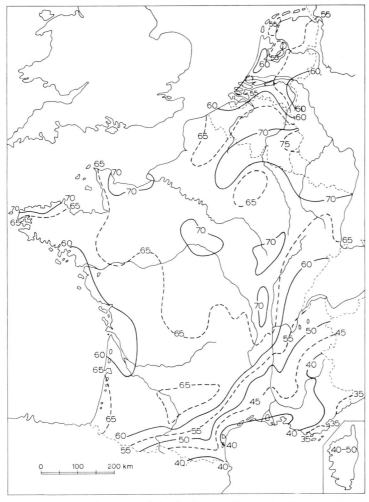

Fig.10. Percentage frequency of cloudy days in winter—cloudiness equal to or greater than 8/10 (1931–1960).

the limits of which are given in the first and second lines of the table, whilst the six highest values are spread over the range limited by the figures given in the fifth and sixth lines.

*Insolation.* Because of large inaccuracies in insolation measurements, Fig.7 shows only the annual duration in broad outlines. It was therefore thought desirable to draw a similar chart showing the duration of sunshine during the six warmest months of the year, or, to be more exact, during those months which best coincide with the growing season—April to September (Fig.8).

*Cloud cover.* The representation of this element on maps, which is very often a convenient method of distinguishing climatic differences between regions having similar temperatures and rainfall, presents problems as to the times of observation to be considered and the choice of limits of the degree of cloudiness. Two extreme cases have been selected: the frequency of clear days in the summer (Fig.9) and that of cloudy days in winter (Fig.10). However, the definitions of a "clear day" and a "cloudy day" are very arbitrary, and the lines on the maps are very generalized.

It has been decided not to give maps of other important climatic elements, such as humidity, numbers of days with rain, thunderstorm, snow, etc. It may be mentioned, however, that maps based on periods other than 1931–1960 may be consulted in various national atlases. For France, for example, plates 12–19 of the *Atlas de France*, edited by the Comité National Français de Géographie with the collaboration of Météorologie Nationale, contain more information than can be given here.

## Climatic tables

The climatic tables which appear on pp.162–193 are in conformity with others given in this collection. This makes it unnecessary to repeat the explanations. It is however necessary to give details about certain points where departures occur from general principles. Values relating to periods other than 1931–1960 are indicated in footnotes, or the length of the period is stated. Thus unless there is no note to the contrary, the basic period is 1931–1960. This does not, however, mean that a continuous series of observations for thirty years was available for the station in question. In many cases, notably on account of the closure of many stations during the second World War, it was necessary to reduce the means by the classical methods, meteorological elements being suitable for this type of treatment. In addition, the sites of some stations may have been slightly changed; in such cases the coordinates and height refer to the position of the station at the end of the period (1960) for which the data are given.

The material for the French stations has been prepared by the author with the active collaboration of Marcel Garnier, Ingenieur de la Météorologie (Division de Climatologie du Service Météorologique Métropolitain, Paris). Pressure, mean temperature, mean diurnal variation of temperature, amount of precipitation, and number of days with precipitation have been reduced to the period 1931–1960. Extreme values and insolation data were generally available for the 15 years 1946–1960. The insolation data were derived from measurements made with Jordan sunshine recorders at all the French stations. The results relating to evaporation were derived from measurements with Piche evaporimeters placed in the meteorological enclosure. These measurements were usually only made from April to October. Evaporation data, as also data on the mean vapour pressure of water, numbers of days with thunderstorms, fog, and strong wind, prevailing wind direction, mean wind speed and mean cloudiness, were only available for the decade 1951–1960.

The data for total mean daily radiation (direct + diffused), are of particular importance, but they were only obtainable from a very limited number of recording stations. It was thought best to assemble these data in one separate table (Table XXXV), where the values shown represent total mean daily radiation in Langleys (cal./cm²) daily. These values represent the period 1931–1960, but in the case of six of the French stations (Nancy, Mâcon, Limoges, Millau, Nice and Ajaccio), data was only available for three complete years. These values, thus, can only be considered as a first approximation of the 30-year averages.

The author wishes to express his gratitude to his esteemed colleagues L. J. L. Deij and L. Poncelet, who provided the tables for stations in The Netherlands and in Belgium. In the tables for The Netherlands, it is to be noted that on account of the low altitude of the stations it was considered preferable to give pressures reduced to the sea-level, and that the evaporation data have been calculated using Penman's classical methods.

Data relating to the two stations in the Grand Duchy of Luxembourg were compiled by the author from data published by LAHR (1950) and supplemented by summarizing data for the eleven years 1950–1960 contained in meteorological and hydrological year-books published by the Administration des Services Agricoles of Luxembourg.

A greater volume of data, referring to different periods and usually for restricted areas within the territories in question, appear in various publications referred to in the bibliography, but not mentioned in the text.

Tables II and III for prevailing wind direction and mean speed give only an imperfect idea of the distribution of this important element, for which it was not possible to include maps. It is therefore necessary to conclude with a brief review of what is shown on the wind maps.

During the autumn and winter the most frequent wind directions are between *south* and *west* in the northern half of the area, and between *west* and *north* in the southern parts of France. The Mediterranean regions are subject to markedly *north* to *northwest* winds (mistral). In all regions the greatest wind speeds occur in winter.

During spring and summer winds from between *west* and *north* prevail over almost all of the area except southeastern France, and the lowest wind speeds usually occur from July to September.

The mean wind speed is of the order of 2–7 m/sec, depending on the region and the year. The highest values occur either in the coastal regions (North Sea, English Channel, Atlantic Ocean, Mediterranean) or in valleys, which may be broad or narrow, subject to local winds (*bise* in the north and northeast, Jura and Alps; *cers*, *marin*, *autan*, *mistral* and *tramontane* in the Mediterranean south; *traverse* in the Massif Central).

The regions which are least subject to high winds lie mainly in a belt 200–300 km wide, running southwest–northeast, from the Bassin d'Arcachon to Franche Comté. Except in some localities which are particularly exposed to the prevailing west or southwest winds, more than half of all the observed speeds are less than 5 m/sec at stations in this belt, whilst the frequency of mean speeds equal to or greater than 16 m/sec seldom exceeds 3 or 4 per 1,000 observations.

**References**

ANGOT, A., 1895. Premier catalogue des observations météorologiques faites en France depuis l'origine jusqu'en 1850. *Ann. Bur. Central Météorol., B*, pp.89–146.

ANGOT, A., 1897. Études sur le climat de la France. Température. 1. Stations de comparaison. *Ann. Bur. Central Météorol., Tome 1, B*, pp.93–170.

ANGOT, A., 1900. Études sur le climat de la France. Température. 1. Stations de comparaison (fin). *Ann. Bur. Central Météorol., Tome 1, B*, pp.33–118.

ANGOT, A., 1902. Études sur le climat de la France. Température. 2. Variation diurne de la température. *Ann. Bur. Central Météorol., Tome 1*, pp.41–130.

ANGOT, A., 1903. Études sur le climat de la France. Température. 3. Températures moyennes. *Ann. Bur. Central Météorol., Tome 1*, pp.119–232.

ANGOT, A., 1904. Études sur le climat de la France. Température. 4. Températures extrêmes; jours de gelée. *Ann. Bur. Central Météorol., Tome 1*, pp.157–366.

ANGOT, A., 1906. Études sur le climat de la France. Pression atmosphérique. *Ann. Bur. Central Météorol., Tome 1*, pp.83–250.

ANGOT, A., 1907. Études sur le climat de la France. Régime des vents. *Ann. Bur. Central Météorol., Tome 1*, pp.33–100.

ANGOT, A., 1911. Études sur le climat de la France. Régime des pluies. 1. Considérations générales. Région Nord-Ouest. *Ann. Bur. Central Météorol., Tome 1*, pp.109–236.

ANGOT, A., 1912. Études sur le climat de la France. Régime des pluies. 2. Régions du Sud-Ouest et du Sud. *Ann. Bur. Central Météorol., Tome 1*, pp.101–214.

ANGOT, A., 1913. Études sur le climat de la France. Régime des pluies. 3. Régions du Nord et de l'Est. *Ann. Bur. Central Météorol., Tome 1*, pp.71–216.

ANGOT, A., 1914. Études sur le climat de la France. Régime des pluies. 4. Régions du Sud-Ouest; résumé général; jours de pluie. *Ann. Bur. Central Météorol., Tome 1*, pp.67–156.

ARLÉRY, R., 1961. La durée d'insolation (1946–1960) en France. *Monograph. Météorol. Natl.*, 24: 24 pp.

ARLÉRY, R., GARNIER, M. et LANGLOIS, R., 1954. Application des méthodes de C. W. Thornthwaite à l'esquisse d'une description agronomique des climats de la France. *La Météorologie*, 35: 345–367.

BÉNÉVENT, E., 1926. Le climat des Alpes Françaises. *Mém. Office Natl. Météorol. France*, 14: 435 pp.

BIGOURDAN, G., 1923. Le climat de la France. L'eau atmosphérique. *Ann. Bur. Longitudes*, 1923: A. 1–118.

BRAAK, C., 1929. Het klimaat van Nederland. C. Luchtdrukking. D. Wind. *Mededeel. Verhandel. K.N.M.I.*, 32: 158 pp.

BRAAK, C., 1930. Het klimaat van Nederland. B (vervolg). Lucht- en grondtemperatuur. *Mededeel. Verhandel. K.N.M.I.*, 33: 78 pp.

BRAAK, C., 1933. Het klimaat van Nederland. A (vervolg). Neerslag. 1. Neerslag volgens zelfregistreerende en gewone regenmeters, regenkaarten. *Mededeel. Verhandel. K.N.M.I.*, 34a: 101 pp.

BRAAK, C., 1934. Het klimaat van Nederland. A (vervolg). Neerslag. 2. Nieuwe bewerking der tabellen van no.15 (Hartman). *Mededeel. Verhandel. K.N.M.I.*, 34b: 53 pp.

BRAAK, C., 1936. Het klimaat van Nederland. E. Verdamping. *Mededeel. Verhandel. K.N.M.I.*, 39: 50 pp.

BRAAK, C., 1937. Het klimaat van Nederland. F. Zonneschijn en bewolking. *Mededeel. Verhandel. K.N.M.I.*, 40: 51 pp.

BRAAK, C., 1938. Het klimaat van Nederland. G. Vochtigheid. *Mededeel. Verhandel. K.N.M.I.*, 41: 42 pp.

BRAAK, C., 1939. Het klimaat van Nederland. H. Mist. *Mededeel. Verhandel. K.N.M.I.*, 42: 49 pp.

BRAAK, C., 1940. Het klimaat van Nederland. B (vervolg). Luchttemperatuur. *Mededeel. Verhandel. K.N.M.I.*, 43, 44 pp.

BRAAK, C., 1942. Het klimaat van Nederland. D (vervolg). Wind. *Mededeel. Verhandel. K.N.M.I.*, 46: 109 pp.

BRAAK, C., 1943. Het klimaat van Nederland. B (vervolg). Grondtemperatuur, minimum temperatuur nabij den grond en nachtvorst. *Mededeel. Verhandel. K.N.M.I.*, 47: 109 pp.

DUFOUR, L., 1950. *Esquisse d'une Histoire de la Météorologie en Belgique*. Inst. Roy. Météorol. Belg., Bruxelles, 55 pp.

ESTIENNE, P., 1956. Recherches sur le climat du Massif Central français. *Mém. Météorol. Natl.*, 43: 242 pp.

GARENC, P., 1957. Contribution à l'étude du climat d'entre Loire inférieure et Gironde d'après des archives climatologiques inédites. *Mém. Météorol. Natl.*, 44: 197 pp.

GARNIER, M., 1954. Contribution à l'étude des gelées en France. *La Météorologie*, 35: 369–378.

GARNIER, M., 1963. Nombre moyen de jours de précipitations en France (périodes 1921–1950 et 1931–1960). *Monograph. Météorol. Natl.*, 29: 42 pp.

GARNIER, M., 1964. Valeurs normales des températures en France (1921–1950). *Monographies Météorol. Natl.*, 30: 48 pp.

GARNIER, M., 1966. Climatologie de la France; sélection de données statistiques et éléments de la variation diurne. *Mém. Météorol. Natl.*, 50: 294 pp.; 51: 148 pp.

GODARD, A., 1951. Contribution à l'étude du climat lorrain. *Rev. Géograph. Lyon*, 26: 297–310.

GRISOLLET, H., 1958. Climatologie de Paris et de la région parisienne. *Mém. Météorol. Natl.*, 45: 78 pp.

HARTMAN, CH. M. A., 1913. Het klimaat van Nederland. A. Neerslag. *Mededeel. Verhandel. K.N.M.I.*, 15: 64 pp.

HARTMAN, CH. M. A., 1918. Het klimaat van Nederland. B. Luchttemperatuur. *Mededeel. Verhandel. K.N.M.I.*, 24: 53 pp.

LABRIJN, A., 1945. Het klimaat van Nederland gedurende de laatste twee en een halve eeuw. *Mededeel. Verhandel. K.N.M.I.*, 49: 114 pp.

LAHR, E., 1950. *Un Siècle d'Observations Météorologiques Appliquées á l'étude du Climat Luxembourgeois.* Bourg-Bourger, Luxembourg, 287 pp.

MAURAIN, CH., 1947. *Le Climat Parisien.* Presses Universitaires de France, Paris, 160 pp.

MÉZIN, M., 1945. Évolution des types de temps en Europe occidentale. *Assoc. Franc. l.Avan. Sci. Paris, 64th Session*, 64/2: 171–183.

PÉDELABORDE, P., 1957. *Le climat du Bassin Parisien: Essai d'une Méthode Rationnelle de Climatologie Physique.* Librairie de Médicis, Paris, 1: 539, pp., 2: 116 pp.

PIÉRY, M., et al., 1934. *Traité de Climatologie Biologique et Médicale.* Masson, Paris, 1–3: 2715 pp.

PIÉRY, M., et al., 1946. *Le Climat de Lyon et de la Région Lyonnaise.* Cartier, Lyon, 389 pp.

PONCELET, L. et MARTIN, H., 1947. Esquisse climatographique de la Belgique. *Inst. Roy. Météorol. Belg. Bruxelles, Mém.*, 27: 265 pp.

QUENEY, P., 1954. Les grands mouvements de l'atmosphère. *La Météorologie*, 34: 195–207.

QUETELET, A., 1849. *Sur le Climat de la Belgique.* Hayez, Bruxelles, 1: 358 pp.

QUETELET, A., 1857. *Sur le Climat de la Belgique.* Hayez, Bruxelles, 2: 285 pp.

QUETELET, A., 1867. *Météorologie de la Belgique comparée à celle du Globe.* Hayez, Bruxelles, 505 pp.

ROULLEAU, J., 1954. L'évolution du temps. *La Météorologie*, 34: 209–220.

SANSON, J., 1945. Recueil de données statistiques relatives au climat de la France. *Mém. Météorol. Natl.* 30: 148 pp.

VIAUT, A., 1947. *Les Aspects du Temps en Europe Occidentale.* Blondel La Rougery, Paris, 104 pp.

WORLD METEOROLOGICAL ORGANIZATION, 1962. Climatological Normals—CLINO—for climat and climat ship stations for the period 1931–1960. *W.M.O./O.M.M., Publ.* 117, TP. 52.

TABLE IV

Latitude 53°08′N, longitude 6°35′E, elevation 4.9 m

| Month | Mean sta. press.[1] (mbar) | Mean daily temp. (°C) | Mean daily temp. range (°C) | Temp. extremes[2](°C) highest | lowest | Mean vapor press.[3] (mbar) | Mean precip.[4] (mm) | Max. precip.[2,4] 24 h (mm) |
|---|---|---|---|---|---|---|---|---|
| Jan. | 1014.1 | 0.9 | 5.0 | 12.2 | −18.7 | 6.2 | 67 | 59.5 |
| Feb. | 1014.9 | 1.3 | 5.7 | 16.0 | −22.9 | 6.2 | 49 | 41.0 |
| Mar. | 1015.7 | 3.9 | 7.6 | 20.2 | −15.6 | 7.0 | 41 | 25.2 |
| Apr. | 1014.6 | 7.6 | 8.9 | 24.4 | − 3.9 | 8.5 | 48 | 31.9 |
| May | 1016.0 | 11.6 | 10.0 | 29.7 | − 3.3 | 10.8 | 53 | 25.7 |
| June | 1015.9 | 14.7 | 10.0 | 33.3 | 1.5 | 13.2 | 56 | 38.0 |
| July | 1014.5 | 16.5 | 9.3 | 33.2 | 3.7 | 15.3 | 91 | 35.1 |
| Aug. | 1014.4 | 16.4 | 9.4 | 31.5 | 3.6 | 15.6 | 87 | 75.6 |
| Sept. | 1015.7 | 13.8 | 8.9 | 31.8 | 0.8 | 13.6 | 73 | 53.7 |
| Oct. | 1015.0 | 9.4 | 7.4 | 24.5 | − 3.7 | 10.7 | 74 | 55.1 |
| Nov. | 1013.7 | 5.4 | 5.6 | 15.5 | − 7.6 | 8.4 | 73 | 41.7 |
| Dec. | 1014.0 | 2.5 | 4.8 | 13.4 | −14.6 | 7.0 | 65 | 37.5 |
| Annual | 1014.9 | 8.7 | 7.7 | 33.3 | −22.9 | 10.2 | 776 | 75.6 |

| Month | Mean evap. (mm) | Number of days with precip. ⩾0.1 mm | thunder-storm | Mean cloudiness (oktas) | Mean sunshine (h) | Most. freq. wind dir. | Mean wind speed (m/sec) |
|---|---|---|---|---|---|---|---|
| Jan. | 5 | 20 | 0 | 7.6 | 47 | SW | 6.3 |
| Feb. | 17 | 18 | 0 | 7.4 | 64 | SW | 6.0 |
| Mar. | 40 | 16 | 0 | 6.9 | 111 | E | 5.9 |
| Apr. | 77 | 16 | 1 | 6.6 | 158 | SW, W | 6.3 |
| May | 111 | 14 | 3 | 6.3 | 209 | NE | 5.3 |
| June | 123 | 14 | 4 | 6.3 | 208 | W | 4.7 |
| July | 115 | 17 | 6 | 6.9 | 187 | W | 4.7 |
| Aug. | 97 | 17 | 5 | 6.6 | 179 | SW | 4.4 |
| Sept. | 62 | 18 | 2 | 6.4 | 139 | SW | 4.1 |
| Oct. | 29 | 20 | 1 | 6.8 | 94 | SW | 5.1 |
| Nov. | 10 | 21 | 1 | 7.7 | 46 | SW | 5.4 |
| Dec. | 3 | 21 | 0 | 7.8 | 38 | SW | 5.5 |
| Annual | 690 | 212 | 23 | 6.9 | 1479 | SW, W | 5.3 |

[1] Reduced to mean sea level.
[2] Extreme temperatures since 1945; max. precipitation in 24 h since 1850.
[3] Calculated from observations at 08h00, 14h00 and 19h00.
[4] Groningen.

TABLE V

CLIMATIC TABLE FOR DEN HELDER
Latitude 52°58′N, longitude 4°45′E, elevation 5.8 m

| Month | Mean sta. press.[1] (mbar) | Mean daily temp. (°C) | Mean daily temp. range(°C) | Temp. extremes[2](°C) | | Mean vapor press.[3] (mbar) | Mean precip. (mm) | Max. precip.[2] 24 h (mm) |
|---|---|---|---|---|---|---|---|---|
| | | | | highest | lowest | | | |
| Jan. | 1013.7 | 2.5 | 3.9 | 12.4 | −14.4 | 6.6 | 65 | 83.3 |
| Feb. | 1014.6 | 2.4 | 3.9 | 14.9 | −18.5 | 6.5 | 46 | 26.8 |
| Mar. | 1015.4 | 4.4 | 4.7 | 20.0 | −14.5 | 7.2 | 39 | 35.6 |
| Apr. | 1014.8 | 7.6 | 4.9 | 24.3 | − 3.7 | 8.7 | 40 | 51.7 |
| May | 1015.9 | 11.3 | 5.5 | 28.5 | − 2.1 | 10.8 | 38 | 29.6 |
| June | 1016.0 | 14.5 | 5.5 | 30.8 | 2.0 | 13.3 | 37 | 57.0 |
| July | 1014.4 | 16.7 | 5.3 | 33.9 | 6.8 | 15.4 | 64 | 49.4 |
| Aug. | 1014.2 | 17.1 | 5.4 | 32.8 | 6.4 | 15.7 | 71 | 79.7 |
| Sept. | 1015.4 | 15.2 | 5.2 | 32.6 | 3.0 | 13.9 | 75 | 55.4 |
| Oct. | 1014.6 | 11.2 | 4.7 | 24.8 | − 4.6 | 11.2 | 90 | 67.4 |
| Nov. | 1013.2 | 7.2 | 4.0 | 17.6 | −10.6 | 8.8 | 83 | 52.0 |
| Dec. | 1013.5 | 4.3 | 3.8 | 13.0 | −12.3 | 7.5 | 65 | 36.5 |
| Annual | 1014.6 | 9.5 | 4.7 | 33.9 | −18.5 | 10.5 | 714 | 83.3 |

| Month | Mean evap. (mm) | Number of days with | | Mean cloudiness (oktas) | Mean sunshine (h) | Most. freq. wind dir. | Mean wind speed (m/sec) |
|---|---|---|---|---|---|---|---|
| | | precip. ⩾0.1 mm | thunder- storm | | | | |
| Jan. | 11 | 21 | 0 | 7.2 | 54 | S | 7.7 |
| Feb. | 22 | 18 | 0 | 7.0 | 74 | SW | 7.3 |
| Mar. | 43 | 16 | 0 | 6.4 | 127 | E | 6.6 |
| Apr. | 78 | 15 | 0 | 6.2 | 181 | SW | 7.0 |
| May | 111 | 13 | 2 | 5.9 | 227 | NE | 6.4 |
| June | 129 | 12 | 2 | 5.8 | 238 | SW | 6.2 |
| July | 126 | 15 | 4 | 6.4 | 217 | SW | 6.3 |
| Aug. | 110 | 15 | 5 | 6.2 | 207 | SW | 6.1 |
| Sept. | 77 | 18 | 3 | 6.3 | 151 | SW | 6.5 |
| Oct. | 43 | 20 | 2 | 6.6 | 102 | S, SW | 7.0 |
| Nov. | 19 | 21 | 2 | 7.6 | 48 | S | 7.0 |
| Dec. | 10 | 22 | 1 | 7.4 | 40 | S | 7.3 |
| Annual | 779 | 206 | 21 | 6.6 | 1665 | SW | 6.8 |

[1] Reduced to mean sea level.
[2] Extreme values since 1850.
[3] Calculated from observations at 08h00, 14h00 and 19h00.

TABLE VI

CLIMATIC TABLE FOR DE BILT
Latitude 52°06′N, longitude 5°11′E, elevation 2.9 m

| Month | Mean sta. press.[1] (mbar) | Mean daily temp. (°C) | Mean daily temp. range(°C) | Temp. extr.[2](°C) highest | lowest | Mean vapor press.[3] (mbar) | Mean precip. (mm) | Max. precip.[2] 24 h (mm) | Global radiation (cal./cm² month) |
|---|---|---|---|---|---|---|---|---|---|
| Jan. | 1014.7 | 1.7 | 5.1 | 13.0 | −24.8 | 6.4 | 68 | 46.2 | 1,797 |
| Feb. | 1015.4 | 2.0 | 6.0 | 18.5 | −21.6 | 6.4 | 52 | 53.6 | 3,078 |
| Mar. | 1015.8 | 5.0 | 8.2 | 22.5 | −12.3 | 7.2 | 45 | 29.8 | 6,237 |
| Apr. | 1015.1 | 8.5 | 9.1 | 27.4 | − 6.0 | 8.6 | 49 | 33.4 | 9,458 |
| May | 1015.9 | 12.4 | 10.2 | 33.6 | − 3.7 | 10.9 | 52 | 46.3 | 12,255 |
| June | 1016.4 | 15.5 | 10.3 | 36.8 | 0.7 | 13.2 | 58 | 51.4 | 12,724 |
| July | 1015.1 | 17.0 | 9.4 | 35.6 | 3.4 | 15.3 | 77 | 66.2 | 11,417 |
| Aug. | 1014.8 | 16.8 | 9.4 | 35.8 | 3.8 | 15.7 | 88 | 74.3 | 9,823 |
| Sept. | 1016.1 | 14.3 | 9.0 | 34.2 | − 0.8 | 13.8 | 71 | 43.7 | 7,251 |
| Oct. | 1015.5 | 10.0 | 7.5 | 26.6 | − 7.8 | 10.9 | 72 | 56.2 | 4,353 |
| Nov. | 1014.1 | 5.9 | 5.5 | 19.3 | −14.4 | 8.5 | 70 | 39.1 | 1,967 |
| Dec. | 1014.6 | 3.0 | 4.8 | 14.4 | −20.8 | 7.1 | 63 | 42.0 | 1,288 |
| Annual | 1015.3 | 9.3 | 7.9 | 36.8 | −24.8 | 10.3 | 765 | 74.3 | 81,648 |

| Month | Mean evap. (mm) | Number of days with precip. ⩾0.1 mm | thunder- storm | Mean cloudiness (oktas) | Mean sunshine (h) | Most. freq. wind dir. | Mean wind speed (m/sec) |
|---|---|---|---|---|---|---|---|
| Jan. | 4 | 21 | 0 | 7.2 | 56 | SW | 3.8 |
| Feb. | 17 | 19 | 1 | 7.0 | 69 | SW | 3.7 |
| Mar. | 42 | 16 | 1 | 6.3 | 127 | NE | 3.5 |
| Apr. | 78 | 16 | 2 | 6.3 | 164 | SW | 3.6 |
| May | 109 | 14 | 4 | 6.0 | 211 | NE | 3.1 |
| June | 126 | 14 | 5 | 6.0 | 223 | SW | 2.9 |
| July | 118 | 17 | 6 | 6.5 | 199 | SW | 2.9 |
| Aug. | 96 | 18 | 6 | 6.3 | 186 | SW | 2.8 |
| Sept. | 61 | 19 | 3 | 6.1 | 146 | SW | 2.9 |
| Oct. | 28 | 20 | 2 | 6.4 | 102 | SW | 3.1 |
| Nov. | 9 | 21 | 1 | 7.4 | 50 | S, SW | 3.4 |
| Dec. | 3 | 21 | 1 | 7.4 | 41 | S, SW | 3.6 |
| Annual | 691 | 216 | 32 | 6.6 | 1572 | SW | 3.3 |

[1] Reduced to mean sea level.
[2] Extreme values since 1848.
[3] Calculated from observations at 08h00, 14h00 and 19h00.

TABLE VII

CLIMATIC TABLE FOR VLISSINGEN
Latitude 51°27′N, longitude 3°36′E, elevation 12.8 m

| Month | Mean sta. press.[1] (mbar) | Mean daily temp. (°C) | Mean daily temp. range(°C) | Temp. extremes[2](°C) highest | Temp. extremes[2](°C) lowest | Mean vapor press.[3] (mbar) | Mean precip. (mm) | Max. precip.[2] 24 h (mm) |
|---|---|---|---|---|---|---|---|---|
| Jan. | 1014.6 | 2.9 | 3.8 | 14.2 | −15.3 | 6.9 | 62 | 38.8 |
| Feb. | 1015.4 | 3.1 | 4.3 | 15.5 | −19.6 | 6.7 | 45 | 41.0 |
| Mar. | 1015.5 | 5.1 | 5.6 | 21.5 | − 6.5 | 7.6 | 39 | 33.0 |
| Apr. | 1015.2 | 8.3 | 6.2 | 29.9 | − 3.3 | 8.9 | 42 | 25.3 |
| May | 1015.7 | 11.9 | 6.8 | 34.9 | 0.8 | 11.1 | 44 | 40.7 |
| June | 1016.4 | 15.1 | 6.6 | 34.4 | 4.3 | 13.6 | 46 | 48.7 |
| July | 1015.2 | 17.1 | 6.1 | ca. 35.0 | 6.3 | 15.6 | 69 | 53.9 |
| Aug. | 1014.9 | 17.4 | 6.0 | 33.4 | 7.0 | 16.0 | 66 | 80.0 |
| Sept. | 1016.1 | 15.4 | 5.6 | 33.2 | 3.9 | 14.4 | 73 | 70.8 |
| Oct. | 1015.3 | 11.4 | 4.8 | 24.4 | − 2.8 | 11.8 | 70 | 37.5 |
| Nov. | 1013.9 | 7.3 | 3.9 | 17.2 | − 7.5 | 9.1 | 75 | 57.2 |
| Dec. | 1014.4 | 4.3 | 3.7 | 14.5 | −13.5 | 7.6 | 58 | 63.0 |
| Annual | 1015.2 | 9.9 | 5.3 | ca. 35.0 | −19.6 | 10.8 | 689 | 80.0 |

| Month | Mean evap. (mm) | Number of days with precip. ⩾0.1 mm | Number of days with thunder- storm | Mean cloudiness (oktas) | Mean sunshine (h) | Most. freq. wind dir. | Mean wind speed (m/sec) |
|---|---|---|---|---|---|---|---|
| Jan. | 8 | 20 | 0 | 7.3 | 54 | S | 7.0 |
| Feb. | 20 | 16 | 0 | 7.1 | 70 | SW | 6.7 |
| Mar. | 43 | 16 | 0 | 6.4 | 134 | NE | 5.6 |
| Apr. | 78 | 15 | 2 | 6.3 | 169 | N | 6.0 |
| May | 111 | 13 | 3 | 6.1 | 216 | N | 5.2 |
| June | 130 | 13 | 3 | 5.8 | 232 | SW, W | 5.6 |
| July | 124 | 16 | 4 | 6.2 | 210 | W | 5.6 |
| Aug. | 106 | 16 | 4 | 6.2 | 200 | SW, W | 4.9 |
| Sept. | 71 | 17 | 3 | 6.4 | 152 | SW | 5.9 |
| Oct. | 37 | 19 | 1 | 6.4 | 104 | S, SW | 5.4 |
| Nov. | 14 | 20 | 1 | 7.6 | 51 | S | 6.1 |
| Dec. | 7 | 21 | 0 | 7.8 | 39 | S | 6.5 |
| Annual | 749 | 202 | 21 | 6.6 | 1629 | SW | 5.9 |

[1] Reduced to mean sea level.
[2] Extreme values since 1855.
[3] Calculated from observations at 08h00, 14h00 and 19h00.

TABLE VIII

CLIMATIC TABLE FOR OSTEND
Latitude 51°14′N, longitude 2°55′E, elevation 10 m.

| Month | Mean daily temp. (°C) | Mean daily temp. range(°C) | Temp. extremes (°C) | | Mean vapor press. (mbar) | Mean precip. (mm) | Max. precip. 24 h (mm) |
|---|---|---|---|---|---|---|---|
| | | | highest | lowest | | | |
| Jan. | 3.5 | 4.1 | 12.8 | −13.9 | 7.4 | 52 | 35 |
| Feb. | 3.1 | 4.6 | 17.2 | −14.8 | 7.2 | 44 | 35 |
| Mar. | 5.6 | 6.5 | 19.1 | −10.2 | 7.8 | 28 | 25 |
| Apr. | 8.3 | 5.9 | 26.8 | − 5.2 | 9.2 | 31 | 29 |
| May | 11.4 | 6.5 | 29.6 | 0.4 | 12.0 | 32 | 31 |
| June | 14.3 | 6.7 | 34.8 | 2.1 | 14.7 | 32 | 42 |
| July | 16.5 | 6.6 | 34.8 | 5.3 | 16.8 | 61 | 45 |
| Aug. | 17.0 | 7.1 | 34.2 | 7.9 | 16.6 | 56 | 46 |
| Sept. | 15.1 | 7.2 | 30.0 | 4.3 | 14.7 | 58 | 42 |
| Oct. | 11.4 | 6.6 | 24.8 | − 5.0 | 11.8 | 74 | 36 |
| Nov. | 7.4 | 4.8 | 17.8 | − 9.7 | 9.5 | 60 | 40 |
| Dec. | 4.5 | 4.1 | 15.0 | −13.5 | 8.0 | 51 | 32 |
| Annual | 9.9 | 5.9 | 34.8 | −14.8 | 11.3 | 579 | 46 |

| Month | Number of days with | | | | | Most freq. wind dir. | Mean wind speed (m/sec) |
|---|---|---|---|---|---|---|---|
| | precip. ≥0.1 mm | thunder- storm | fog | cloudy sky ≥$\frac{3}{4}$ | clear sky ≤$\frac{1}{4}$ | | |
| Jan. | 14 | 0 | 3 | 13 | 3 | S | 7.1 |
| Feb. | 13 | 0 | 7 | 14 | 2 | NE | 6.7 |
| Mar. | 10 | 0.1 | 5 | 10 | 4 | WSW | 6.4 |
| Apr. | 10 | 0.3 | 3 | 11 | 2 | NNE | 6.6 |
| May | 9 | 0.7 | 1 | 9 | 2 | NNE | 6.7 |
| June | 9 | 0.4 | 1 | 8 | 5 | NE | 6.3 |
| July | 12 | 1.6 | 1 | 10 | 3 | WSW | 6.2 |
| Aug. | 13 | 1.0 | 1 | 11 | 2 | WSW | 6.7 |
| Sept. | 10 | 1.0 | 2 | 10 | 3 | NE | 6.2 |
| Oct. | 14 | 0.3 | 4 | 13 | 3 | S | 6.0 |
| Nov. | 14 | 0.1 | 4 | 16 | 2 | S | 6.2 |
| Dec. | 15 | 0.1 | 5 | 15 | 3 | S | 6.5 |
| Annual | 143 | 5.6 | 37 | 140 | 34 | | 6.5 |

TABLE IX

CLIMATIC TABLE FOR GERDINGEN-BREE
Latitude 51°7′N, longitude 5°35′E, elevation 63 m

| Month | Mean daily temp. (°C) | Mean daily temp. range(°C) | Temp. extremes(°C) | | Mean vapor press. (mbar) | Mean precip. (mm) | Max. precip. 24 h (mm) |
|---|---|---|---|---|---|---|---|
| | | | highest | lowest | | | |
| Jan. | 1.6 | 5.6 | 16.0 | −19.5 | 6.4 | 83 | 46 |
| Feb. | 1.8 | 7.2 | 20.1 | −21.5 | 6.2 | 62 | 39 |
| Mar. | 5.3 | 9.5 | 22.3 | − 9.9 | 6.8 | 46 | 31 |
| Apr. | 8.9 | 11.1 | 29.6 | − 4.8 | 8.6 | 51 | 30 |
| May | 12.6 | 11.7 | 35.0 | − 3.5 | 10.8 | 60 | 58 |
| June | 15.9 | 12.0 | 38.0 | − 0.2 | 13.2 | 54 | 52 |
| July | 17.3 | 11.5 | 37.5 | 3.1 | 15.3 | 82 | 49 |
| Aug. | 17.2 | 11.0 | 37.1 | 4.8 | 15.4 | 87 | 46 |
| Sept. | 14.4 | 11.2 | 35.6 | 1.5 | 13.2 | 77 | 48 |
| Oct. | 10.0 | 9.2 | 24.8 | − 7.8 | 10.2 | 74 | 43 |
| Nov. | 5.9 | 6.8 | 20.5 | − 8.2 | 8.1 | 52 | 65 |
| Dec. | 3.2 | 4.9 | 16.9 | −18.6 | 9.1 | 75 | 36 |
| Annual | 9.5 | 9.3 | 38.0 | −21.5 | 10.8 | 803 | 65 |

| Month | Number of days with | | | | | Most freq. wind dir. | Mean wind speed (m/sec) |
|---|---|---|---|---|---|---|---|
| | precip. ⩾0.1 mm | thunder-storms | fog | cloudy sky ⩾$\frac{3}{4}$ | clear sky ⩽$\frac{1}{4}$ | | |
| Jan. | 23 | 0.1 | 6 | 14 | 2 | SW | 4.9 |
| Feb. | 16 | 0 | 6 | 12 | 4 | SW | 4.6 |
| Mar. | 17 | 0.3 | 5 | 13 | 2 | SW | 4.4 |
| Apr. | 18 | 0.1 | 5 | 14 | 1 | SW | 4.3 |
| May | 17 | 1.4 | 3 | 11 | 2 | NE | 3.9 |
| June | 12 | 1.9 | 2 | 11 | 4 | SW | 3.8 |
| July | 17 | 3.0 | 3 | 15 | 2 | SW | 3.9 |
| Aug. | 19 | 3.6 | 4 | 12 | 2 | SW | 3.9 |
| Sept. | 17 | 0.7 | 6 | 10 | 4 | SW | 3.7 |
| Oct. | 21 | 0.4 | 7 | 13 | 3 | SW | 4.0 |
| Nov. | 19 | 0 | 6 | 17 | 2 | SW | 4.0 |
| Dec. | 21 | 0 | 6 | 16 | 3 | SW | 4.7 |
| Annual | 217 | 11.7 | 59 | 158 | 31 | SW | 4.2 |

TABLE X

Latitude 50°55′N, longitude 5°46′E, elevation 115,7 m

| Month | Mean sta. press.[1] (mbar) | Mean daily temp. (°C) | Mean daily temp. range(°C) | Temp. extremes[2](°C) highest | lowest | Mean vapor press.[3] (mbar) | Mean precip.[4] (mm) | Max. precip.[2,4] 24 h (mm) |
|---|---|---|---|---|---|---|---|---|
| Jan. | 1016.0 | 1.6 | 5.0 | 15.9 | −18.0 | 6.3 | 63 | 28.8 |
| Feb. | 1016.2 | 1.9 | 5.8 | 19.1 | −18.5 | 6.3 | 49 | 33.7 |
| Mar. | 1016.0 | 5.3 | 7.9 | 21.6 | − 9.0 | 7.3 | 44 | 38.0 |
| Apr. | 1015.4 | 8.6 | 9.0 | 28.4 | − 3.7 | 8.5 | 54 | 32.6 |
| May | 1015.8 | 12.5 | 10.2 | 31.0 | − 1.0 | 10.9 | 51 | 47.3 |
| June | 1016.8 | 15.6 | 10.0 | 37.3 | 3.7 | 13.3 | 67 | 62.7 |
| July | 1015.7 | 17.2 | 9.6 | 36.0 | 6.5 | 15.1 | 70 | 67.6 |
| Aug. | 1015.4 | 17.1 | 9.4 | 34.1 | 5.7 | 15.1 | 75 | 74.6 |
| Sept. | 1016.8 | 14.4 | 8.7 | 33.1 | 2.8 | 13.5 | 64 | 47.5 |
| Oct. | 1016.3 | 9.8 | 7.2 | 24.2 | − 4.6 | 10.7 | 52 | 40.1 |
| Nov. | 1015.3 | 5.8 | 5.2 | 20.4 | − 6.1 | 8.3 | 59 | 34.3 |
| Dec. | 1015.7 | 2.8 | 4.5 | 16.7 | −16.7 | 7.0 | 53 | 40.2 |
| Annual | 1016.0 | 9.4 | 7.7 | 37.3 | −18.5 | 10.2 | 701 | 74.6 |

| Month | Mean evap. (mm) | Number of days with precip. ⩾0.1 mm | thunder- storm | Mean cloudiness (oktas) | Mean sunshine (h) | Most. freq. wind dir. |
|---|---|---|---|---|---|---|
| Jan. | 7 | 19 | 0 | 7.2 | 50 | SW |
| Feb. | 19 | 17 | 0 | 7.2 | 63 | SW |
| Mar. | 45 | 14 | 1 | 6.4 | 118 | S |
| Apr. | 79 | 14 | 2 | 6.5 | 150 | SW |
| May | 112 | 14 | 4 | 6.3 | 197 | NE |
| June | 125 | 14 | 4 | 6.3 | 211 | SW |
| July | 121 | 15 | 5 | 6.5 | 197 | SW |
| Aug. | 100 | 15 | 5 | 6.4 | 183 | S, SW |
| Sept. | 66 | 15 | 2 | 6.2 | 143 | S |
| Oct. | 31 | 16 | 1 | 6.6 | 102 | S |
| Nov. | 13 | 18 | 0 | 7.6 | 51 | S |
| Dec. | 6 | 18 | 0 | 7.5 | 42 | S |
| Annual | 724 | 189 | 24 | 6.7 | 1505 | S, SW |

[1] Reduced to mean sea level.

[2] Extreme temperatures since 1946: max. precipitation in 24 h since 1852.

[3] Calculated from observations at 08h00, 14h00 and 19h00.

[4] Maastricht.

## TABLE XI

CLIMATIC TABLE FOR UCCLE-BRUSSELS
Latitude 50°48′N, longitude 4°21′E, elevation 100 m

| Month | Mean sta. press. (mbar) | Mean daily temp. (°C) | Mean daily temp. range(°C) | Temp. extremes(°C) highest | lowest | Mean vapor press. (mbar) | Mean precip. (mm) | Max. precip. 24 h (mm) |
|---|---|---|---|---|---|---|---|---|
| Jan. | 1002.9 | 2.2 | 5.7 | 15.3 | −18.7 | 6.5 | 83 | 38 |
| Feb. | 1003.3 | 2.6 | 6.8 | 20.0 | −16.7 | 6.6 | 67 | 24 |
| Mar. | 1003.2 | 6.0 | 8.7 | 23.1 | − 9.4 | 7.6 | 47 | 24 |
| Apr. | 1003.0 | 9.2 | 9.9 | 28.7 | − 3.8 | 9.0 | 53 | 38 |
| May | 1003.4 | 13.0 | 11.0 | 33.8 | − 2.2 | 11.4 | 42 | 39 |
| June | 1004.5 | 16.0 | 11.2 | 38.8 | 0.3 | 14.0 | 55 | 42 |
| July | 1003.6 | 17.5 | 11.0 | 37.1 | 5.2 | 15.8 | 97 | 75 |
| Aug. | 1003.2 | 17.3 | 10.5 | 36.5 | 4.8 | 15.9 | 83 | 43 |
| Sept. | 1004.5 | 14.7 | 10.0 | 31.7 | 0.2 | 14.1 | 69 | 48 |
| Oct. | 1003.9 | 10.3 | 8.2 | 26.0 | − 6.8 | 11.1 | 90 | 34 |
| Nov. | 1002.3 | 6.2 | 6.2 | 20.4 | − 6.4 | 8.6 | 64 | 62 |
| Dec. | 1002.7 | 3.3 | 5.3 | 16.1 | −16.0 | 7.2 | 67 | 34 |
| Annual | 1003.4 | 9.9 | 8.7 | 38.8 | −18.7 | 10.7 | 817 | 75 |

| Month | Number of days with precip. ⩾0.1 mm | thunder- storms | fog | cloudy sky ⩾¾ | clear sky ⩽¼ | Mean cloudi- ness (oktas) | Mean sun- shine (h) | Most freq. wind dir. | Mean wind speed (m/sec) |
|---|---|---|---|---|---|---|---|---|---|
| Jan. | 23 | 0.7 | 6 | 18 | 2 | 7.7 | 54 | SW | 4.3 |
| Feb. | 17 | 0.6 | 7 | 15 | 4 | 7.6 | 77 | SW | 4.1 |
| Mar. | 15 | 0.3 | 3 | 14 | 3 | 7.4 | 119 | SW | 4.0 |
| Apr. | 16 | 0.9 | 2 | 13 | 1 | 7.3 | 158 | SW | 3.8 |
| May | 15 | 2.7 | 1 | 12 | 2 | 6.7 | 219 | SW | 3.6 |
| June | 14 | 2.7 | 1 | 11 | 4 | 7.2 | 196 | SW | 3.4 |
| July | 17 | 3.3 | 0.4 | 14 | 4 | 7.2 | 198 | SW | 3.4 |
| Aug. | 18 | 3.4 | 2 | 13 | 2 | 7.1 | 198 | SW | 3.4 |
| Sept. | 14 | 1.1 | 4 | 11 | 5 | 6.7 | 160 | SW | 3.4 |
| Oct. | 19 | 1.1 | 7 | 15 | 3 | 7.2 | 110 | SW | 3.6 |
| Nov. | 19 | 0.3 | 8 | 19 | 1 | 7.7 | 60 | SW | 3.9 |
| Dec. | 19 | 0 | 6 | 20 | 3 | 8.1 | 36 | SW | 4.2 |
| Annual | 206 | 17.1 | 47 | 175 | 34 | 7.3 | 1,585 | SW | 3.8 |

TABLE XII

CLIMATIC TABLE FOR LILLE (LESQUIN)
Latitude 50°44′N, longitude 3°6′E Gr., elevation 44 m

| Month | Mean sta. press. (mbar) | Mean daily temp. (°C) | Mean daily temp. range(°C) | Temp. extremes(°C)* highest | Temp. extremes(°C)* lowest | Mean vapor press.** (mbar) | Mean precip. (mm) | Max. precip. 24 h* (mm) |
|---|---|---|---|---|---|---|---|---|
| Jan. | 1010.0 | 2.4 | 4.9 | 13.3 | −17.6 | 6.7 | 53 | 21 |
| Feb. | 1010.9 | 2.9 | 5.7 | 18.9 | −17.8 | 6.5 | 44 | 18 |
| Mar. | 1010.3 | 6.0 | 7.8 | 22.0 | − 8.7 | 7.7 | 37 | 17 |
| Apr. | 1010.5 | 8.9 | 9.1 | 27.6 | − 3.0 | 8.5 | 42 | 29 |
| May | 1010.2 | 12.4 | 9.9 | 30.7 | − 1.0 | 10.9 | 50 | 34 |
| June | 1011.3 | 15.3 | 10.3 | 34.8 | 1.8 | 13.2 | 51 | 49 |
| July | 1010.8 | 17.1 | 10.0 | 36.1 | 4.0 | 14.9 | 63 | 38 |
| Aug. | 1010.2 | 17.1 | 10.0 | 35.6 | 3.9 | 14.9 | 63 | 24 |
| Sept. | 1011.2 | 14.7 | 9.1 | 33.8 | 1.2 | 13.3 | 57 | 48 |
| Oct. | 1010.7 | 10.4 | 7.4 | 26.4 | − 4.4 | 11.0 | 63 | 23 |
| Nov. | 1009.8 | 6.1 | 5.3 | 18.5 | − 7.6 | 8.6 | 60 | 22 |
| Dec. | 1009.9 | 3.5 | 4.7 | 15.7 | −14.0 | 7.8 | 54 | 22 |
| Annual | 1010.5 | 9.7 | 7.8 | 36.1 | −17.8 | 10.3 | 637 | 49 |

| Month | Mean evap.** (mm) | Number of days with precip. ≥0.1 mm | Number of days with thunder storms** | Number of days with fog** | Mean cloudi-ness** (oktas) | Mean sun-shine* (h) | Most freq. wind dir.** | Mean wind speed** (m/sec) | No. days gale** |
|---|---|---|---|---|---|---|---|---|---|
| Jan. | | 18 | 0.3 | 8.4 | 7.2 | 59 | W | 5.5 | 9 |
| Feb. | | 14 | 0 | 9.2 | 6.8 | 74 | SSW | 5.0 | 6 |
| Mar. | | 13 | 0.4 | 6.8 | 6.2 | 130 | NE | 4.8 | 5 |
| Apr. | 67 | 14 | 1.6 | 3.7 | 5.6 | 176 | NE | 5.0 | 5 |
| May | 86 | 13 | 3.4 | 3.3 | 6.1 | 198 | W | 4.4 | 4 |
| June | 79 | 12 | 3.7 | 4.1 | 6.1 | 205 | W | 4.0 | 2 |
| July | 85 | 13 | 4.2 | 3.5 | 6.3 | 201 | W | 4.1 | 3 |
| Aug. | 79 | 13 | 5.6 | 5.0 | 6.2 | 179 | W | 3.9 | 3 |
| Sept. | 70 | 14 | 2.8 | 6.8 | 5.8 | 150 | W | 4.0 | 4 |
| Oct. | 42 | 14 | 0.6 | 10.5 | 6.4 | 112 | W | 3.9 | 4 |
| Nov. | | 16 | 0.1 | 8.8 | 7.4 | 51 | S | 4.5 | 4 |
| Dec. | | 17 | 0.1 | 10.6 | 7.7 | 39 | SSW | 4.9 | 6 |
| Annual | | 171 | 22.8 | 80.7 | 6.5 | 1574 | W | 4.5 | 55 |

  * Period 1946–1960
  ** Period 1951–1960

TABLE XIII

CLIMATIC TABLE FOR BOTRANGE-ROBERTVILLE
Latitude 50°30′N, longitude 6°6′E, elevation 694 m

| Month | Mean daily temp. (°C) | Mean daily temp. range(°C) | Temp. extremes(°C) | | Mean vapor press. (mbar) | Mean precip. (mm) | Max. precip. 24 h (mm) |
|---|---|---|---|---|---|---|---|
| | | | highest | lowest | | | |
| Jan. | − 2.1 | 3.3 | 15.9 | −23.6 | 5.0 | 145 | 61 |
| Feb. | − 1.6 | 4.3 | 16.4 | −23.1 | 5.2 | 141 | 48 |
| Mar. | 1.4 | 6.5 | 19.8 | −14.4 | 5.8 | 85 | 61 |
| Apr. | 4.8 | 7.0 | 26.2 | − 7.4 | 7.6 | 93 | 56 |
| May | 8.4 | 8.0 | 29.2 | − 6.4 | 9.8 | 114 | 81 |
| June | 11.6 | 8.2 | 33.0 | − 1.5 | 12.2 | 116 | 77 |
| July | 13.1 | 7.4 | 31.4 | 1.2 | 13.7 | 161 | 101 |
| Aug. | 13.2 | 7.5 | 33.3 | 2.9 | 13.5 | 150 | 73 |
| Sept. | 11.0 | 7.4 | 33.0 | − 0.5 | 11.3 | 133 | 77 |
| Oct. | 6.6 | 5.8 | 22.7 | − 9.8 | 8.6 | 128 | 128 |
| Nov. | 2.2 | 3.8 | 16.4 | −10.0 | 6.9 | 96 | 63 |
| Dec. | − 0.6 | 3.3 | 14.8 | −20.7 | 5.6 | 148 | 73 |
| Annual | 5.7 | 6.0 | 33.3 | −23.6 | 8.8 | 1510 | 128 |

| Month | Number of days with | | | | | Mean wind speed (m/sec) |
|---|---|---|---|---|---|---|
| | precip. ≥0.1 mm | thunder-storms | fog | cloudy sky ≥$\frac{3}{4}$ | clear sky ≤$\frac{1}{4}$ | |
| Jan. | 21 | 0.1 | 19 | 5 | 16 | 5.2 |
| Feb. | 16 | 0 | 15 | 6 | 13 | 5.1 |
| Mar. | 16 | 0 | 14 | 5 | 13 | 4.5 |
| Apr. | 18 | 0.6 | 16 | 3 | 13 | 4.5 |
| May | 18 | 1.3 | 15 | 4 | 12 | 4.2 |
| June | 16 | 1.3 | 13 | 6 | 12 | 3.9 |
| July | 19 | 0.7 | 17 | 4 | 13 | 4.0 |
| Aug. | 20 | 0.9 | 16 | 4 | 14 | 4.3 |
| Sept. | 15 | 0.6 | 13 | 6 | 14 | 4.3 |
| Oct. | 19 | 0.3 | 16 | 5 | 16 | 4.6 |
| Nov. | 17 | 0 | 21 | 2 | 16 | 4.6 |
| Dec. | 18 | 0 | 20 | 4 | 16 | 5.2 |
| Annual | 213 | 5.8 | 195 | 54 | 168 | 4.5 |

TABLE XIV

CLIMATIC TABLE FOR WERBOMONT
Latitude 50°23′N, longitude 5°40′E, elevation 460 m

| Month | Mean daily temp. (°C) | Mean daily temp. range(°C) | Temp. extremes(°C) highest | lowest | Mean vapor press. (mbar) | Mean precip. (mm) | Max. precip. 24 h (mm) |
|-------|------|------|------|------|------|------|------|
| Jan. | − 0.5 | 4.8 | 13.2 | −23.8 | 5.7 | 110 | 42 |
| Feb. | 0.1 | 6.3 | 20.1 | −20.9 | 5.6 | 94 | 31 |
| Mar. | 3.3 | 7.5 | 22.6 | −14.0 | 6.4 | 62 | 33 |
| Apr. | 6.7 | 8.3 | 28.6 | − 6.5 | 7.7 | 74 | 39 |
| May | 10.3 | 9.1 | 34.6 | − 3.6 | 9.8 | 95 | 32 |
| June | 13.4 | 9.2 | 37.1 | − 1.0 | 12.2 | 89 | 41 |
| July | 14.9 | 8.9 | 33.0 | 4.0 | 13.8 | 125 | 61 |
| Aug. | 14.8 | 8.8 | 36.6 | 1.5 | 13.5 | 120 | 60 |
| Sept. | 12.5 | 8.9 | 32.6 | 1.0 | 11.8 | 106 | 94 |
| Oct. | 8.0 | 7.7 | 28.1 | − 8.3 | 9.1 | 89 | 40 |
| Nov. | 3.8 | 5.2 | 19.6 | −10.3 | 7.6 | 73 | 47 |
| Dec. | 1.2 | 4.4 | 14.0 | −17.8 | 6.4 | 101 | 35 |
| Annual | 7.4 | 7.4 | 37.1 | −23.8 | 9.1 | 1138 | 94 |

| Month | Number of days with precip. ⩾0.1 mm | thunder- storms | fog | cloudy sky ⩾$\frac{3}{4}$ | clear sky ⩽$\frac{1}{4}$ | Mean wind speed (m/sec) |
|-------|------|------|------|------|------|------|
| Jan. | 22 | 0 | 7 | 20 | 4 | 5.2 |
| Feb. | 18 | 0.1 | 8 | 14 | 5 | 5.0 |
| Mar. | 17 | 0.1 | 6 | 15 | 4 | 4.6 |
| Apr. | 19 | 1.0 | 8 | 16 | 2 | 4.4 |
| May | 19 | 2.3 | 6 | 15 | 3 | 4.1 |
| June | 16 | 2.3 | 5 | 12 | 3 | 3.9 |
| July | 19 | 2.4 | 4 | 17 | 3 | 4.1 |
| Aug. | 19 | 1.7 | 6 | 16 | 2 | 4.1 |
| Sept. | 17 | 0.6 | 8 | 13 | 5 | 4.1 |
| Oct. | 19 | 0.4 | 10 | 13 | 4 | 4.7 |
| Nov. | 19 | 0 | 11 | 20 | 2 | 4.8 |
| Dec. | 20 | 0.4 | 9 | 21 | 3 | 5.3 |
| Annual | 224 | 11.4 | 88 | 192 | 40 | 4.5 |

TABLE XV

CLIMATIC TABLE FOR CLERVAUX
Latitude 50°3′N, longitude 6°1′E, elevation 454 m

| Month | Mean sta. press. (mbar) | Mean daily temp. (°C) | Mean daily temp. range(°C) | Temp. extremes(°C) highest | lowest | Mean vapor press. (mbar) | Mean precip. (mm) | Max. precip. 24 h (mm) |
|-------|------|------|------|------|------|------|------|------|
| Jan.   | 960.4 | − 0.0 | 5.6  | 11.0 | −18.2 | 5.4  | 85  | 27 |
| Feb.   | 960.3 | 0.6   | 7.0  | 16.0 | −23.5 | 5.4  | 61  | 29 |
| Mar.   | 960.3 | 4.3   | 9.2  | 20.0 | −11.5 | 6.5  | 51  | 41 |
| Apr.   | 960.3 | 7.4   | 10.5 | 26.5 | − 7.0 | 7.7  | 66  | 27 |
| May    | 960.5 | 11.8  | 11.2 | 31.5 | − 3.5 | 9.9  | 64  | 25 |
| June   | 962.2 | 15.1  | 11.0 | 35.5 | 0.5   | 12.5 | 82  | 45 |
| July   | 961.8 | 16.6  | 10.6 | 35.5 | 1.6   | 14.2 | 88  | 67 |
| Aug.   | 960.5 | 15.9  | 10.1 | 31.5 | 1.0   | 14.0 | 99  | 39 |
| Sept.  | 962.5 | 13.3  | 9.8  | 29.0 | − 3.0 | 12.4 | 69  | 31 |
| Oct.   | 961.6 | 8.7   | 8.6  | 22.8 | − 4.0 | 9.5  | 64  | 29 |
| Nov.   | 960.3 | 4.3   | 5.4  | 17.5 | −10.0 | 7.5  | 77  | 26 |
| Dec.   | 959.0 | 1.1   | 5.0  | 13.5 | −10.5 | 6.0  | 75  | 28 |
| Annual | 960.8 | 8.3   | 8.7  | 35.5 | −23.5 | 9.3  | 881 | 67 |

| Month | Number of days precip. ≥0.1 mm | Mean cloudiness (oktas) | Mean sunshine (h) | Most freq. wind dir.* |
|-------|------|------|------|------|
| Jan.   | 17  | 7.5 | 53   | S |
| Feb.   | 16  | 6.7 | 82   | S |
| Mar.   | 13  | 6.1 | 141  | S |
| Apr.   | 15  | 6.6 | 174  | N |
| May    | 13  | 6.3 | 218  | N |
| June   | 14  | 6.2 | 237  | N |
| July   | 15  | 6.6 | 237  | N |
| Aug.   | 15  | 6.6 | 210  | S |
| Sept.  | 14  | 6.2 | 173  | S |
| Oct.   | 17  | 6.5 | 121  | S |
| Nov.   | 18  | 7.7 | 67   | S |
| Dec.   | 17  | 7.8 | 41   | S |
| Annual | 184 | 6.7 | 1754 | S |

* 1950–1960

TABLE XVI

CLIMATIC TABLE FOR LUXEMBOURG-CITY
Latitude 49°37′N, longitude 6°3′E, elevation 334 m

| Month | Mean sta. press. (mbar) | Mean daily temp. (°C) | Mean daily temp. range(°C) | Temp. extremes(°C)* | | Mean vapor press. (mbar) | Mean precip. (mm) | Max. precip. 24 h (mm) |
|---|---|---|---|---|---|---|---|---|
| | | | | highest | lowest | | | |
| Jan. | 975.8 | 0.3 | 3.8 | 11.7 | −16.0 | 5.5 | 73 | 28 |
| Feb. | 975.9 | 1.0 | 5.7 | 17.2 | −19.6 | 5.6 | 56 | 23 |
| Mar. | 975.6 | 4.9 | 8.0 | 22.5 | −10.5 | 6.7 | 43 | 23 |
| Apr. | 975.4 | 8.5 | 9.7 | 29.4 | − 3.9 | 8.1 | 54 | 28 |
| May | 975.3 | 12.8 | 10.3 | 30.2 | − 1.6 | 10.8 | 60 | 67 |
| June | 977.0 | 15.7 | 9.6 | 36.8 | 3.7 | 13.0 | 64 | 33 |
| July | 976.5 | 17.4 | 9.3 | 36.6 | 5.3 | 15.2 | 66 | 50 |
| Aug. | 975.4 | 16.7 | 9.1 | 33.8 | 4.3 | 14.4 | 74 | 51 |
| Sept. | 977.5 | 13.8 | 8.6 | 33.0 | 1.0 | 12.6 | 63 | 44 |
| Oct. | 976.8 | 9.0 | 6.8 | 23.5 | − 4.5 | 9.8 | 55 | 28 |
| Nov. | 975.6 | 4.6 | 4.0 | 17.0 | − 7.2 | 7.6 | 64 | 42 |
| Dec. | 974.7 | 1.3 | 3.3 | 14.0 | −15.2 | 6.0 | 68 | 42 |
| Annual | 976.0 | 8.8 | 7.4 | 36.8 | −19.6 | 9.6 | 740 | 67 |

| Month | Number of days with | | | Mean cloudi- ness (oktas) | Mean sun- shine (h) | Most freq. wind dir.* |
|---|---|---|---|---|---|---|
| | precip. ≤0.1 mm | thunder- storms* | fog** | | | |
| Jan. | 19 | 0.1 | 9 | 7.5 | 50 | SW |
| Feb. | 15 | 0 | 6 | 7.0 | 75 | SW |
| Mar. | 13 | 0.1 | 5 | 5.9 | 145 | E |
| Apr. | 13 | 1.2 | 5 | 5.9 | 165 | E |
| May | 14 | 2.8 | 3 | 5.8 | 210 | E |
| June | 13 | 3.1 | 4 | 5.9 | 215 | S |
| July | 15 | 3.6 | 3 | 6.0 | 225 | SW |
| Aug. | 14 | 4.0 | 4 | 5.9 | 200 | SW |
| Sept. | 14 | 2.0 | 6 | 5.8 | 160 | S |
| Oct. | 16 | 0.1 | 9 | 6.6 | 105 | E |
| Nov. | 18 | 0 | 10 | 7.8 | 45 | S |
| Dec. | 18 | 0.1 | 7 | 7.9 | 35 | SW |
| Annual | 182 | 17.1 | 71 | 6.5 | 1,630 | SW |

  * 1946–1960
** 1930–1949

TABLE XVII

CLIMATIC TABLE FOR VIRTON
Latitude 49°33′N, longitude 5°34′E, elevation 242 m

| Month | Mean daily temp. (°C) | Mean daily temp. range(°C) | Temp. extremes(°C) | | Mean vapor press. (mbar) | Mean precip. (mm) | Max. precip. 24 h (mm) |
|---|---|---|---|---|---|---|---|
| | | | highest | lowest | | | |
| Jan. | 0.9 | 5.6 | 13.3 | −22.8 | 6.3 | 97 | 38 |
| Feb. | 1.4 | 7.5 | 16.5 | −21.2 | 6.5 | 82 | 33 |
| Mar. | 5.1 | 10.1 | 23.5 | −15.5 | 7.0 | 60 | 31 |
| Apr. | 9.0 | 11.1 | 30.5 | −10.5 | 8.2 | 67 | 44 |
| May | 12.6 | 12.3 | 34.4 | − 5.5 | 11.2 | 74 | 42 |
| June | 15.7 | 11.9 | 36.8 | − 3.4 | 13.7 | 59 | 60 |
| July | 17.2 | 11.7 | 35.8 | 1.9 | 15.4 | 79 | 44 |
| Aug. | 16.9 | 11.3 | 37.0 | 1.2 | 15.5 | 88 | 45 |
| Sept. | 14.1 | 11.7 | 33.8 | − 1.3 | 12.9 | 76 | 36 |
| Oct. | 9.0 | 10.2 | 26.1 | − 9.5 | 9.7 | 72 | 39 |
| Nov. | 5.3 | 5.9 | 20.8 | −10.8 | 8.1 | 61 | 41 |
| Dec. | 2.3 | 5.3 | 18.0 | −19.1 | 7.4 | 87 | 46 |
| Annual | 9.1 | 9.5 | 37.0 | −22.8 | 9.7 | 902 | 60 |

| Month | Number of days with | | | | | Mean wind speed (m/sec) |
|---|---|---|---|---|---|---|
| | precip. ≥0.1 mm | thunder-storms | fog | cloudy sky ≥$\frac{3}{4}$ | clear sky ≤$\frac{1}{4}$ | |
| Jan. | 20 | 0 | 1 | 18 | 4 | 3.4 |
| Feb. | 16 | 0 | 3 | 14 | 5 | 3.1 |
| Mar. | 15 | 0.3 | 1 | 13 | 6 | 2.9 |
| Apr. | 15 | 0.6 | 1 | 12 | 3 | 2.8 |
| May | 16 | 1.9 | 1 | 11 | 4 | 2.4 |
| June | 13 | 0.7 | 0 | 11 | 5 | 2.3 |
| July | 15 | 0.9 | 1 | 13 | 3 | 2.4 |
| Aug. | 18 | 0.4 | 2 | 14 | 3 | 2.4 |
| Sept. | 14 | 0.7 | 3 | 13 | 6 | 2.4 |
| Oct. | 19 | 0.1 | 5 | 13 | 4 | 2.4 |
| Nov. | 17 | 0 | 2 | 22 | 2 | 2.8 |
| Dec. | 18 | 0 | 3 | 20 | 4 | 3.3 |
| Annual | 196 | 5.6 | 23 | 174 | 49 | 2.7 |

TABLE XVIII

CLIMATIC TABLE FOR LE HAVRE (LA HÈVE),
Latitude 49°31′N, longitude 0°4′E, elevation 101 m

| Month | Mean sta. press. (mbar) | Mean daily temp. (°C) | Mean daily temp. range(°C) | Temp. extremes(°C)* | | Mean vapor press.** (mbar) | Mean precip. (mm) | Max. precip. 24 h* (mm) |
|---|---|---|---|---|---|---|---|---|
| | | | | highest | lowest | | | |
| Jan. | 1002.4 | 4.6 | 4.3 | 14.6 | −11.0 | 7.5 | 73 | 21 |
| Feb. | 1003.4 | 4.8 | 4.8 | 20.0 | −12.2 | 7.2 | 55 | 26 |
| Mar. | 1002.4 | 7.0 | 6.6 | 21.2 | − 4.8 | 8.3 | 45 | 18 |
| Apr. | 1003.3 | 9.1 | 6.3 | 24.4 | − 0.8 | 9.1 | 46 | 30 |
| May | 1003.4 | 12.3 | 7.0 | 30.2 | 2.8 | 11.2 | 50 | 26 |
| June | 1004.7 | 14.6 | 6.2 | 32.4 | 5.6 | 13.5 | 47 | 41 |
| July | 1004.0 | 16.8 | 6.2 | 36.1 | 8.8 | 15.4 | 62 | 18 |
| Aug. | 1003.8 | 17.1 | 6.0 | 32.4 | 9.8 | 15.8 | 71 | 30 |
| Sept. | 1004.3 | 15.8 | 5.9 | 32.0 | 6.6 | 14.1 | 74 | 35 |
| Oct. | 1003.6 | 12.1 | 5.4 | 25.4 | − 1.2 | 12.1 | 85 | 50 |
| Nov. | 1002.0 | 8.1 | 4.6 | 18.4 | − 3.4 | 9.5 | 89 | 30 |
| Dec. | 1002.3 | 5.5 | 4.3 | 15.0 | − 7.8 | 8.8 | 67 | 32 |
| Annual | 1003.3 | 10.6 | 5.6 | 36.1 | −12.2 | 11.0 | 764 | 50 |

| Month | Number of days with | | | Mean cloudiness** (oktas) | Most freq. wind dir.** |
|---|---|---|---|---|---|
| | precip. ⩾0.1 mm | thunder-storms** | fog** | | |
| Jan. | 18 | 0.2 | 5.6 | 7.2 | SW |
| Feb. | 14 | 0.1 | 5.7 | 7.1 | SW |
| Mar. | 12 | 0.5 | 5.5 | 6.4 | E |
| Apr. | 12 | 0.4 | 4.1 | 5.9 | NE |
| May | 13 | 2.3 | 2.6 | 5.8 | NE |
| June | 11 | 2.0 | 3.9 | 6.2 | W |
| July | 11 | 1.6 | 4.0 | 6.1 | W |
| Aug. | 12 | 2.2 | 2.4 | 6.4 | W |
| Sept. | 14 | 1.6 | 3.4 | 6.3 | W |
| Oct. | 15 | 0.8 | 3.8 | 6.9 | SW |
| Nov. | 17 | 0.4 | 4.4 | 7.5 | SW |
| Dec. | 17 | 0.1 | 6.2 | 7.6 | SW |
| Annual | 166 | 12.2 | 51.6 | 6.6 | SW |

 * Period 1946–1960
** Period 1951–1960

## TABLE XIX

CLIMATIC TABLE FOR PARIS (LE BOURGET)
Latitude 48°58′N, longitude 2°27′E, Gr., elevation 52 m

| Month | Mean sta. press. (mbar) | Mean daily temp. (°C) | Mean daily temp. range(°C) | Temp. extremes[1](°C) highest | lowest | Mean vapor press.[2] (mbar) | Mean precip. (mm) | Max. precip. 24 h[1] (mm) | Mean daily global radiation (ly/day)[3] |
|---|---|---|---|---|---|---|---|---|---|
| Jan. | 1010.3 | 3.1 | 5.5 | 15.6 | −17.0 | 6.9 | 54 | 18 | 74 |
| Feb. | 1010.4 | 3.8 | 6.4 | 20.8 | −16.8 | 6.7 | 43 | 17 | 128 |
| Mar. | 1009.4 | 7.2 | 9.2 | 24.7 | − 7.8 | 7.9 | 32 | 16 | 244 |
| Apr. | 1009.5 | 10.3 | 10.2 | 31.9 | − 3.7 | 8.4 | 38 | 26 | 353 |
| May | 1009.2 | 14.0 | 10.8 | 33.1 | − 1.6 | 11.0 | 52 | 29 | 439 |
| June | 1010.8 | 17.1 | 10.9 | 36.2 | 1.7 | 13.4 | 50 | 35 | 478 |
| July | 1010.1 | 19.0 | 10.7 | 39.6 | 4.9 | 14.8 | 55 | 56 | 454 |
| Aug. | 1009.7 | 18.5 | 10.5 | 36.6 | 5.1 | 14.9 | 62 | 37 | 383 |
| Sept. | 1010.9 | 15.9 | 9.8 | 34.8 | 0.1 | 13.4 | 51 | 31 | 289 |
| Oct. | 1010.5 | 11.1 | 8.3 | 27.2 | − 4.6 | 11.0 | 49 | 31 | 174 |
| Nov. | 1009.3 | 6.8 | 5.9 | 20.3 | − 8.3 | 8.8 | 50 | 27 | 84 |
| Dec. | 1009.9 | 4.1 | 5.0 | 16.2 | −13.2 | 8.0 | 49 | 25 | 58 |
| Annual | 1010.0 | 10.9 | 8.6 | 39.6 | −17.0 | 10.4 | 585 | 56 | 263 |

| Month | Mean evapor.[1] (mm) | Number of days with precip. ⩾0.1 mm | thunder- storm[2] | fog[2] | gale[2] | Mean cloudi- ness[2] (oktas) | Mean sun- shine (h)[1,3] | Most freq. wind dir.[2] | Mean wind speed[2] (m/sec) |
|---|---|---|---|---|---|---|---|---|---|
| Jan. | | 17 | 0.3 | 8.8 | 6 | 7.0 | 64 | W | 4.4 |
| Feb. | | 14 | 0.5 | 7.1 | 4 | 6.5 | 83 | W | 4.3 |
| Mar. | | 12 | 0.6 | 3.9 | 4 | 5.7 | 152 | E | 4.2 |
| Apr. | 100 | 13 | 0.7 | 0.9 | 3 | 5.1 | 185 | NNE | 4.5 |
| May | 120 | 13 | 3.6 | 0.3 | 4 | 5.6 | 223 | NNE | 3.9 |
| June | 111 | 12 | 4.5 | 0.6 | 2 | 5.7 | 233 | W | 3.4 |
| July | 133 | 12 | 4.4 | 1.3 | 3 | 5.7 | 231 | W | 3.7 |
| Aug. | 115 | 13 | 3.6 | 1.5 | 3 | 5.5 | 204 | W | 3.3 |
| Sept. | 89 | 13 | 2.0 | 3.4 | 3 | 5.4 | 166 | W | 3.4 |
| Oct. | 55 | 14 | 0.5 | 9.3 | 3 | 5.9 | 122 | W | 3.3 |
| Nov. | | 15 | 0.2 | 8.4 | 4 | 7.2 | 63 | S | 3.7 |
| Dec. | | 16 | 0 | 9.6 | 6 | 7.5 | 53 | S | 4.1 |
| Annual | | 164 | 20.9 | 55.1 | 45 | 6.1 | 1779 | W | 3.9 |

[1] Period 1946–1960.
[2] Period 1951–1960.
[3] Observatory of Parc-Saint-Maur (48°49′N 2°30′E, 50 m).

TABLE XX

CLIMATIC TABLE FOR STRASBOURG (ENTZHEIM),
Latitude 48°33′N, longitude 7°38′E, elevation 149 m

| Month | Mean sta. press. (mbar) | Mean daily temp. (°C) | Mean daily temp. range(°C) | Temp. extremes(°C) highest | lowest | Mean vapor press.[2] (mbar) | Mean precip. (mm) | Max. precip. 24 h[1] (mm) | Mean evapor. (mm) |
|-------|------|------|------|------|------|------|------|------|------|
| Jan. | 999.5 | 0.4 | 5.7 | 15.1 | −22.0 | 5.7 | 39 | 25 | |
| Feb. | 999.0 | 1.5 | 7.1 | 20.8 | −22.2 | 5.8 | 33 | 36 | |
| Mar. | 998.3 | 5.6 | 10.1 | 23.5 | −15.7 | 7.1 | 30 | 26 | |
| Apr. | 997.7 | 9.8 | 11.0 | 29.7 | − 5.6 | 8.2 | 39 | 28 | 87 |
| May | 997.5 | 14.0 | 11.5 | 31.5 | − 2.4 | 11.3 | 60 | 33 | 102 |
| June | 999.1 | 17.2 | 11.5 | 37.0 | 2.8 | 14.4 | 77 | 54 | 89 |
| July | 998.8 | 19.0 | 11.4 | 37.4 | 5.6 | 16.1 | 77 | 44 | 104 |
| Aug. | 998.4 | 18.3 | 11.2 | 35.3 | 5.1 | 15.7 | 80 | 46 | 90 |
| Sept. | 999.9 | 15.1 | 10.5 | 33.4 | 0.4 | 13.6 | 58 | 22 | 68 |
| Oct. | 999.9 | 9.5 | 8.7 | 25.6 | − 7.6 | 10.2 | 42 | 24 | 41 |
| Nov. | 999.0 | 4.9 | 5.8 | 21.0 | −10.0 | 7.7 | 41 | 30 | |
| Dec. | 999.4 | 1.3 | 4.9 | 16.5 | −21.0 | 6.6 | 31 | 23 | |
| Annual | 998.9 | 9.7 | 9.1 | 37.4 | −22.2 | 10.2 | 607 | 54 | |

| Month | Number of days with precip. ⩾0.1 mm | thunder- storm[2] | fog[2] | gale[2] | Mean cloudi- ness[2] (oktas) | Mean sun- shine[1] (h) | Most freq. wind dir.[2] | Mean wind speed[2] (m/sec) |
|-------|------|------|------|------|------|------|------|------|
| Jan. | 15 | 0.1 | 6.5 | 2 | 7.7 | 49 | S | 2.8 |
| Feb. | 13 | 0.2 | 6.8 | 2 | 7.1 | 68 | S | 2.6 |
| Mar. | 12 | 0.2 | 3.7 | 1 | 5.9 | 148 | NNE | 2.6 |
| Apr. | 13 | 1.0 | 1.2 | 1 | 5.8 | 188 | NNE | 2.8 |
| May | 13 | 4.4 | 1.6 | 2 | 5.9 | 211 | NNE | 2.4 |
| June | 14 | 7.1 | 2.3 | 0.3 | 6.6 | 206 | S | 1.8 |
| July | 14 | 6.9 | 2.0 | 1 | 5.9 | 225 | W | 1.9 |
| Aug. | 13 | 5.2 | 3.5 | 1 | 5.9 | 216 | S | 1.8 |
| Sept. | 12 | 2.4 | 7.2 | 1 | 5.6 | 168 | S | 1.9 |
| Oct. | 12 | 0.4 | 12.7 | 1 | 6.3 | 121 | S | 1.7 |
| Nov. | 13 | 0.1 | 10.8 | 1 | 7.9 | 49 | S | 1.9 |
| Dec. | 14 | 0 | 11.3 | 2 | 8.0 | 36 | S | 2.3 |
| Annual | 158 | 28.0 | 69.6 | 15 | 6.5 | 1685 | S | 2.2 |

[1] Period 1946–1960.
[2] Period 1951–1960.

TABLE XXI

CLIMATIC TABLE FOR BREST (GUIPAVAS),
Latitude 48°27′N, longitude 4°25′W, elevation 98 m

| Month | Mean sta. press. (mbar) | Mean daily temp. (°C) | Mean daily temp. range(°C) | Temp. extremes[1](°C) highest | lowest | Mean vapor press.[2] (mbar) | Mean precip. (mm) | Max. precip. 24 h[1] (mm) | Mean evapor.[2] (mm) |
|---|---|---|---|---|---|---|---|---|---|
| Jan. | 1003.3 | 6.1 | 4.9 | 14.5 | −14.0 | 8.5 | 133 | 42 | |
| Feb. | 1004.1 | 5.8 | 5.5 | 19.3 | −13.4 | 8.0 | 96 | 34 | |
| Mar. | 1002.0 | 7.8 | 7.9 | 20.1 | − 4.4 | 9.1 | 83 | 47 | |
| Apr. | 1004.0 | 9.2 | 7.3 | 26.5 | − 0.7 | 9.4 | 69 | 55 | 68 |
| May | 1003.7 | 11.6 | 7.4 | 28.6 | − 0.0 | 11.5 | 68 | 38 | 72 |
| June | 1005.5 | 14.4 | 7.5 | 28.8 | 4.3 | 13.6 | 56 | 39 | 67 |
| July | 1005.3 | 15.6 | 7.4 | 35.2 | 6.3 | 15.0 | 62 | 26 | 69 |
| Aug. | 1004.5 | 16.0 | 7.3 | 33.7 | 6.6 | 15.2 | 80 | 48 | 67 |
| Sept. | 1004.9 | 14.7 | 6.9 | 29.0 | 4.0 | 14.3 | 87 | 37 | 57 |
| Oct. | 1004.0 | 12.0 | 6.5 | 25.3 | − 0.8 | 12.4 | 104 | 33 | 50 |
| Nov. | 1002.1 | 9.0 | 5.2 | 18.8 | − 4.0 | 10.1 | 138 | 48 | |
| Dec. | 1002.8 | 7.0 | 5.1 | 16.2 | − 5.0 | 9.6 | 150 | 57 | |
| Annual | 1003.9 | 10.8 | 6.5 | 35.2 | −14.0 | 11.4 | 1126 | 57 | |

| Month | Number of days with precip. ⩾0.1 mm | thunder- storm[2] | fog[2] | gale[2] | Mean cloudi- ness[2] (oktas) | Mean sun- shine[1] (h) | Most freq. wind dir.[2] | Mean wind speed[2] (m/sec) |
|---|---|---|---|---|---|---|---|---|
| Jan. | 22 | 0.8 | 4.5 | 10 | 7.2 | 66 | SW | 5.5 |
| Feb. | 16 | 0.9 | 3.9 | 10 | 6.9 | 85 | W | 5.4 |
| Mar. | 15 | 1.3 | 6.5 | 8 | 7.0 | 142 | NE | 5.7 |
| Apr. | 15 | 0.2 | 6.5 | 7 | 5.9 | 189 | NE | 5.4 |
| May | 14 | 1.1 | 7.7 | 5 | 6.2 | 220 | NE | 5.2 |
| June | 13 | 1.9 | 9.0 | 2 | 6.5 | 209 | NE | 4.5 |
| July | 14 | 1.2 | 11.0 | 3 | 6.5 | 210 | W | 4.8 |
| Aug. | 15 | 0.9 | 9.6 | 3 | 6.4 | 207 | SW | 4.5 |
| Sept. | 16 | 2.2 | 7.5 | 4 | 6.3 | 156 | SW | 4.5 |
| Oct. | 19 | 0.9 | 7.4 | 5 | 6.6 | 120 | SW | 4.2 |
| Nov. | 20 | 1.3 | 3.5 | 8 | 7.3 | 69 | W | 4.7 |
| Dec. | 22 | 2.0 | 4.6 | 11 | 7.3 | 56 | W | 5.2 |
| Annual | 201 | 14.7 | 81.7 | 76 | 6.7 | 1729 | SW | 5.0 |

[1] Period 1946–1960.
[2] Period 1951–1960

TABLE XXII

CLIMATIC TABLE FOR TOURS (SAINT-SYMPHORIEN)
Latitude 47°25′N, longitude 0°46′E, elevation 98 m

| Month | Mean sta. press. (mbar) | Mean daily temp. (°C) | Mean daily temp. range(°C) | Temp. extremes(°C) highest | lowest | Mean vapor press.[2] (mbar) | Mean precip. (mm) | Max. precip. 24 h[1] (mm) | Mean evapor. (mm) |
|---|---|---|---|---|---|---|---|---|---|
| Jan. | 1005.2 | 3.5 | 5.8 | 15.3 | −14.6 | 7.3 | 64 | 24 | |
| Feb. | 1005.0 | 4.4 | 6.7 | 20.4 | −13.2 | 7.1 | 55 | 25 | |
| Mar. | 1003.3 | 7.7 | 9.2 | 26.4 | − 6.0 | 8.4 | 49 | 41 | |
| Apr. | 1003.6 | 10.6 | 10.0 | 29.3 | − 2.0 | 8.9 | 48 | 28 | 75 |
| May | 1003.5 | 13.9 | 10.3 | 32.6 | − 2.7 | 11.3 | 63 | 23 | 83 |
| June | 1005.3 | 17.3 | 10.8 | 37.3 | 2.6 | 13.8 | 49 | 27 | 82 |
| July | 1005.1 | 19.1 | 11.1 | 41.4 | 7.1 | 15.1 | 49 | 48 | 98 |
| Aug. | 1004.6 | 18.7 | 10.8 | 36.8 | 6.7 | 15.3 | 61 | 57 | 82 |
| Sept. | 1005.2 | 16.2 | 9.9 | 34.8 | 2.0 | 14.1 | 59 | 74 | 60 |
| Oct. | 1004.9 | 11.7 | 8.5 | 28.0 | − 4.4 | 11.6 | 60 | 26 | 38 |
| Nov. | 1003.9 | 7.2 | 6.3 | 22.6 | − 9.0 | 9.0 | 64 | 53 | |
| Dec. | 1004.3 | 4.3 | 5.4 | 18.9 | −18.0 | 8.4 | 68 | 42 | |
| Annual | 1004.5 | 11.2 | 8.8 | 41.4 | −18.0 | 10.9 | 689 | 74 | |

| Month | Number of days with precip. ≥0.1 mm | thunder- storm[2] | fog[2] | gale[2] | Mean cloudi- ness[2] (oktas) | Mean sun- shine[1] (h) | Most freq. wind dir.[2] | Mean wind speed[2] (m/sec) |
|---|---|---|---|---|---|---|---|---|
| Jan. | 16 | 0.1 | 8.6 | 5 | 7.0 | 68 | W | 4.2 |
| Feb. | 13 | 0.4 | 6.0 | 7 | 6.5 | 86 | W | 4.4 |
| Mar. | 12 | 0.6 | 3.8 | 4 | 6.1 | 154 | E | 4.1 |
| Apr. | 12 | 1.6 | 1.2 | 3 | 5.4 | 208 | NE | 4.2 |
| May | 13 | 4.2 | 2.1 | 2 | 5.8 | 212 | W | 3.6 |
| June | 11 | 4.5 | 1.3 | 2 | 6.1 | 218 | W | 3.2 |
| July | 11 | 3.8 | 1.4 | 2 | 5.7 | 235 | W | 3.6 |
| Aug. | 12 | 2.7 | 1.6 | 3 | 5.8 | 213 | W | 3.2 |
| Sept. | 13 | 1.8 | 4.8 | 2 | 5.5 | 176 | W | 3.0 |
| Oct. | 13 | 0.5 | 8.3 | 2 | 6.2 | 133 | W | 2.9 |
| Nov. | 15 | 0.4 | 9.9 | 3 | 7.1 | 66 | E | 3.6 |
| Dec. | 16 | 0.2 | 9.8 | 5 | 7.6 | 50 | W | 4.0 |
| Annual | 157 | 20.8 | 58.8 | 40 | 6.4 | 1879 | W | 3.7 |

[1] Period 1946–1960.
[2] Period 1951–1960.

TABLE XXIII

CLIMATIC TABLE FOR DIJON
Latitude 47°16′N, longitude 5°6′E, elevation 220 m

| Month | Mean sta. press. (mbar) | Mean daily temp. (°C) | Mean daily temp. range(°C) | Temp. extremes[1](°C) | | Mean vapor press.[2] (mbar) | Mean precip. (mm) | Max. precip. 24 h[1] (mm) | Mean evapor. (mm) |
|---|---|---|---|---|---|---|---|---|---|
| | | | | highest | lowest | | | | |
| Jan. | 990.1 | 1.3 | 6.0 | 16.5 | −16.6 | 6.1 | 64 | 26 | |
| Feb. | 989.6 | 2.6 | 7.4 | 19.7 | −19.6 | 6.2 | 42 | 21 | |
| Mar. | 988.6 | 6.9 | 10.3 | 23.5 | −15.3 | 7.5 | 42 | 33 | |
| Apr. | 988.5 | 10.4 | 10.7 | 29.0 | − 5.3 | 8.1 | 46 | 27 | 98 |
| May | 988.3 | 14.3 | 11.1 | 33.0 | − 3.3 | 10.9 | 64 | 30 | 105 |
| June | 990.3 | 17.7 | 10.9 | 36.0 | 0.8 | 13.9 | 81 | 110 | 95 |
| July | 990.4 | 19.6 | 11.2 | 37.2 | 5.3 | 15.1 | 58 | 62 | 115 |
| Aug. | 989.9 | 19.0 | 11.2 | 36.0 | 5.4 | 15.1 | 77 | 63 | 92 |
| Sept. | 991.0 | 15.9 | 10.4 | 33.0 | − 0.0 | 13.8 | 72 | 57 | 67 |
| Oct. | 990.5 | 10.5 | 8.9 | 26.0 | − 3.8 | 10.5 | 60 | 53 | 41 |
| Nov. | 989.5 | 5.7 | 6.4 | 21.6 | − 9.4 | 8.0 | 76 | 52 | |
| Dec. | 990.0 | 2.1 | 5.1 | 16.0 | −20.4 | 7.1 | 57 | 38 | |
| Annual | 989.7 | 10.5 | 9.2 | 37.2 | −20.4 | 10.2 | 739 | 110 | |

| Month | Number of days with | | | | Mean cloudi- ness[2] (oktas) | Mean sun- shine[1] (h) | Most freq. wind dir.[2] | Mean wind speed[2] (m/sec) |
|---|---|---|---|---|---|---|---|---|
| | precip. ⩾0.1 mm | thunder- storm[2] | fog[2] | gale[2] | | | | |
| Jan. | 16 | 0 | 9.4 | 4 | 7.6 | 69 | S | 3.7 |
| Feb. | 13 | 0.1 | 7.3 | 4 | 6.8 | 93 | S | 3.9 |
| Mar. | 10 | 0.9 | 4.7 | 2 | 6.0 | 185 | S | 3.5 |
| Apr. | 11 | 1.7 | 0.9 | 3 | 5.8 | 212 | N | 4.0 |
| May | 12 | 3.4 | 1.2 | 2 | 5.7 | 249 | N | 3.2 |
| June | 12 | 5.7 | 1.6 | 2 | 6.1 | 252 | N | 2.6 |
| July | 11 | 4.4 | 1.2 | 1 | 5.3 | 269 | W | 2.8 |
| Aug. | 11 | 5.1 | 1.8 | 2 | 5.7 | 244 | S | 2.6 |
| Sept. | 11 | 2.3 | 5.6 | 1 | 5.6 | 207 | S | 2.5 |
| Oct. | 12 | 0.6 | 10.1 | 1 | 6.2 | 148 | S | 2.5 |
| Nov. | 14 | 0.3 | 12.0 | 2 | 7.5 | 71 | S | 2.9 |
| Dec. | 14 | 0 | 12.8 | 3 | 8.0 | 45 | S | 3.2 |
| Annual | 147 | 24.5 | 68.6 | 27 | 6.4 | 2044 | S | 3.1 |

[1] Period 1946–1960.

[2] Period 1951–1960.

TABLE XXIV

CLIMATIC TABLE FOR NANTES (CHATEAU-BOUGON)
Latitude 47°10′N, longitude 1°37′W, elevation 26 m

| Month | Mean sta. press. (mbar) | Mean daily temp. (°C) | Mean daily temp. range(°C) | Temp.extremes[1](°C) highest | lowest | Mean vapor press.[2] (mbar) | Mean precip. (mm) | Max. precip. 24 h[1] (mm) | Mean evapor.[2] (mm) |
|---|---|---|---|---|---|---|---|---|---|
| Jan. | 1014.2 | 5.0 | 6.5 | 15.4 | −11.4 | 8.0 | 81 | 28 | |
| Feb. | 1014.5 | 5.3 | 7.5 | 21.4 | −15.6 | 7.7 | 66 | 35 | |
| Mar. | 1012.3 | 8.4 | 9.1 | 22.6 | − 5.7 | 9.1 | 57 | 38 | |
| Apr. | 1013.4 | 10.8 | 9.9 | 27.5 | − 1.5 | 9.7 | 45 | 22 | 70 |
| May | 1012.6 | 13.9 | 10.5 | 32.7 | 0.1 | 12.2 | 58 | 57 | 81 |
| June | 1014.8 | 17.2 | 10.9 | 36.8 | 4.2 | 14.4 | 44 | 36 | 84 |
| July | 1014.5 | 18.8 | 10.7 | 40.3 | 5.8 | 16.0 | 48 | 32 | 89 |
| Aug. | 1013.7 | 18.6 | 11.2 | 36.0 | 5.6 | 16.0 | 63 | 40 | 83 |
| Sept. | 1014.3 | 16.4 | 10.1 | 33.8 | 2.8 | 15.0 | 73 | 38 | 62 |
| Oct. | 1014.0 | 12.2 | 9.0 | 27.4 | − 2.5 | 12.5 | 75 | 40 | 42 |
| Nov. | 1012.9 | 8.2 | 7.1 | 21.1 | − 5.3 | 9.8 | 83 | 46 | |
| Dec. | 1013.8 | 5.5 | 6.2 | 18.4 | −10.8 | 9.3 | 89 | 36 | |
| Annual | 1013.8 | 11.7 | 9.1 | 40.3 | −15.6 | 11.6 | 782 | 57 | |

| Month | Number of days with precip. ⩾0.1 mm | thunder- storm[2] | fog[2] | gale[2] | Mean cloudi- ness[2] (oktas) | Mean sun- shine[1] (h) | Most freq. wind dir.[2] | Mean wind speed[2] (m/sec) |
|---|---|---|---|---|---|---|---|---|
| Jan. | 18 | 0.4 | 7.3 | 7 | 6.6 | 75 | NE | 4.2 |
| Feb. | 14 | 1.1 | 7.4 | 7 | 6.0 | 94 | NE | 4.2 |
| Mar. | 14 | 0.7 | 4.4 | 5 | 6.2 | 151 | NE | 4.1 |
| Apr. | 11 | 0.5 | 2.4 | 3 | 5.3 | 203 | NE | 4.1 |
| May | 13 | 1.6 | 2.5 | 3 | 5.5 | 218 | W | 3.7 |
| June | 11 | 2.3 | 2.6 | 1 | 5.7 | 223 | W | 3.2 |
| July | 12 | 1.8 | 2.0 | 2 | 5.5 | 228 | W | 3.5 |
| Aug. | 12 | 2.2 | 4.0 | 2 | 5.4 | 218 | W | 3.3 |
| Sept. | 14 | 1.6 | 6.9 | 3 | 5.6 | 171 | W | 3.1 |
| Oct. | 15 | 0.5 | 7.7 | 2 | 6.0 | 140 | W | 3.1 |
| Nov. | 16 | 0.5 | 8.9 | 3 | 6.7 | 77 | NE | 3.4 |
| Dec. | 18 | 0.9 | 8.4 | 7 | 7.2 | 54 | W | 3.8 |
| Annual | 168 | 14.1 | 64.5 | 45 | 6.0 | 1852 | W | 3.6 |

[1] Period 1946–1960
[2] Period 1951–1960.

TABLE XXV

CLIMATIC TABLE FOR LIMOGES,
Latitude 45°49′N, longitude 1°17′E, elevation 282 m

| Month | Mean sta. press. (mbar) | Mean daily temp. (°C) | Mean daily temp. range(°C) | Temp.extremes(°C)[1] | | Mean vapor. press.[2] (mbar) | Mean precip. (mm) | Max. precip. 24 h[1] (mm) | Mean evapor. (mm) |
|---|---|---|---|---|---|---|---|---|---|
| | | | | highest | lowest | | | | |
| Jan. | 983.8 | 3.1 | 7.5 | 17.4 | −18.2 | 6.7 | 89 | 30 | |
| Feb. | 983.5 | 3.9 | 8.5 | 22.3 | −21.7 | 6.4 | 76 | 33 | |
| Mar. | 981.6 | 7.4 | 11.1 | 24.8 | − 9.4 | 7.8 | 66 | 31 | |
| Apr. | 982.2 | 9.9 | 11.6 | 29.7 | − 5.6 | 8.1 | 65 | 28 | 84 |
| May | 982.0 | 13.3 | 12.1 | 32.2 | − 3.9 | 10.6 | 80 | 43 | 100 |
| June | 984.3 | 16.8 | 12.5 | 36.7 | 0.2 | 13.3 | 67 | 48 | 96 |
| July | 984.7 | 18.4 | 13.0 | 38.8 | 3.2 | 14.7 | 71 | 38 | 113 |
| Aug. | 984.0 | 17.8 | 12.7 | 38.2 | 2.7 | 14.8 | 74 | 46 | 97 |
| Sept. | 984.6 | 15.3 | 11.9 | 32.3 | − 0.6 | 13.6 | 84 | 62 | 71 |
| Oct. | 983.9 | 10.7 | 11.0 | 27.0 | − 8.2 | 10.6 | 80 | 50 | 50 |
| Nov. | 982.9 | 6.7 | 8.3 | 25.3 | −11.2 | 8.3 | 88 | 48 | |
| Dec. | 983.5 | 3.8 | 6.9 | 19.6 | −17.0 | 7.7 | 94 | 30 | |
| Annual | 983.4 | 10.6 | 10.6 | 38.8 | −21.7 | 10.2 | 934 | 62 | |

| Month | Number of days with | | | | Mean cloudi-ness[2] (oktas) | Mean sun-shine[1] (h) | Most freq. wind dir.[2] | Mean wind speed[2] (m/sec) |
|---|---|---|---|---|---|---|---|---|
| | precip. ⩾0.1 mm | thunder-storm[2] | fog[2] | gale[2] | | | | |
| Jan. | 17 | 0 | 4.4 | 4 | 7.2 | 75 | SSW | 2.9 |
| Feb. | 14 | 0.5 | 1.9 | 5 | 6.6 | 96 | W | 3.2 |
| Mar. | 13 | 1.1 | 1.9 | 3 | 6.4 | 151 | S | 3.0 |
| Apr. | 13 | 1.6 | 2.3 | 2 | 6.0 | 188 | W | 2.9 |
| May | 14 | 4.0 | 2.6 | 2 | 6.1 | 207 | W | 2.5 |
| June | 12 | 5.8 | 2.9 | 2 | 6.3 | 225 | W | 2.1 |
| July | 12 | 4.0 | 3.1 | 1 | 5.8 | 240 | W | 2.3 |
| Aug. | 12 | 4.9 | 4.0 | 1 | 5.7 | 221 | W | 2.1 |
| Sept. | 12 | 2.5 | 5.8 | 1 | 5.9 | 185 | W | 1.9 |
| Oct. | 14 | 0.8 | 7.9 | 1 | 6.0 | 154 | W | 1.9 |
| Nov. | 15 | 0.6 | 5.0 | 3 | 7.1 | 84 | S | 2.6 |
| Dec. | 17 | 0.4 | 6.2 | 4 | 7.5 | 59 | S | 2.9 |
| Annual | 165 | 26.2 | 48.0 | 29 | 6.8 | 1885 | W | 2.5 |

[1] Period 1946–1960.
[2] Period 1951–1960.

TABLE XXVI

CLIMATIC TABLE FOR CLERMONT-FERRAND (AULNAT),
Latitude 45°48′N, longitude 3°9′E, elevation 329 m

| Month | Mean sta. press. (mbar) | Mean daily temp. (°C) | Mean daily temp. range(°C) | Temp. extremes(°C) highest | Temp. extremes(°C) lowest | Mean vapor. press.[2] (mbar) | Mean precip. (mm) | Max. precip. 24 h[1] (mm) | Mean evapor. (mm) |
|---|---|---|---|---|---|---|---|---|---|
| Jan. | 987.0 | 2.7 | 7.8 | 19.5 | −22.0 | 6.0 | 25 | 17 | |
| Feb. | 977.6 | 3.5 | 9.1 | 25.9 | −19.6 | 5.8 | 25 | 22 | |
| Mar. | 976.1 | 7.3 | 11.2 | 26.3 | −21.3 | 7.0 | 29 | 23 | |
| Apr. | 976.5 | 10.1 | 11.4 | 31.3 | − 6.4 | 7.8 | 43 | 28 | 103 |
| May | 976.3 | 13.7 | 11.8 | 33.0 | − 4.2 | 10.4 | 67 | 46 | 117 |
| June | 978.5 | 17.2 | 11.8 | 37.4 | 1.8 | 13.3 | 72 | 45 | 101 |
| July | 978.7 | 19.2 | 12.5 | 39.3 | 3.8 | 14.4 | 51 | 45 | 129 |
| Aug. | 978.1 | 18.8 | 12.5 | 39.6 | 2.8 | 14.1 | 68 | 55 | 118 |
| Sept. | 979.1 | 16.1 | 11.4 | 34.6 | − 2.8 | 13.0 | 61 | 41 | 91 |
| Oct. | 978,4 | 11.0 | 10.4 | 29.1 | − 8.2 | 10.0 | 49 | 40 | 67 |
| Nov. | 977.3 | 6.7 | 8.1 | 23.4 | −11.6 | 7.7 | 40 | 32 | |
| Dec. | 977.7 | 3.5 | 7.2 | 19.4 | −25.8 | 6.8 | 33 | 43 | |
| Annual | 977.7 | 10.9 | 10.5 | 39.6 | −25.8 | 9.7 | 563 | 55 | |

| Month | Number of days with precip. ⩾0.1 mm | Number of days with thunder- storm[2] | Number of days with fog[2] | Number of days with gale[2] | Mean cloudi- ness[2] (oktas) | Mean sun- shine[1] (h) | Most freq. wind dir.[2] | Mean wind speed[2] (m/sec) |
|---|---|---|---|---|---|---|---|---|
| Jan. | 12 | 0 | 4.2 | 6 | 7.2 | 82 | S | 3.7 |
| Feb. | 11 | 0.1 | 2.4 | 5 | 6.7 | 108 | S | 3.9 |
| Mar. | 9 | 0.1 | 0.6 | 4 | 6.2 | 167 | S | 3.8 |
| Apr. | 12 | 1.1 | 0.4 | 3 | 6.3 | 190 | N | 3.6 |
| May | 12 | 4.1 | 0.4 | 2 | 6.1 | 215 | N | 3.0 |
| June | 12 | 5.6 | 0.5 | 1 | 6.4 | 221 | N | 2.5 |
| July | 9 | 4.9 | 0.4 | 1 | 5.5 | 259 | N | 2.6 |
| Aug. | 10 | 5.9 | 0.8 | 2 | 5.6 | 235 | N | 2.6 |
| Sept. | 10 | 2.3 | 2.0 | 2 | 5.7 | 197 | S | 2.5 |
| Oct. | 11 | 0.4 | 3.6 | 2 | 6.2 | 158 | S | 2.7 |
| Nov. | 12 | 0.4 | 5.2 | 4 | 7.3 | 86 | S | 3.3 |
| Dec. | 12 | 0.2 | 4.7 | 6 | 7.3 | 72 | S | 3.5 |
| Annual | 132 | 25.1 | 25.2 | 38 | 6.4 | 1990 | S | 3.1 |

[1] Period 1946–1960.
[2] Period 1951–1960.

TABLE XXVII

CLIMATIC TABLE FOR LYON (BRON)
Latitude 45°43′N, longitude 4°57′E, elevation 200 m

| Month | Mean sta. press. (mbar) | Mean daily temp. (°C) | Mean daily temp. range(°C) | Temp. extremes(°C) highest | lowest | Mean vapor. press.[2] (mbar) | Mean precip. (mm) | Max. precip. 24 h[1] (mm) | Mean daily global radiation[3] (ly/day) |
|---|---|---|---|---|---|---|---|---|---|
| Jan. | 994.2 | 2.1 | 6.3 | 17.7 | −20.7 | 6.4 | 52 | 37 | 86 |
| Feb. | 993.3 | 3.3 | 7.7 | 21.9 | −21.4 | 6.4 | 46 | 44 | 159 |
| Mar. | 991.8 | 7.7 | 9.9 | 23.0 | −10.0 | 8.1 | 53 | 43 | 276 |
| Apr. | 991.7 | 10.9 | 10.6 | 30.1 | − 4.4 | 8.8 | 56 | 56 | 390 |
| May | 991.3 | 14.9 | 11.0 | 34.2 | − 3.8 | 11.5 | 69 | 47 | 458 |
| June | 993.2 | 18.5 | 11.3 | 36.8 | 2.3 | 14.6 | 85 | 71 | 513 |
| July | 993.4 | 20.7 | 11.8 | 39.5 | 6.4 | 15.8 | 56 | 40 | 545 |
| Aug. | 992.7 | 20.1 | 11.6 | 39.7 | 4.6 | 16.0 | 89 | 59 | 438 |
| Sept. | 994.1 | 16.9 | 10.7 | 35.8 | 0.7 | 14.7 | 93 | 89 | 316 |
| Oct. | 993.8 | 11.4 | 8.9 | 27.9 | − 4.5 | 11.1 | 77 | 47 | 193 |
| Nov. | 993.1 | 6.7 | 6.6 | 22.9 | − 8.0 | 8.4 | 80 | 81 | 95 |
| Dec. | 993.8 | 3.1 | 5.7 | 18.9 | −24.6 | 7.4 | 57 | 81 | 61 |
| Annual | 993.0 | 11.4 | 9.3 | 39.7 | −24.6 | 10.8 | 813 | 89 | 294 |

| Month | Mean evapor. (mm) | Number of days with precip. ≥0.1 mm | thunder- storm[2] | fog[2] | gale[2] | Mean cloudi- ness[2] (oktas) | Mean sun- shine[1] (h) | Most freq. wind dir.[2] | Mean wind speed[2] (m/sec) |
|---|---|---|---|---|---|---|---|---|---|
| Jan. | | 15 | 0.2 | 9.2 | 5 | 7.4 | 64 | N | 3.1 |
| Feb. | | 12 | 0.3 | 5.8 | 6 | 6.7 | 94 | N, S | 3.5 |
| Mar. | | 11 | 0.6 | 2.8 | 6 | 5.8 | 181 | S | 3.4 |
| Apr. | 97 | 11 | 1.5 | 0.6 | 6 | 5.5 | 213 | N | 4.0 |
| May | 118 | 13 | 3.4 | 0.2 | 4 | 5.4 | 250 | N | 3.5 |
| June | 115 | 11 | 6.8 | 0.6 | 3 | 5.7 | 262 | N | 2.9 |
| July | 152 | 10 | 4.9 | 0.7 | 2 | 4.7 | 293 | N | 2.8 |
| Aug. | 133 | 11 | 7.2 | 0.9 | 3 | 5.0 | 257 | N | 2.6 |
| Sept. | 81 | 11 | 3.7 | 3.8 | 2 | 5.3 | 205 | S | 2.4 |
| Oct. | 53 | 12 | 1.7 | 10.2 | 2 | 6.1 | 144 | S | 2.4 |
| Nov. | | 14 | 0.4 | 8.9 | 5 | 7.6 | 61 | S | 3.0 |
| Dec. | | 14 | 0.3 | 9.9 | 5 | 7.8 | 48 | S | 2.8 |
| Annual | | 145 | 31.0 | 53.6 | 49 | 6.1 | 2072 | N | 3.0 |

[1] Period 1946–1960.
[2] Period 1951–1960.
[3] Saint-Genis-Laval (45°42′N 4°47′E; 299 m).

TABLE XXVIII

CLIMATIC TABLE FOR BORDEAUX (MÉRIGNAC)
Latitude 44°50′N, longitude 0°42′W, elevation 47 m

| Month | Mean sta. press. (mbar) | Mean daily temp. (°C) | Mean daily temp. range(°C) | Temp. extremes(°C) highest | Temp. extremes(°C) lowest | Mean vapor. press.[2] (mbar) | Mean precip. (mm) | Max. precip. 24 h[1] (mm) | Mean evapor.[2] (mm) |
|---|---|---|---|---|---|---|---|---|---|
| Jan. | 1012.4 | 5.2 | 7.5 | 19.7 | −14.4 | 8.1 | 90 | 39 | |
| Feb. | 1012.1 | 5.9 | 8.5 | 22.4 | −15.2 | 7.7 | 75 | 34 | |
| Mar. | 1009.6 | 9.3 | 10.7 | 25.5 | − 6.1 | 9.2 | 63 | 32 | |
| Apr. | 1010.3 | 11.7 | 11.2 | 30.9 | − 4.8 | 9.7 | 48 | 33 | 90 |
| May | 1009.5 | 14.7 | 11.2 | 33.6 | − 1.8 | 12.4 | 61 | 40 | 97 |
| June | 1011.7 | 18.0 | 11.5 | 38.4 | 2.5 | 15.0 | 65 | 49 | 95 |
| July | 1012.1 | 19.6 | 11.7 | 38.6 | 5.1 | 16.5 | 56 | 42 | 102 |
| Aug. | 1011.1 | 19.5 | 12.1 | 37.1 | 4.7 | 16.6 | 70 | 54 | 89 |
| Sept. | 1011.7 | 17.1 | 11.1 | 36.2 | − 1.7 | 15.9 | 84 | 44 | 67 |
| Oct. | 1011.3 | 12.7 | 9.9 | 30.4 | − 5.3 | 12.6 | 83 | 41 | 48 |
| Nov. | 1010.8 | 8.4 | 8.0 | 23.9 | − 6.4 | 9.8 | 96 | 47 | |
| Dec. | 1011.7 | 5.7 | 6.8 | 21.2 | −13.4 | 9.1 | 109 | 36 | |
| Annual | 1011.2 | 12.3 | 10.0 | 38.6 | −15.2 | 11.9 | 900 | 54 | |

| Month | Number of days with precip. ⩾0.1 mm | Number of days with thunder-storm[2] | Number of days with fog[2] | gale[2] | Mean cloudi-ness[2] (oktas) | Mean sun-shine[1] (h) | Most freq. wind dir.[2] | Mean wind speed[2] (m/sec) |
|---|---|---|---|---|---|---|---|---|
| Jan. | 16 | 0.7 | 8.5 | 4 | 6.5 | 81 | W | 3.3 |
| Feb. | 13 | 1.3 | 6.4 | 5 | 5.9 | 103 | W | 3.5 |
| Mar. | 13 | 0.9 | 5.8 | 4 | 5.9 | 174 | W | 3.8 |
| Apr. | 13 | 1.6 | 3.6 | 3 | 5.4 | 210 | W | 3.6 |
| May | 14 | 4.3 | 5.4 | 2 | 5.6 | 229 | W | 3.2 |
| June | 11 | 5.4 | 3.9 | 2 | 5.6 | 253 | W | 3.1 |
| July | 11 | 5.9 | 4.2 | 2 | 5.1 | 262 | W | 3.1 |
| Aug. | 12 | 6.0 | 3.9 | 3 | 5.0 | 243 | W | 2.8 |
| Sept. | 13 | 3.3 | 9.0 | 2 | 5.3 | 195 | W | 2.5 |
| Oct. | 14 | 1.2 | 11.2 | 2 | 5.4 | 158 | W | 2.5 |
| Nov. | 15 | 0.7 | 10.4 | 2 | 6.4 | 84 | W | 2.9 |
| Dec. | 17 | 1.7 | 10.6 | 5 | 7.3 | 60 | WSW | 3.3 |
| Annual | 162 | 33.0 | 82.9 | 36 | 5.8 | 2052 | W | 3.1 |

[1] Period 1946–1960.
[2] Period 1951–1960.

TABLE XXIX

CLIMATIC TABLE FOR NICE (AIRPORT)
Latitude 43°40′N, longitude 7°12′E, elevation 5 m

| Month | Mean sta. press. (mbar) | Mean daily temp. (°C) | Mean daily temp. range(°C) | Temp. extremes[1](°C) highest | lowest | Mean vapor press.[2] (mbar) | Mean precip. (mm) | Max. precip. 24 h[1] (mm) | Mean evapor. (mm) |
|---|---|---|---|---|---|---|---|---|---|
| Jan. | 1013.7 | 7.5 | 8.3 | 22.2 | − 1.6 | 7.5 | 68 | 52 | |
| Feb. | 1012.7 | 8.5 | 8.7 | 21.0 | − 4.6 | 7.8 | 61 | 75 | |
| Mar. | 1013.1 | 10.8 | 8.1 | 21.2 | − 1.5 | 9.4 | 73 | 58 | |
| Apr. | 1012.5 | 13.3 | 8.1 | 26.0 | 3.2 | 11.2 | 73 | 117 | 94 |
| May | 1012.5 | 16.7 | 7.9 | 29.9 | 5.1 | 14.4 | 68 | 53 | 109 |
| June | 1013.7 | 20.1 | 7.9 | 31.2 | 7.6 | 17.9 | 35 | 45 | 106 |
| July | 1013.1 | 22.7 | 8.3 | 34.0 | 12.8 | 20.3 | 20 | 91 | 134 |
| Aug. | 1013.0 | 22.5 | 8.6 | 35.8 | 11.4 | 20.3 | 27 | 60 | 125 |
| Sept. | 1014.6 | 20.3 | 8.5 | 32.0 | 10.0 | 17.6 | 77 | 88 | 113 |
| Oct. | 1014.1 | 16.0 | 8.7 | 28.6 | 4.2 | 13.2 | 124 | 96 | 109 |
| Nov. | 1013.6 | 11.5 | 8.5 | 22.8 | 1.2 | 10.2 | 129 | 147 | |
| Dec. | 1013.6 | 8.2 | 8.2 | 22.0 | − 2.2 | 8.5 | 107 | 126 | |
| Annual | 1013.3 | 14.8 | 8.3 | 35.8 | − 4.6 | 13.2 | 862 | 147 | |

| Month | Number of days with precip. ⩾0.1 mm | thunder- storm[2] | fog[2] | gale[2] | Mean cloudi- ness[2] (oktas) | Mean sun- shine[1] (h) | Most freq. wind dir.[2] | Mean wind speed[2] (m/sec) |
|---|---|---|---|---|---|---|---|---|
| Jan. | 9 | 1.3 | 0 | 5 | 5.0 | 148 | NW | 4.6 |
| Feb. | 7 | 1.6 | 0 | 5 | 5.2 | 165 | NNW | 4.3 |
| Mar. | 8 | 1.2 | 0 | 6 | 5.7 | 196 | E | 4.2 |
| Apr. | 9 | 2.5 | 0 | 4 | 5.2 | 243 | E | 3.7 |
| May | 8 | 2.8 | 0.4 | 2 | 5.3 | 272 | E | 3.3 |
| June | 5 | 3.3 | 0 | 2 | 4.7 | 312 | E | 3.1 |
| July | 2 | 3.3 | 0 | 2 | 2.8 | 363 | E | 3.0 |
| Aug. | 4 | 3.2 | 0 | 3 | 3.4 | 324 | E | 3.3 |
| Sept. | 7 | 3.9 | 0.2 | 3 | 4.3 | 263 | E | 3.3 |
| Oct. | 9 | 4.1 | 0 | 2 | 4.8 | 200 | NW | 3.7 |
| Nov. | 9 | 2.8 | 0 | 2 | 5.4 | 155 | NNW | 4.4 |
| Dec. | 9 | 2.1 | 0 | 4 | 5.2 | 137 | NNW | 4.9 |
| Annual | 86 | 32.1 | 0.6 | 40 | 4.7 | 2778 | E | 3.8 |

[1] Period 1946–1960.
[2] Period 1951–1960.

TABLE XXX

CLIMATIC TABLE FOR TOULOUSE (BLAGNAC)
Latitude 43°37'N, longitude 1°22'E, elevation 151 m

| Month | Mean sta. press. (mbar) | Mean daily temp. (°C) | Mean daily temp. range(°C) | Temp. extremes[1](°C) highest | lowest | Mean vapor press.[2] (mbar) | Mean precip. (mm) | Max. precip. 24 h[1] (mm) | Mean evapor.[2] (mm) |
|-------|------|------|------|------|------|------|------|------|------|
| Jan. | 1000.4 | 4.5 | 7.7 | 21.2 | −17.0 | 7.7 | 49 | 24 | |
| Feb. | 1000.0 | 5.4 | 8.8 | 22.2 | −19.2 | 7.3 | 46 | 57 | |
| Mar. | 997.6 | 9.0 | 10.3 | 24.0 | − 5.3 | 9.1 | 53 | 38 | |
| Apr. | 998.0 | 11.4 | 10.8 | 30.0 | − 3.0 | 9.6 | 50 | 32 | 94 |
| May | 997.5 | 14.8 | 10.9 | 32.8 | − 0.8 | 12.4 | 75 | 43 | 106 |
| June | 999.8 | 18.6 | 11.4 | 39.8 | 6.2 | 15.0 | 61 | 62 | 115 |
| July | 1000.1 | 20.8 | 12.1 | 39.0 | 8.2 | 16.7 | 44 | 78 | 134 |
| Aug. | 999.0 | 20.7 | 11.9 | 40.2 | 7.3 | 16.7 | 54 | 47 | 129 |
| Sept. | 999.9 | 18.0 | 10.8 | 33.0 | 3.6 | 15.7 | 64 | 40 | 99 |
| Oct. | 999.6 | 13.0 | 9.9 | 31.5 | − 3.0 | 12.3 | 45 | 39 | 67 |
| Nov. | 999.2 | 8.3 | 8.1 | 22.9 | − 6.3 | 9.6 | 51 | 36 | |
| Dec. | 1000.0 | 5.3 | 6.6 | 20.3 | −10.5 | 8.7 | 67 | 40 | |
| Annual | 999.3 | 12.5 | 9.9 | 40.2 | −19.2 | 11.7 | 659 | 78 | |

| Month | Number of days with precip. ≥0.1 mm | thunder- storm[2] | fog[2] | gale[2] | Mean cloudi- ness[2] (oktas) | Mean sun- shine[1] (h) | Most freq. wind dir.[2] | Mean wind speed[2] (m/sec) |
|-------|------|------|------|------|------|------|------|------|
| Jan. | 14 | 0.2 | 9.2 | 5 | 6.5 | 80 | W | 3.5 |
| Feb. | 12 | 0.6 | 4.1 | 5 | 6.0 | 117 | W | 3.8 |
| Mar. | 11 | 1.0 | 2.7 | 7 | 5.9 | 183 | SE | 4.5 |
| Apr. | 12 | 1.4 | 0.9 | 5 | 5.9 | 199 | W | 4.6 |
| May | 13 | 2.6 | 1.1 | 3 | 6.0 | 223 | W | 3.8 |
| June | 10 | 4.1 | 0.6 | 3 | 6.1 | 234 | W | 3.5 |
| July | 9 | 3.9 | 0.8 | 3 | 5.2 | 262 | W | 3.5 |
| Aug. | 9 | 5.0 | 0.7 | 3 | 5.0 | 254 | W | 3.3 |
| Sept. | 10 | 2.1 | 2.7 | 2 | 5.4 | 206 | SE | 3.1 |
| Oct. | 11 | 0.6 | 6.5 | 2 | 5.9 | 168 | W | 3.1 |
| Nov. | 12 | 0.1 | 9.6 | 3 | 6.7 | 92 | W | 3.1 |
| Dec. | 15 | 0.2 | 9.1 | 5 | 7.7 | 62 | W | 3.7 |
| Annual | 138 | 21.8 | 48.0 | 46 | 6.0 | 2080 | W | 3.6 |

[1] Period 1946–1960.

[2] Period 1951–1960.

TABLE XXXI

CLIMATIC TABLE FOR MARSEILLES-MARIGNANE (AIRPORT)
Latitude 43°27′N, longitude 5°13′E, elevation 3 m

| Month | Mean sta. press. (mbar) | Mean daily temp. (°C) | Mean daily temp. range(°C) | Temp. extremes(°C) highest | lowest | Mean vapor press.[2] (mbar) | Mean precip. (mm) | Max. precip. 24 h[1] (mm) | Mean evapor.[2] (mm) |
|-------|------|------|------|------|------|------|------|------|------|
| Jan. | 1016.2 | 5.5 | 8.5 | 20.0 | −10.7 | 7.0 | 43 | 48 | |
| Feb. | 1015.3 | 6.6 | 9.4 | 21.9 | −16.8 | 7.1 | 32 | 81 | |
| Mar. | 1014.6 | 10.0 | 9.9 | 24.0 | −10.0 | 8.7 | 43 | 80 | |
| Apr. | 1014.1 | 13.0 | 10.3 | 28.5 | − 2.4 | 9.6 | 42 | 47 | 147 |
| May | 1013.8 | 16.8 | 10.7 | 33.0 | − 0.0 | 12.3 | 46 | 45 | 176 |
| June | 1015.3 | 20.8 | 11.4 | 36.0 | 5.4 | 15.2 | 24 | 55 | 223 |
| July | 1015.1 | 23.3 | 11.8 | 39.0 | 7.8 | 16.6 | 11 | 26 | 276 |
| Aug. | 1014.8 | 22.8 | 11.3 | 36.6 | 8.6 | 17.0 | 34 | 46 | 241 |
| Sept. | 1016.3 | 19.9 | 10.4 | 34.3 | 1.0 | 15.6 | 60 | 68 | 175 |
| Oct. | 1016.0 | 15.0 | 9.4 | 30.2 | − 2.2 | 12.3 | 76 | 86 | 125 |
| Nov. | 1015.5 | 10.2 | 8.7 | 22.8 | − 5.4 | 9.6 | 69 | 44 | |
| Dec. | 1015.7 | 6.9 | 7.9 | 20.2 | −12.8 | 8.3 | 66 | 46 | |
| Annual | 1015.2 | 14.2 | 10.0 | 39.0 | −16.8 | 11.6 | 546 | 86 | |

| Month | Number of days with precip. ⩾0.1 mm | thunder-storm[2] | fog[2] | gale[2] | Mean cloudi-ness[2] (oktas) | Mean sun-shine[1] (h) | Most freq. wind dir.[2] | Mean wind speed[2] (m/sec) |
|-------|------|------|------|------|------|------|------|------|
| Jan. | 8 | 0.6 | 2.0 | 6 | 4.6 | 134 | NNW | 4.0 |
| Feb. | 6 | 0.8 | 2.1 | 7 | 4.5 | 157 | NW | 4.3 |
| Mar. | 7 | 0.7 | 1.2 | 8 | 5.1 | 208 | NW | 4.6 |
| Apr. | 6 | 0.9 | 0.1 | 8 | 4.4 | 251 | NNW | 5.2 |
| May | 7 | 1.3 | 0.1 | 6 | 4.4 | 281 | W | 4.6 |
| June | 4 | 1.9 | 0.1 | 4 | 4.1 | 323 | NW | 4.8 |
| July | 2 | 1.5 | 0.3 | 5 | 2.5 | 368 | W | 5.0 |
| Aug. | 4 | 3.2 | 0.3 | 4 | 3.3 | 324 | NW | 4.4 |
| Sept. | 6 | 3.0 | 0.5 | 5 | 4.0 | 253 | W | 4.0 |
| Oct. | 8 | 2.5 | 0.9 | 6 | 4.8 | 191 | NW | 3.9 |
| Nov. | 8 | 0.6 | 1.9 | 6 | 4.9 | 151 | NW | 4.1 |
| Dec. | 10 | 1.3 | 2.5 | 8 | 5.1 | 123 | E | 4.2 |
| Annual | 76 | 18.3 | 12.0 | 73 | 4.3 | 2764 | NW | 4.4 |

[1] Period 1946–1960.
[2] Period 1951–1960.

TABLE XXXII

CLIMATIC TABLE FOR PERPIGNAN (LLABANÈRE)
Latitude 42°44′N, longitude 2°52′E, elevation 43 m

| Month | Mean sta. press. (mbar) | Mean daily temp. (°C) | Mean daily temp. range(°C) | Temp. extremes(°C) highest | lowest | Mean vapor press.[2] (mbar) | Mean precip. (mm) | Max. precip. 24 h[1] (mm) | Mean evapor.[2] (mm) |
|-------|------|------|------|------|------|------|------|------|------|
| Jan. | 1011.4 | 7.5 | 8.1 | 22.6 | − 6.9 | 7.4 | 39 | 59 | |
| Feb. | 1010.8 | 8.4 | 8.4 | 26.4 | −11.0 | 7.3 | 52 | 109 | |
| Mar. | 1009.4 | 11.3 | 8.8 | 26.2 | − 2.7 | 9.3 | 66 | 42 | |
| Apr. | 1009.4 | 13.9 | 9.2 | 32.4 | 1.7 | 9.6 | 39 | 39 | 137 |
| May | 1008.8 | 17.1 | 9.1 | 33.1 | 5.5 | 12.5 | 51 | 59 | 148 |
| June | 1010.7 | 21.1 | 9.6 | 36.8 | 9.1 | 15.3 | 38 | 91 | 150 |
| July | 1010.8 | 23.8 | 10.0 | 38.0 | 12.0 | 16.6 | 24 | 45 | 187 |
| Aug. | 1009.9 | 23.3 | 9.7 | 38.7 | 11.6 | 17.3 | 31 | 37 | 152 |
| Sept. | 1011.2 | 20.5 | 8.7 | 36.8 | 7.4 | 16.0 | 82 | 186 | 126 |
| Oct. | 1010.8 | 15.9 | 8.2 | 29.1 | 1.2 | 12.5 | 74 | 97 | 98 |
| Nov. | 1010.3 | 11.5 | 7.9 | 25.4 | − 2.1 | 9.6 | 56 | 55 | |
| Dec. | 1010.7 | 8.6 | 7.3 | 25.0 | − 5.0 | 8.7 | 87 | 114 | |
| Annual | 1010.4 | 15.2 | 8.7 | 38.7 | −11.0 | 11.8 | 639 | 186 | |

| Month | Number of days with precip. ⩾0.1 mm | thunder-storm[2] | fog[2] | gale[2] | Mean cloudi-ness[2] (oktas) | Mean sun-shine[1] (h) | Most freq. wind dir.[2] | Mean wind speed[2] (m/sec) |
|-------|------|------|------|------|------|------|------|------|
| Jan. | 7 | 0.1 | 0.5 | 14 | 5.8 | 161 | NW | 5.4 |
| Feb. | 6 | 0.4 | 1.3 | 13 | 5.1 | 172 | NW | 5.6 |
| Mar. | 8 | 1.1 | 1.2 | 11 | 5.8 | 209 | NW | 4.9 |
| Apr. | 7 | 1.7 | 0.6 | 14 | 5.4 | 245 | NW | 6.3 |
| May | 9 | 2.6 | 1.1 | 9 | 5.6 | 255 | NW | 5.0 |
| June | 7 | 4.9 | 0.6 | 7 | 5.4 | 279 | NW | 4.6 |
| July | 5 | 5.9 | 0.2 | 8 | 4.1 | 322 | NW | 5.0 |
| Aug. | 6 | 6.7 | 1.2 | 8 | 4.7 | 282 | NW | 4.0 |
| Sept. | 7 | 3.6 | 1.8 | 7 | 5.3 | 236 | NW | 3.9 |
| Oct. | 8 | 1.7 | 1.6 | 10 | 5.8 | 189 | NW | 4.3 |
| Nov. | 6 | 0.1 | 1.1 | 12 | 5.6 | 158 | NW | 4.7 |
| Dec. | 9 | 0.7 | 0.9 | 14 | 5.8 | 136 | NW | 5.4 |
| Annual | 85 | 29.5 | 12.1 | 127 | 5.3 | 2644 | NW | 4.9 |

[1] Period 1946–1960.

[2] Period 1951–1960.

TABLE XXXIII

CLIMATIC TABLE FOR BASTIA (PORETTA), CORSICA
Latitude 42°33′N, longitude 9°29′E, elevation 10 m

| Month | Mean sta. press. (mbar) | Mean daily temp. (°C) | Mean daily temp. range(°C) | Temp.extremes[1](°C) highest | lowest | Mean vapor press.[2] (mbar) | Mean precip. (mm) | Max. precip. 24 h[1] (mm) | Mean evapor.[2] (mm) |
|---|---|---|---|---|---|---|---|---|---|
| Jan. | 1013.4 | 7.9 | 9.4 | 23.6 | − 4.6 | 8.0 | 75 | 75 | |
| Feb. | 1012.5 | 8.6 | 9.6 | 22.0 | − 5.0 | 8.2 | 65 | 120 | |
| Mar. | 1013.0 | 10.3 | 9.8 | 23.8 | − 3.8 | 9.6 | 60 | 44 | |
| Apr. | 1012.4 | 12.7 | 10.0 | 24.2 | 0.5 | 11.1 | 65 | 47 | 72 |
| May | 1012.3 | 16.2 | 10.5 | 29.7 | 1.3 | 14.1 | 50 | 32 | 77 |
| June | 1013.6 | 20.2 | 11.3 | 32.6 | 8.2 | 17.6 | 20 | 44 | 99 |
| July | 1012.9 | 23.0 | 11.7 | 35.8 | 10.2 | 19.1 | 10 | 21 | 136 |
| Aug. | 1012.8 | 23.0 | 11.6 | 36.0 | 11.8 | 19.4 | 25 | 201 | 133 |
| Sept. | 1014.4 | 20.4 | 11.1 | 34.0 | 7.8 | 17.5 | 65 | 156 | 101 |
| Oct. | 1013.9 | 16.0 | 9.7 | 27.6 | 3.0 | 13.8 | 110 | 135 | 83 |
| Nov. | 1013.3 | 11.9 | 9.2 | 23.8 | 1.2 | 11.3 | 95 | 66 | |
| Dec. | 1013.2 | 9.1 | 9.6 | 24.0 | − 1.8 | 9.5 | 95 | 57 | |
| Annual | 1013.1 | 14.9 | 10.3 | 36.0 | − 5.0 | 13.3 | 735 | 201 | |

| Month | Number of days with precip. ≥0.1 mm | thunder- storm[2] | fog[2] | gale[2] | Mean cloudi- ness[2] (oktas) | Mean sun- shine[1] (h) | Most freq. wind dir.[2] | Mean wind speed[2] (m/sec) |
|---|---|---|---|---|---|---|---|---|
| Jan. | 10 | 1.0 | 0 | 2 | 5.4 | 137 | SW | 2.5 |
| Feb. | 8 | 0.7 | 0.1 | 2 | 6.0 | 132 | W | 2.7 |
| Mar. | 10 | 0.7 | 0.4 | 1 | 5.7 | 188 | SW | 2.4 |
| Apr. | 9 | 1.8 | 0 | 1 | 5.3 | 223 | SE | 2.2 |
| May | 9 | 2.4 | 0.3 | 1 | 4.9 | 258 | SE | 1.9 |
| June | 4 | 3.0 | 0.1 | 0.2 | 4.3 | 306 | SE | 2.1 |
| July | 1 | 3.8 | 0 | 0 | 2.3 | 364 | SE | 2.2 |
| Aug. | 2 | 3.1 | 0 | 1 | 3.1 | 313 | SE, E | 2.1 |
| Sept. | 6 | 3.2 | 0.1 | 1 | 4.0 | 249 | SE | 2.0 |
| Oct. | 10 | 3.1 | 0 | 1 | 4.9 | 197 | SE | 2.1 |
| Nov. | 11 | 2.3 | 0 | 1 | 5.7 | 126 | SW | 2.2 |
| Dec. | 11 | 1.7 | 0 | 2 | 5.8 | 110 | SW | 2.4 |
| Annual | 91 | 26.8 | 1.0 | 13 | 4.8 | 2603 | SE | 2.2 |

[1] Period 1946–1960.
[2] Period 1951–1960.

TABLE XXXIV

CLIMATIC TABLE FOR AJACCIO (CAMPO-DELL'ORO), CORSICA
Latitude 41°55′N, longitude 8°48′E, elevation 4 m

| Month | Mean sta. press. (mbar) | Mean daily temp. (°C) | Mean daily temp. range(°C) | Temp.extremes[1](°C) | | Mean vapor press.[2] (mbar) | Mean precip. (mm) | Max. precip. 24 h[1] (mm) | Mean evapor.[2] (mm) |
|---|---|---|---|---|---|---|---|---|---|
| | | | | highest | lowest | | | | |
| Jan. | 1014.6 | 7.7 | 9.9 | 20.6 | −5.0 | 8.7 | 76 | 66 | |
| Feb. | 1014.0 | 8.7 | 9.7 | 23.3 | −6.0 | 9.0 | 65 | 32 | |
| Mar. | 1014.0 | 10.5 | 10.3 | 26.0 | −3.8 | 10.1 | 53 | 32 | |
| Apr. | 1013.5 | 12.6 | 10.5 | 29.3 | −1.7 | 11.2 | 48 | 34 | 80 |
| May | 1013.5 | 15.9 | 10.7 | 32.8 | 3.0 | 14.2 | 50 | 41 | 82 |
| June | 1015.0 | 19.8 | 11.3 | 37.0 | 7.0 | 17.6 | 21 | 61 | 106 |
| July | 1014.6 | 22.0 | 11.9 | 35.7 | 9.2 | 19.3 | 10 | 39 | 132 |
| Aug. | 1014.3 | 22.2 | 12.3 | 38.6 | 9.1 | 19.4 | 16 | 52 | 137 |
| Sept. | 1015.6 | 20.3 | 11.3 | 36.0 | 7.6 | 17.7 | 50 | 50 | 108 |
| Oct. | 1015.0 | 16.3 | 10.9 | 29.7 | 2.0 | 14.0 | 88 | 75 | 87 |
| Nov. | 1014.5 | 11.8 | 10.5 | 28.0 | −2.0 | 11.2 | 97 | 68 | |
| Dec. | 1014.4 | 8.7 | 10.3 | 22.2 | −3.6 | 9.9 | 98 | 43 | |
| Annual | 1014.4 | 14.7 | 10.8 | 38.6 | −6.0 | 13.5 | 672 | 75 | |

| Month | Number of days with | | | | Mean cloudi- ness[2] (oktas) | Mean sun- shine[1] (h) | Most freq. wind dir.[2] | Mean wind speed[2] (m/sec) |
|---|---|---|---|---|---|---|---|---|
| | precip. ≥0.1 mm | thunder- storm[2] | fog[2] | gale[2] | | | | |
| Jan. | 12 | 2.0 | 0 | 1 | 5.8 | 132 | ENE | 2.4 |
| Feb. | 10 | 2.5 | 0.3 | 2 | 6.3 | 137 | ENE | 2.7 |
| Mar. | 9 | 1.7 | 0 | 1 | 5.8 | 200 | ENE | 2.3 |
| Apr. | 9 | 3.0 | 0.4 | 1 | 5.4 | 240 | SW | 2.7 |
| May | 8 | 2.3 | 0.8 | 0.2 | 4.9 | 286 | SW | 2.5 |
| June | 4 | 2.6 | 0.3 | 0 | 4.0 | 336 | SW | 2.8 |
| July | 1 | 1.8 | 0.1 | 0.1 | 2.1 | 382 | SW | 3.0 |
| Aug. | 2 | 2.7 | 0 | 0.1 | 2.8 | 342 | SW | 2.9 |
| Sept. | 6 | 5.1 | 0 | 1 | 4.0 | 275 | SW | 2.5 |
| Oct. | 10 | 5.2 | 0 | 1 | 5.2 | 205 | ENE | 2.4 |
| Nov. | 11 | 4.3 | 1.0 | 1 | 5.6 | 140 | ENE | 2.3 |
| Dec. | 13 | 3.8 | 0 | 2 | 6.0 | 115 | ENE | 2.4 |
| Annual | 95 | 37.0 | 20.0 | 10.5 | 4.8 | 2790 | ENE | 2.6 |

[1] Period 1946–1960.

[2] Period 1951–1960.

TABLE XXXV

MEAN DAILY GLOBAL RADIATION (ly./day)

| | Jan. | Feb. | Mar. | Apr. | May | June | July | Aug. | Sept. | Oct. | Nov. | Dec. | Annual |
|---|---|---|---|---|---|---|---|---|---|---|---|---|---|
| *De Bilt (52°6′N 5°11′E, elevation 2.9 m)* | | | | | | | | | | | | | |
| | 58 | 110 | 201 | 315 | 395 | 424 | 368 | 317 | 242 | 140 | 66 | 42 | 223 |
| *Uccle (50°48′N 4°21′E, elevation 100 m)* | | | | | | | | | | | | | |
| | 62 | 122 | 208 | 317 | 400 | 429 | 400 | 338 | 260 | 158 | 73 | 49 | 235 |
| *Paris-Saint-Maur (48°58′N 2°27′E, elevation 52 m)* | | | | | | | | | | | | | |
| | 74 | 128 | 244 | 353 | 439 | 478 | 454 | 383 | 289 | 174 | 84 | 58 | 263 |
| *Trappes (48°46′N 2°4′E, elevation 168 m)* | | | | | | | | | | | | | |
| | 75 | 131 | 238 | 333 | 432 | 501 | 480 | 398 | 300 | 195 | 92 | 58 | 269 |
| *Nancy (48°41′N 6°13′E, elevation 212 m)* | | | | | | | | | | | | | |
| | 80 | 145 | 246 | 347 | 441 | 472 | 458 | 382 | 283 | 185 | 83 | 60 | 265 |
| *Baugé (47°32′N 0°7′W, elevation 51 m)* | | | | | | | | | | | | | |
| | 92 | 163 | 292 | 380 | 456 | 508 | 485 | 411 | 311 | 194 | 104 | 71 | 289 |
| *Mâcon (46°18′N 4°48′E, elevation 216 m)* | | | | | | | | | | | | | |
| | 87 | 162 | 282 | 396 | 480 | 525 | 506 | 430 | 315 | 205 | 90 | 64 | 295 |
| *Limoges (45°49′N 1°17′E, elevation 282 m)* | | | | | | | | | | | | | |
| | 100 | 167 | 268 | 362 | 451 | 515 | 490 | 411 | 309 | 206 | 108 | 72 | 288 |
| *Saint-Genis-Laval (45°43′N 4°57′E, elevation 200 m)* | | | | | | | | | | | | | |
| | 86 | 159 | 276 | 390 | 458 | 513 | 545 | 438 | 316 | 193 | 95 | 61 | 294 |
| *Millau (44°7′N 3°31′E, elevation 715 m)* | | | | | | | | | | | | | |
| | 131 | 195 | 305 | 419 | 522 | 603 | 610 | 504 | 358 | 214 | 125 | 110 | 341 |
| *Nice (43°40′N 7°12′E, elevation 5 m)* | | | | | | | | | | | | | |
| | 154 | 221 | 365 | 478 | 556 | 602 | 612 | 521 | 386 | 275 | 188 | 154 | 376 |
| *Ajaccio (41°55′N 8°48′E, elevation 4 m)* | | | | | | | | | | | | | |
| | 142 | 202 | 358 | 475 | 581 | 644 | 638 | 547 | 409 | 277 | 169 | 136 | 382 |

# Chapter 5

# The Climate of the Iberian Peninsula

A. LINÉS ESCARDÓ

## Introduction

The Iberian Peninsula, lying between latitudes 35°59′50″ and 43°47′25″N and between longitudes 3°19′22″E and 9°18′19″W, is the westernmost and the largest of the three Mediterranean peninsulas. It has an area of 581,000 km², and is shaped roughly like a pentagon with one vertex pointing towards Africa from which it is separated by the Strait of Gibraltar.

The Iberian Peninsula is one of the most mountainous countries in Europe, having a mean altitude of more than 500 m above M.S.L. Spain is appreciably more mountainous than Portugal; 34% of peninsular Spain is between 800 and 2,000 m in altitude.

The interior of the peninsula is formed chiefly by the Central Plateau, traversed by an orographic range called the Central Range (Sistema Central) which divides it into two sub-plateaus: an upper one having altitudes between 700 and 900 m, and a lower one, situated to the south, at between 500 and 800 m. The Central Plateau is limited in the north by the Cantabrian Mountains which separate it from the Cantabrian slopes. To the northwest it is limited by the Galicia-Douro massif, to the northeast and east by the Iberian Mountains, and to the south by the Sierra Morena. To the west it has a gentle slope towards Portugal which is broken by the western offshoots of the Central Range. The greatest heights in the peninsula are encountered in the Pyrenees and in the Sierra Nevada, respectively in the north and in the south. The Pyrenees separate the peninsula from France and frequently rise to heights of over 3,000 m in Cataluña and Aragon: the highest peak is Aneto (3,404 m). In the Sierra Nevada, the main core of the Penibetica Range, are found the highest points in the peninsula, namely Mulhacén (3,478 m) and Veleta (3,392 m).

The mountain ranges run generally from east to west, except for the Iberian Range, a complex system whose orientation is roughly from north to south. The Cantabrian Mountains, which constitute an extension of the Pyrenees, are parallel with the Cantabrian coast, and there the descent to the coast is abrupt, with locally steep slopes. Similarly, the Penibetica Range is roughly parallel with the southern Mediterranean coast and rises to great heights near the sea. The climatological contrasts are therefore very large, with a rapid transition from permanent snow to the tropical orchards of the coasts of Granada.

There are two important valleys: that of the Ebro, which is limited by the Pyrenees and the Iberian Mountains, and the Bética Valley which is traversed by the Guadalquivir. The coastal plains of the Mediterranean Sea occupy about 9% of the peninsula. They have a very homogeneous climate characterized by strong influences from the Mediterranean.

The rivers of the peninsula are irregular in their flow due to the fact that they are chiefly fed by rains which reach their maximum during the spring and autumn, and their minimum during the summer. The main rivers flow towards the Atlantic Ocean.

The Cantabrian slopes, which constitute a narrow coastal zone between the Cantabrian Mountains and the sea, have short torrential rivers with abundant flow of water since their source region has considerable rainfall. With the exception of the Ebro, the Atlantic slopes possess the chief rivers of the peninsula, such as the Douro, Tagus, Guadiana, and Guadalquivir, the last being in the Bética Valley. The Tagus is the longest river of the peninsula, being 1,008 km in length. It rises in the Iberian Mountains and has a wide estuary in Portugal known as the "Mar de Palha". The Douro, the second of the Atlantic rivers, rises in the Pico de Urbion in the north of the Iberian Range, and after traversing the northern plateau, discharges into the important harbour of Oporto. The rivers of Galicia, in the northwest, also discharge a considerable quantity of water. On the Mediterranean slopes, apart from the Ebro, which is the most important, we may distinguish northeastern, eastern, and southeastern rivers. The northeastern (or Catalonian) rivers are short, and have a considerable flow of water. The eastern rivers are torrential in character, with long periods when they are nearly dry, and their courses are very irregular. The southeastern rivers have short courses with considerable downslope, and for this reason they exert a strong erosive action. The upper section of the Ebro is torrential in character; later, the river crosses the plain of Aragon and flows into a wide delta. Its most important tributaries are those of the left bank, which are fed by the snows of the Pyrenees.

The coasts of the Iberian Peninsula stretch for some 4,000 km. The northern, and especially the northwestern coasts are more broken and have numerous "rias" similar to the Norwegian fjords. Those of the west-southwest are smoother, with many beaches. The Mediterranean coasts are very varied, with highly indented zones alternating with large beaches.

Near the peninsula, in the Mediterranean, the most important islands are the Balearics, comprising the three main islands of Mallorca (Majorca), Menorca (Minorca) and Ibiza, and other smaller ones. More remote, in the Atlantic, is the Portuguese island of Madeira, some 500 miles to the southwest of the peninsula. To the northwest of Africa, in latitudes of about 28°N, lies the Spanish archipelago of the Canary Islands, with seven principal islands: Lanzarote, Fuerteventura, Gran Canaria, Tenerife, Gomera, La Palma and Hierro. These islands are volcanic in origin. On the island of Tenerife there is a peak called Teide rising to 3,718 m. The orographic variety of this island results in a great diversity of climates. The Portuguese archipelago of the Azores lies in about latitude 39°N between the meridians of 25° and 32°W, and is divided into three groups: eastern, central, and western. These are also of volcanic origin.

## Historical notes

Meteorological studies in the Iberian Peninsula go back to remote times (LORENTE PEREZ, 1941, 1956). In the 1st century A.D., Lucius Annaeus Seneca devoted half of his

*Questiones Naturales* to meteorology and the rest to seismology and other natural sciences. Also in the 1st century, Lucius Junius Moderatus Columela put forward a meteorological calendar in *De Re Rustica* which may have applied to Andalucia where he had possessions or, perhaps, to Italy.

In the 7th century, San Isidoro of Seville, in his works *De Rerum Natura* and *Ethymologiarum sive Originum Libri XX*, compiled what was known in his time about meteorology, and established a nomenclature and classification of winds.

There are also references to meteorological questions, mingled with astronomical matters, in the work of Alfonso X (Alfonso the Wise) of Castille.

Of the Hispano–Arabic culture, mention should be made of Asib ben Said el Kateb (tenth century) who compiled a calendar with a section devoted to meteorology, and of Averroes (twelfth century).

In the thirteenth century, Raimundo Lulio wrote a work describing the causes of the winds and perhaps established the 16-point wind rose.

To Prince Henry the Navigator, who founded the world-famous "School of Navigation" at Sagres (Portugal), goes the credit for having given a strong stimulus to meteorology, with particular devotion to the study of winds and atmospheric circulation: as early as the 15th century Portuguese navigators had a reasonably fair knowledge of the zones with prevailing easterly and westerly winds. Many meteorological references, especially relating to winds, appear in the documents of the great Portuguese navigators such as Vasco da Gama, Alvares Cabral, and Magellan.

A *Cronología y Repertorio de la Razón de los Tiempos* (Chronology and Repertoire of the Causes of Weather) by Rodrigo Zamorano was published in Seville in 1594. This work contained 470 definite rules for weather forecasting, with special applications to agriculture.

In the 16th century, Father José de Acosta wrote his book *Natural and Moral History of the Indies*, which was published in Sevilla and translated into several languages. He established some of the first ideas for a zonal classification of climates, considerably advancing the knowledge of his times by denying the influence of the moon on meteorological forecasts, and explained many meteorological phenomena in modern terms. In the same century, Andrés de Urdaneta determined the circulation of the winds in the Pacific anticyclone.

Thanks to the enthusiasm of certain individuals, meteorological observatories were set up as early as 1654, and in 1737 Francisco Fernandez de Navarrete published a "barometric-medical" diary for Madrid. By 1805 Francisco Salva was making temperature and rainfall observations in Barcelona, only some of which have been preserved. In 1803 observations were being made in Madrid, although for practical purposes, the observations for this capital did not commence until 1841. The Observatory of San Fernando (Cadiz) started to function in 1805 and was fully operative in 1835. In Gibraltar observations were started at the end of the eighteenth century; many of these records have been lost, but monthly data since 1850 and annual data since 1791 have been preserved. Continuous observations are available for Lisbon from 1856 onwards and for Coimbra since 1866, although many observations had already been made earlier. In 1860 an observational network was established in Spain with the help of the universities and the secondary schools.

**Dynamic climatology**

Except in the summer months, the Iberian Peninsula remains within the current of prevailing westerly winds and, in consequence, Atlantic cyclones are the disturbances which contribute the most rainfall to the annual balance of precipitation (MORAN SAMANIEGO, 1944, p. 187).

In the period when westerly winds predominate and the cyclones penetrate the peninsula, these cyclones tend to weaken as they cross the Iberian Mountains separating the Atlantic slopes from those of the Mediterranean. Similarly, disturbances of Mediterranean origin have little effect on the Atlantic river valleys. Moreover, the number of cyclones proceeding from the Mediterranean is far smaller than that coming from the Atlantic.

During the hot months, westerly winds almost disappear from the lower layers of the atmosphere. Most of the precipitation is then of convective origin, and the weather is similar to sub-tropical dry weather.

The intensity and frequency of Atlantic cyclones depend closely on the position of the southern branch of the polar "jet" stream. During autumn, this branch lies between latitudes 40° and 55°N, so that the depressions associated with it tend to affect the Iberian Peninsula (RODRIGUEZ FRANCO, 1955). In winter this "jet" stream branch lies even further south, between latitudes 25°N and 45°N, and precipitation, except in the south, then tends to decrease in comparison to the autumn months. In spring, as the jet stream moves northwards, there is another period in which cyclones seriously affect the Iberian Peninsula and, in the majority of regions, there is a second maximum of precipitation.

The various types of cyclones which generate precipitation over the Iberian Peninsula may be classified as follows (LINÉS ESCARDÓ, 1955–1965):

(*A*) Cyclones of Atlantic origin, formed far from the Iberian Peninsula.

(*B*) Cyclones forming or deepening appreciably in regions adjacent to the Iberian Peninsula (between about 25°W and 10°E).

(*C*) Intense advection of polar air sometimes connected with cyclogenesis over the Iberian Peninsula. Rainfall generally originates over the Cantabrian slopes and/or northern side of the Pyrenean range.

(*D*) Mediterranean cyclones originating not immediately adjacent to the Iberian Peninsula.

The type *A* cyclones amount to roughly 54% of the disturbances creating rainfall over the Iberian Peninsula (LINÉS ESCARDÓ, 1966; Fig. 1).

Large cyclones extended roughly in the Atlantic from 35° to 60°N, normally affect the whole of the Iberian Peninsula, although many fronts are weakened when traversing the peninsula and are often quite weak upon reaching the Mediterranean. Cyclones moving along higher latitudes cause appreciable amounts of rainfall only over Galicia and northern Portugal.

Southern Galicia gets heavy rainfall (80 mm a day is not uncommon) with type *A* cyclones when persistent southwest winds are blowing. These winds are more persistent when a steep gradient is caused between a deep type *A* cyclone and a weak high pressure center over the southeastern parts of the Iberian Peninsula.

Among the first type of cyclones we may distinguish:

(*A1*) Large cyclones, or low pressure systems, covering almost the whole north Atlantic

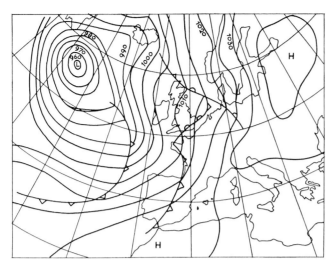

Fig.1. Atlantic type storm. Surface chart for 00h00 on February 25, 1961.

area (Fig.2). These are characteristic of autumn and winter. They often affect the whole peninsula and have an average duration of about eight days, although in stationary situations they frequently may persist for up to nineteen days. This type makes up 18% of the annual total of situations creating rainfall.

These cyclones show up at the surface as deep barometric depressions. At upper levels they have the shape of well conformed troughs and sometimes, in more stationary situations, even of closed depressions. In the case of the upper level troughs, maximum rainfall always occurs ahead of its axis. Fig.3 shows an example of a trough moving slowly eastwards.

(*A*2) Cyclones of medium size, i.e., those that cover only a limited area of the north Atlantic. These cyclones are the most common ones and represent 36% of the total. More than one third of these cyclones are regenerated in the region of the peninsula, either by deepening or by the formation of secondary depressions.

The approximate duration of the effect of these cyclones is eight days, but it is not exceptional for them to persist for up to seventeen days.

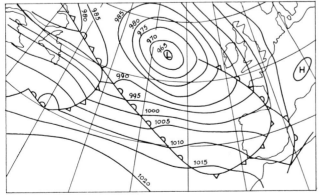

Fig.2. Type *A1* storm. Surface chart for 00h00 on February 15, 1963.

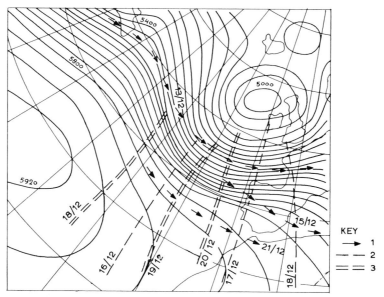

Fig.3. Type *A1* storm. 500-mbar chart for December 13, 1958.
*1* = Jet-stream (on December 13, 15, 21); *2* = trough lines (on December 16, 17, 18); *3* = trough lines (on December 18, 19, 20).

In these cyclones, the maximum precipitation occurs, as with those of type *A1*, in front of the axis of the (upper air) trough.

It is well known that these moving cyclones occur in families: they pass the peninsula when there is a strong westerly zonal circulation. At upper levels a well defined jet stream can be found blowing from the west. Generally there are very strong winds in the core of these jets. The duration of each individual storm is short (3–4 days). The passage of a family of cyclones is usually interrupted by a strong outbreak of polar air, thereby establishing a northerly situation. The most active of these cyclones affect the whole of the Iberian Peninsula. Those that reach the Bay of Biscay frequently pass quickly into the Mediterranean where they deepen (Fig.4). This happens in connection with an active northwesterly jet stream formed in the upper air.

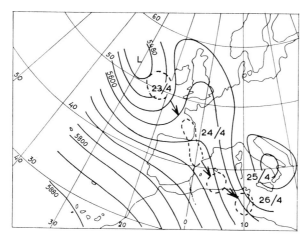

Fig.4. Type *A2* storm. 500 mbar chart for 00h00 on April 20, 1959. Dotted circles indicate positions of the centre of low pressure on subsequent days.

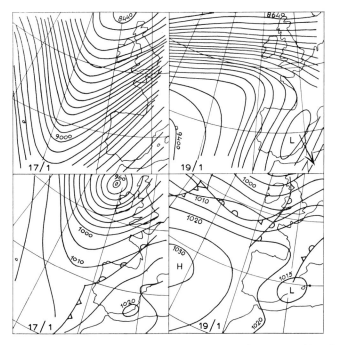

Fig.5. Type *B* storm. Surface and 300 mbar charts for January 17 and 19, 1962. The diagrams indicate decay of the jet-stream and development of a "cut-off" low.

The *B* group of cyclones includes those which may develop or be actively regenerated in the geographical areas adjacent to the Iberian Peninsula; they constitute 25% of the total annual number. The most severe weather conditions in the peninsula are nearly always associated with cyclones of this type.

The characteristics of these cyclones are as follows:

(*1*) They do not usually move with the zonal current of westerly winds. They may remain stationary for several days, and even move from east to west.

(*2*) Their development is difficult to forecast since, because of their special circulation mechanism, the rules of forecasting based on continuity do not always apply.

(*3*) Analysis of the upper air maps is much more significant than that of surface maps.

(*4*) This type also includes cyclones caused by the decay of the jet stream which is usually followed by the formation of a cut-off low (Fig.5).

Commenting particularly on the geographical areas where these disturbances occur, we may establish the following classes:

(*B1*) Disturbances developing or actively re-developing to the northwest of the Iberian Peninsula. These disturbances most resemble the storms of type *A2* and, on occasion, it is difficult to distinguish the one from the other (Fig.6).

(*B2*) Disturbances developing or actively re-developing in the southwestern area (chiefly in the Gulf of Cadiz). They usually cause considerable precipitation in the south of the peninsula and frequently extend to the Mediterranean (Fig.7).

Perturbations of this type begin in the Gulf of Cadiz or north of the Canary Islands on the occasion of a high pressure ridge being centered over the peninsula; in the neighbourhood of the Strait of Gibraltar strong easterly winds in the lower levels start blowing, and at the same time the intensity of the trade winds usually increases over north-

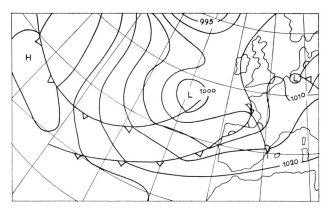

Fig.6. Type *B1* storm. Surface chart for 00h00 on January 26, 1960.

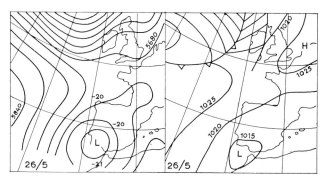

Fig.7. Type *B2* storm. 500 mbar and surface charts for 00h00 on May 26, 1963.

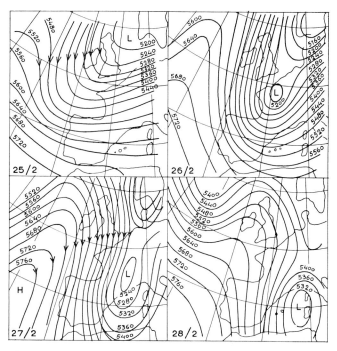

Fig.8. Type *B3* storm. 500 mbar charts for 00h00 on February 25, 26, 27 and 28, 1958.

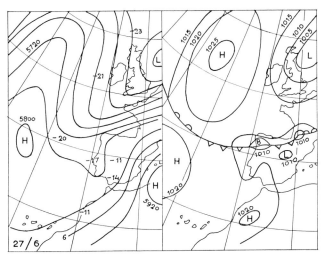

Fig.9. Type *B4* storm. Surface and 500 mbar charts for 00h00 on June 27, 1963.

west Africa and the Canary Islands. After the easterly wind has been blowing for some days over the Strait of Gibraltar, a low pressure center may appear over the Gulf of Cadiz. Usually it is possible to detect it at upper levels earlier than at the surface. This low pressure center usually exhibits little activity with only stratified cloudiness, unless there is advection of cold air at the upper levels occurs, which may intensify the system.

(*B3*) Disturbances developing or actively re-developing in the Mediterranean and the northeast of the peninsula. Most of these arise through decay of the jet stream or simply through the advection of unstable polar air into the Mediterranean in autumn and winter. They usually affect the Mediterranean slopes, especially the northern half of these and the Balearics (Fig.8).

In spring and during the beginning of the summer, thunderstorms with advection of unstable polar air do not usually occur over the western Mediterranean, because this body of water is colder than the surrounding land areas; thunderstorms may build up over the land, but they usually dissipate when moving over the sea. At the end of the summer and the beginning of autumn, the opposite usually occurs, i.e., with polar air advection, strong, unstable phenomena show up over both land and sea. The heaviest and most important precipitation recorded over the peninsula occur with this situation over the Mediterranean coastal zone during the months of September, October and, occasionally, November.

(*B4*) Thundery situations over the peninsula, mainly due to strong surface heating and comparatively cold upper air. Included in this group are periods of thunderstorms or of considerable instability during late spring and summer. The interior regions, especially the mountainous regions, are the ones usually affected (Fig.9) (ZIMMERSCHIED and BAUR, 1949).

The *C* group covers periods characterized by intense advection of polar air, either with or without later deepening, over the Iberian Peninsula. In these situations, and if there is no further cyclogenesis, important precipitation occurs only on the Cantabrian slopes and in the Pyrenees Range. This is mainly due to orographic lifting of unstable air. During autumn and winter these situations may also affect Cataluña and the Balearics (Fig.10).

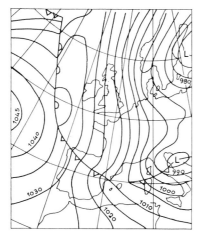

Fig.10. Type *C* storm. Surface chart for February 14, 1962.


Periods of polar air invasion usually occur also in the rear of an active cold front connected with a great Atlantic cyclone. Quite often the cyclone moves along latitudes to the north of the Iberian Peninsula, not therefore fully affecting the situation over that land area. Only occasionally does the cold front cross the peninsula bringing with it a short period of rainfall. These situations are not included in group *C* unless they are connected with deepening, or intense and persistent polar air advection.

Northerly perturbations of this kind make up 17% of the annual total. Of these, only a third affects the whole of the peninsula.

(*D*) Mediterranean storms. These storms are the rarest, hardly amounting to one per annum. They are distinguished from those of type *B3* first of all by their larger size, and secondly because they may originate in regions more remote from the Iberian Peninsula. In any case, the distinction will not always be simple (Fig.11).

The foregoing may be summarized in Table I which tabulates the classifications described.

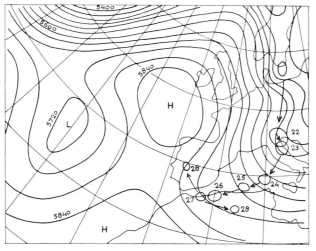

Fig.11. Type *D* storm. 500 mbar chart for 00h00 on October 21, 1958. The circles with dates indicate the positions of the 500 mbar low centre on subsequent days.

TABLE I

TYPES OF CYCLONES WHICH GENERATE PRECIPITATION OVER THE IBERIAN PENINSULA

| Type | Sub-type | Percentage of total | Mean no. of days in which the peninsula is affected | Percentage affecting the entire peninsula | Seasons of highest frequency |
|---|---|---|---|---|---|
| *A* (Atlantic) Total: 54% | *A1*, large storms | 18 | 8 | 13 | autumn winter early spring |
| | *A2*, medium storms | 36 | 8 days in groups of 3–4 days | 21 | autumn winter spring |
| *B* (Peninsular area) Total: 25% | *B1*, NW of the peninsula | 8 | 6 | 5 | autumn winter spring |
| | *B2*, SW (Gulf of Cadiz) | 6.5 | 7 | 3 | autumn winter spring |
| | *B3*, Mediterranean and NE | 3.5 | 6 | 3 | autumn winter |
| | *B4*, thundery | 7 | 6 | 2 | summer |
| *C* (Northerly) | — | 17 | 4 | 6 | autumn winter spring summer |
| *D* (Mediterranean) | — | 4 | 6 | 0 | autumn winter |

## Radiation

The clarity of the air over the Iberian Peninsula is such that values of solar radiation measured near ground level are usually very high. In general, on the southern (Mediterranean) coast, in the Gulf of Cadiz, and on the southwestern coast of Portugal the mean annual total radiation exceeds 400 cal./cm² day. For the most part, the maximum radiation is recorded in July. At some stations, such as at Sevilla, the maximum occurs in June. The minimum occurs in December.

In general, the maximum transparency of the air occurs in spring. At Madrid, the mean radiation measured at midday assumes a maximum of 1.33 cal./cm² min in the first twenty days of April (CASALS MARCÉN, 1956; GONCALVES, and MATA, 1962). The reason is that in spring, the advection of cold air masses over actively heated ground, produces instability which readily disperses impurities in the air. Of measurements taken at midday, the lowest mean values occur at the beginning of December. Another, secondary minimum is observed at the end of summer, due to the atmospheric obscurity resulting from the frequent periods of hazy days.

Observations made at Cercedilla (Madrid) at a height of 1,300 m, at midday in July, show 10% more radiation than do similar observations made at el Retiro Observatory (Madrid) at 667 m.

At Izaña (Tenerife, Canary Islands) at 2,367 m, the following high values of radiation have been recorded (in cal./cm²):

|  | Jan. | Feb. | Mar. | Apr. | May | Jun. | Jul. | Aug. | Sept. | Oct. | Nov. | Dec. |
|---|---|---|---|---|---|---|---|---|---|---|---|---|
| Mean value | 1.62 | 1.65 | 1.64 | 1.62 | 1.61 | 1.57 | 1.51 | 1.52 | 1.55 | 1.60 | 1.61 | 1.62 |
| Absolute max. | 1.69 | 1.71 | 1.69 | 1.70 | 1.67 | 1.67 | 1.65 | 1.65 | 1.65 | 1.67 | 1.69 | 1.67 |

These values rank among the highest recorded anywhere in the world.

## Insolation

The Iberian Peninsula is one of the most highly insolated regions in Europe, having an average of some 2,500 h of sunshine per annum. Only the northernmost borders have an annual average of less than 2,000 h (Fig.12).

The regions with more than 3,000 h per annum are the central region, the southeast, a large part of the south of Portugal, and the Gulf of Cadiz, the latter being the region of maximum sunshine, amounting on average to 3,243 h.

The least insolated region is Alava, in South Vascongadas, with 1,680 h.

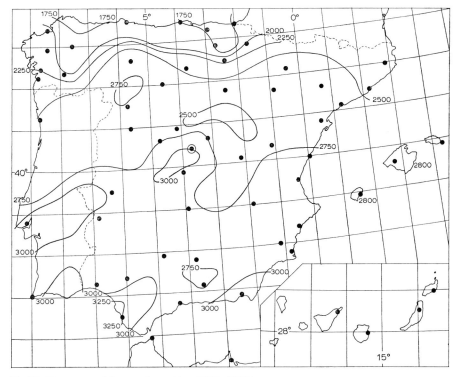

Fig.12. Mean annual hours of sunshine (1931–1960).

The insolation is also high in the Balearics, reaching 2,795 h per annum in Palma (Majorca) and 2,640 h in Mahón (Minorca). It is also high in the Canaries with 2,895 h at Tenerife and 2,664 h at Gando. At Izaña (Tenerife), at a height of 2,367 m, the mean annual insolation is 3,377 h.

The number of clear days is some four times greater in the south than in the north. The maximum number of clear days (meaning less than 2/8 of cloud) occurs in Granada, with 157 clear days. In Huelva, and in the lower basin of the Guadiana, the number of clear days is about 150 per annum. In practically all of the southern half of the peninsula, there are more than a hundred clear days; in the basins of the Douro and the Ebro and in southern Galicia there are between 75 and 100 clear days, with the exception of the source regions of these rivers. Finally, in the Cantabrian region there are between 30 and 50 clear days per annum.

## Temperature

The distribution of temperature in the peninsula is very complex, being governed mainly by altitude which, as has been indicated, is extraordinarily varied, and by continentality. The maritime influence extends along the entire coast, varying greatly in degree. It is only perceptible along a narrow stretch of country in Cantabria owing to the nearness of the mountain range, and is more widespread in Galicia and in the north of Portugal, owing to the irregularity of the coast and the many deep inlets. In the extreme south, thermal contrasts are greater. In Granada, within horizontal distances of little more than 50 km, one passes from the perpetual snows of the Sierra Nevada to the tropical orchards which abound in the heat of the "Costa del Sol". In the Central Plateau the temperature regime is markedly 'continental', with a large seasonal and annual range. The continentality decreases very slowly from the interior towards the coasts, so that in the valleys of the principal rivers the continental characteristics are well developed even at places relatively near the estuaries.

Examination of the mean annual temperature of the Iberian Peninsula reveals that, with the exception of places at very high altitudes, the values range between 13° and 19°C. Values of 19°C are reached on the southeastern coasts and in isolated places in the south. The lowest mean annual temperatures are recorded in Cantabria and, of course, at high altitudes. Altitude, however, is not the only factor which determines the mean temperature in the interior, since in the valleys an important factor is their orientation towards the north or the south.

The highest temperatures in the peninsula (50.5°C) have been recorded at Riodades (Alto Douro, Portugal). Temperatures of 49°C have been recorded several times in Andalucia. Of the stations in the synoptic network, Sevilla has recorded 47°C. At stations in the interior, the hottest month is July. On the coast, for the most part, the highest temperature usually occurs in August, especially along the Mediterranean coast. The lowest temperatures are always recorded at high altitudes and in the interior. Of the synoptic stations, Calamocha (Teruel) has recorded a minimum of −30°C at an altitude of 905 m. It is very probable that considerably lower temperatures have been reached at higher altitudes, particularly at heights above 3,000 m. The highest minimum temperatures occur on the southern coasts of the peninsula. In Almeria the lowest temperature recorded is 0.2°C, and at some places on the southwest coast of Portugal temperatures below zero have never been recorded. The lowest temperatures of the year are usually observed in January, but at times they have occurred in December or in February; on the Mediterranean coast it is not unusual for the minimum to occur in February (ROL-DÁN FLORES, 1962). In the second half of December, in January and part of February, the period in which the temperatures. are at their lowest, the Iberian Peninsula usually remains under the influence of the European continental high-pressure system.

Referring to mean monthly temperatures, those of January (Fig.13) are highest on the southern coasts, reaching 13°C at Málaga, and 12.5°C at Cabo de San Vicente. At other coastal stations they vary between 9° and 12°C, but near the flanks of the Pyrenees, at Gerona and Guipuzcoa, the January means are somewhat lower. The minima decrease notably in the interior and, with the exception of the high mountain regions, the lowest minima occur in the highlands of southern Aragon and in Castilla la Vieja (Old Castille). Fig.14 shows the mean temperatures for the month of July. The maximum may be seen in the lower and middle Guadalquivir, where it reaches 28°C. The whole Iberian Peninsula may be said to have values exceeding 20°C, except for the high mountain regions and for some parts of the north and northwest. Both the Mediterranean coast and the greater part of the southern half all have values exceeding 25°C.

The diurnal variation of temperature is fairly small in the coastal regions, but increases rapidly towards the interior. At stations near the coast, the variation is practically constant throughout the year. In Cantabria, the diurnal variation is some 5°–6°C in all months of the year. On the Mediterranean coasts it is slightly higher. In Cataluña and in the Balearics, the variation is somewhat greater in summer than in winter. In the southeast it is uniform throughout the year, amounting to some 7.5°C. On the Atlantic coast, the

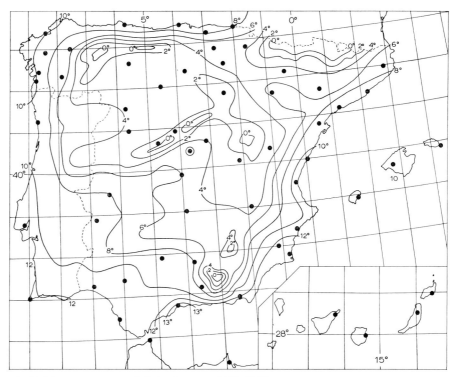

Fig.13. Mean temperature (°C) for January (1931–1960).

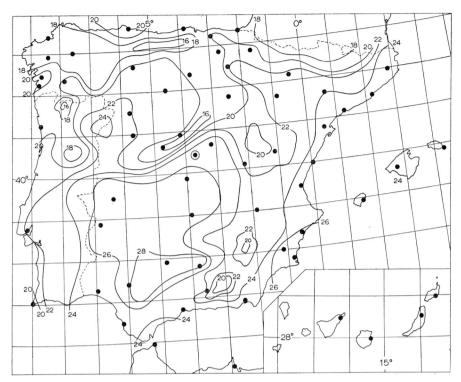

Fig.14. Mean temperatures (°C) for July (1931–1960).

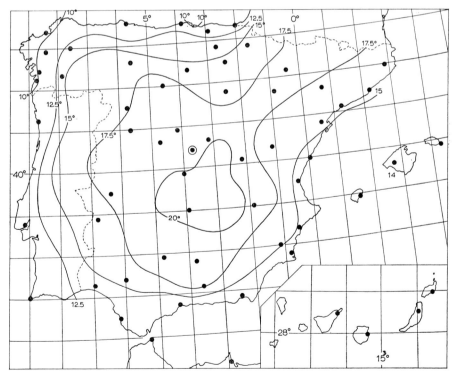

Fig.15. Difference between the mean temperatures in the warmest and the coldest months.

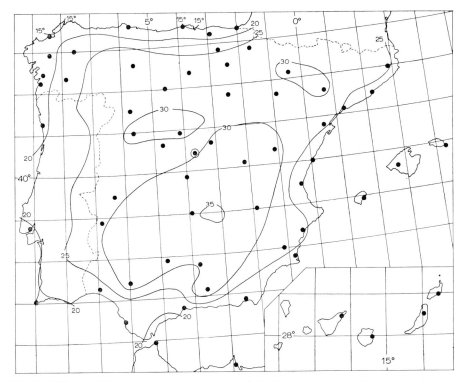

Fig.16. Difference (°C) between mean maximum of the hottest month and the mean minimum of the coldest month.

diurnal variation is 2°–4°C greater in summer than in winter, and is about 6°–10°C. In the interior the diurnal variation in the summer is very different from that in winter: on the Central Plateau it is about 7°C in winter, and is roughly double this amount in the summer. Thus, July is nearly always the month in which the diurnal variation is greatest. Fig.15 shows the annual range, or the difference between the mean temperatures of the warmest and coldest months. The greatest range is that corresponding to the southern plateau, amounting to 20°C. Almost all the interior is shown as having values of over 15°C for the annual range. The smallest range corresponds to the extreme northwest, being less than 10°C. Altitude does not have any very great effect on the annual range; the main factor is undoubtedly the distance from the sea.

Fig.16 shows the difference between the mean maximum temperature in the warmest month and the mean minimum temperature in the coldest month. The lines of equal difference are very similar to those of the previous figure, except that the values are very much greater. The highest value again appears on the southern plateau, to the east of Ciudad Real. Another secondary maximum is found on the northern plateau. The greater value in the south is due to the higher summer temperatures, the smaller amount of cloudiness, the brilliant sunshine, and the fact that the cooling effects of the marine winds do not succeed in penetrating to the plateau. As a result of the strong heating, the formation of a thermal low pressure center over the southern half of Spain is extremely frequent during the summer. During the night this low fills or persists in a much weakened form.

Another temperature maximum appears in the middle reaches of the Ebro Valley. This maximum originates in the relatively low-lying parts of the river basin which, however, are sufficiently remote from the coast to remain outside the maritime influence of the Mediterranean.

The total annual range, i.e., the difference between the maximum and the minimum in the same year, is greatest in the interior of the southern half of the peninsula and has reached 57.7°C in Badajoz. It is smallest on the northwest coast and in the islands, especially in the Canaries, reaching only about 17°C at Santa Cruz de Tenerife.

The highest daily ranges, i.e., the absolute maxima of the daily range, occur in the interior. It is not exceptional for them to reach 20°C and even more on the plateau, in the middle Ebro Valley, in Andalucia and in Murcia, whereas on the Mediterranean coast they seldom exceed 14°C and are smaller on the other coasts.

Within a given month the range of temperature, meaning the difference between the absolute maximum and the absolute minimum, reaches about 30°C or more during the hottest months. Values of 25° to 30°C occur in the hottest months on the Mediterranean coasts, in the Balearics, in Madeira, and in the Canaries. The month with the greatest range is usually May or September in the interior and May on the coasts. The lowest monthly range occurs in the coldest months and is very rarely less than 10°C.

### Average frost-free period

By average frost-free period is meant the average period between the date of the last frost in spring, and the date of the first frost of the following autumn or winter (ROLDÁN FLORES, 1956) .In the Iberian Peninsula the frost-free period is governed by the proximity of the sea and by the altitude (Fig.17).

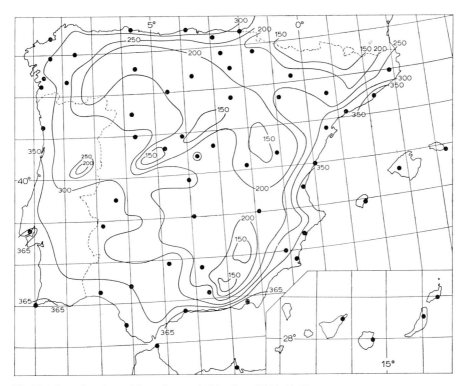

Fig.17. Mean duration of frost-free period in days (1931–1960).

Almeria, on the southeast coast, has a frost-free period lasting throughout the year. The southern Mediterranean coast from Gibraltar to Cabo de Gata has conditions similar to those of Almeria, with a frost-free period of 364 days or more. The southern Atlantic coast is much the same.

In general, all the coasts of the peninsula have a frost-free period of more than 340 days, with the exception of Vascongadas and some places in the northeast. In the interior, the frost-free period is considerably shorter: it is more than 300 days in the Guadalquivir Valley and the lower reaches of the Guadiana and Tagus valleys, about 250 days in the valley of the Ebro, between 215 and 265 days on the southern plateau, and around 200 days on the northern plateau. At high levels, and in the mountain ranges, the altitude becomes very important. In general, at altitudes of more than 1,000 m the frost-free period is always less than 190 days.

Tables II and III show the frequency percentages of the first and last frost for the fifty-year period 1901–1950. In Cuenca, for example, during the first ten day period of October, one finds a frequency percentage of 14, this means that 14% (or 70 days) out of the total number of days, (50 ten-day periods = 500 days), were days on which the first frost occurred.

### Evaporation and evapotranspiration

Potential evaporation reaches very high values in the Iberian Peninsula. In practically the whole territory, annual mean values vary between 1,000 and 2,000 mm. The lowest potential evaporation occurs in the northwest where, in some regions, it only amounts to about 500 mm per annum. The highest occurs in the southeast, towards Alicante, where it reaches 2,500 mm.

In January, the lowest monthly potential evaporation occurs in the interior, amounting to only 20 mm on the northern plateau, in lower Aragon, and in the Pyrenees. The maximum in the said month occurs on the Mediterranean coasts and in the Gulf of Cadiz.

In July, the minimum occurs in the north and northwest of the peninsula, with values varying between 60 and 100 mm.

The maximum occurs in the south, particularly in Murcia, in the lower valley of the Guadiana, and in the middle reaches of the Tagus Valley. These regions have values reaching 400 mm.

Fig.18 shows values of the annual potential evapotranspiration calculated by Thornthwaite's method. Two maxima of over 1,000 mm may be seen, one near Murcia and the other at Córdoba. Areas with 900 mm appear surrounding those just mentioned, and there are also areas having the same value in Estremadura and Baixo Alentejo. There is also another region with 900 mm to the south of Alicante. The lowest values occur in the northwest and, of course, at the summits of the high mountain ranges, as in the Sierra da Estrela, the region of highest rainfall in Portugal, and of lowest evapotranspiration (670 mm).

ELÍAS CASTILLO (1965) has evaluated the potential evapotranspiration by the Thornthwaite, Turc and Penman methods. A comparison between Thornthwaite's and Penman's methods shows that the Penman formula gives values that may be up to 20% higher than

TABLE II

FREQUENCY PERCENTAGES OF THE FIRST DAYS OF FROST, 1901–1950[1]

| Location | September* | | | October | | | November | | | December | | | January | | | February | | | March | | |
|---|---|---|---|---|---|---|---|---|---|---|---|---|---|---|---|---|---|---|---|---|---|
| | 1 | 2 | 3 | 1 | 2 | 3 | 1 | 2 | 3 | 1 | 2 | 3 | 1 | 2 | 3 | 1 | 2 | 3 | 1 | 2 | 3 |
| Almeria | 0 | 0 | 0 | 0 | 0 | 0 | 0 | 0 | 0 | 0 | 0 | 0 | 0 | 0 | 0 | 0 | 0 | 0 | 0 | 0 | 0 |
| Barcelona | 0 | 0 | 0 | 0 | 0 | 0 | 0 | 2 | 0 | 4 | 12 | 8 | 6 | 8 | 15 | 2 | 4 | 6 | 0 | 0 | 0 |
| Córdoba | 0 | 0 | 0 | 0 | 0 | 0 | 0 | 0 | 6 | 0 | 6 | 37 | 27 | 6 | 9 | 6 | 0 | 0 | 0 | 0 | 0 |
| Cuenca | 0 | 0 | 0 | 14 | 8 | 31 | 28 | 14 | 5 | 0 | 0 | 0 | 0 | 0 | 0 | 0 | 0 | 0 | 0 | 0 | 0 |
| La Coruña | 0 | 0 | 0 | 0 | 0 | 2 | 0 | 0 | 4 | 4 | 0 | 9 | 4 | 4 | 7 | 0 | 2 | 7 | 0 | 2 | 0 |
| Madrid | 0 | 0 | 0 | 11 | 13 | 13 | 18 | 11 | 20 | 7 | 7 | 0 | 2 | 0 | 0 | 0 | 0 | 0 | 0 | 0 | 0 |
| Málaga | 0 | 0 | 0 | 0 | 0 | 0 | 0 | 0 | 0 | 0 | 0 | 0 | 3 | 0 | 3 | 3 | 0 | 0 | 0 | 0 | 0 |
| Santander | 0 | 0 | 0 | 0 | 0 | 0 | 0 | 0 | 0 | 0 | 0 | 18 | 4 | 12 | 8 | 0 | 4 | 4 | 0 | 0 | 0 |
| Sevilla | 0 | 0 | 0 | 0 | 0 | 0 | 0 | 0 | 0 | 5 | 25 | 14 | 14 | 2 | 9 | 5 | 0 | 2 | 0 | 0 | 0 |
| Teruel | 0 | 0 | 0 | 14 | 32 | 34 | 11 | 9 | 0 | 0 | 0 | 0 | 0 | 0 | 0 | 0 | 0 | 0 | 0 | 0 | 0 |
| Valladolid | 0 | 0 | 0 | 6 | 4 | 33 | 13 | 36 | 6 | 2 | 0 | 0 | 0 | 0 | 0 | 0 | 0 | 0 | 0 | 0 | 0 |

[1] If each year in the period 1901–1950 had some days of frost at a given location, addition of all numbers in the horizontal column behind that location yields a total of 100%. When at a certain location 100% is not reached, however, this means that at that location there occurred some frost-free years during the period 1901–1950.

* *1, 2, 3* refer to periods of ten days.

TABLE III

FREQUENCY PERCENTAGES OF THE LAST DAYS OF FROST, 1901–1950[1]

| Location | December* | | | January | | | February | | | March | | | April | | | May | | | June | | |
|---|---|---|---|---|---|---|---|---|---|---|---|---|---|---|---|---|---|---|---|---|---|
| | 1 | 2 | 3 | 1 | 2 | 3 | 1 | 2 | 3 | 1 | 2 | 3 | 1 | 2 | 3 | 1 | 2 | 3 | 1 | 2 | 3 |
| Almería | 0 | 0 | 0 | 0 | 0 | 0 | 0 | 0 | 0 | 0 | 0 | 0 | 0 | 0 | 0 | 0 | 0 | 0 | 0 | 0 | 0 |
| Barcelona | 2 | 4 | 2 | 12 | 10 | 8 | 11 | 6 | 8 | 2 | 0 | 0 | 2 | 0 | 0 | 0 | 0 | 0 | 0 | 0 | 0 |
| Córdoba | 0 | 6 | 3 | 12 | 22 | 12 | 6 | 18 | 6 | 6 | 0 | 6 | 0 | 0 | 0 | 0 | 0 | 0 | 0 | 0 | 0 |
| Cuenca | 0 | 0 | 0 | 0 | 0 | 0 | 0 | 0 | 0 | 3 | 3 | 8 | 10 | 13 | 24 | 26 | 8 | 0 | 5 | 0 | 0 |
| La Coruña | 2 | 0 | 0 | 0 | 14 | 9 | 4 | 7 | 7 | 0 | 2 | 0 | 0 | 0 | 0 | 0 | 0 | 0 | 0 | 0 | 0 |
| Madrid | 0 | 0 | 0 | 0 | 2 | 0 | 2 | 7 | 11 | 9 | 11 | 7 | 7 | 20 | 6 | 2 | 2 | 4 | 0 | 0 | 0 |
| Málaga | 0 | 0 | 0 | 3 | 0 | 3 | 3 | 0 | 0 | 0 | 0 | 0 | 0 | 0 | 0 | 0 | 0 | 0 | 0 | 0 | 0 |
| Santander | 0 | 0 | 8 | 0 | 12 | 12 | 0 | 12 | 8 | 0 | 0 | 0 | 0 | 0 | 0 | 0 | 0 | 0 | 0 | 0 | 0 |
| Sevilla | 0 | 7 | 7 | 12 | 7 | 12 | 9 | 7 | 7 | 5 | 0 | 2 | 0 | 0 | 0 | 0 | 0 | 0 | 0 | 0 | 0 |
| Teruel | 0 | 0 | 0 | 0 | 0 | 0 | 0 | 0 | 0 | 0 | 3 | 0 | 6 | 17 | 29 | 20 | 17 | 8 | 0 | 0 | 0 |
| Valladolid | 0 | 0 | 0 | 0 | 0 | 0 | 0 | 0 | 2 | 6 | 6 | 8 | 14 | 23 | 31 | 6 | 2 | 2 | 0 | 0 | 0 |

[1] If each year in the period 1901–1950 had some days of frost at a given location, addition of all numbers in the horizontal column behind that location yields a total of 100%. When at a certain location 100% is not reached, however, this means that at that location there occurred some frost-free years during the period 1901–1950.

* *1, 2, 3* refer to periods of ten days.

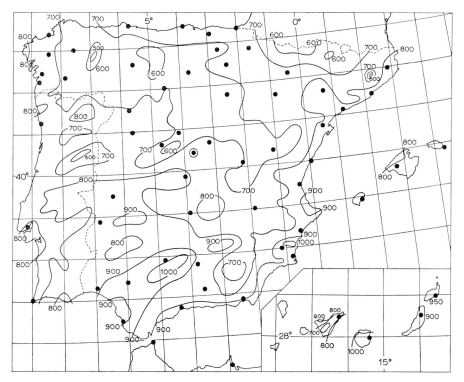

Fig.18. Potential annual evapotranspiration (mm) by Thornthwaite's method.

those calculated with Thornthwaite's. Generally, in the northern region of the Iberian Peninsula the difference is lower than in the southern areas. Using Thornthwaite's method, values higher than 1,000 mm seldom appear. By Penman's method, values of 1,300 mm appear along the coast of the Gulf of Cadiz, and at points on the lower Ebro River. In the south of Portugal, most of Andalucia, Estremadura, Alicante and wide zones along the Ebro, the potential evapotranspiration is, according to Penman's method, calculated to be more than 1,000 mm. The absolute minimum appears in similar locations when both methods are applied, and the minimum values by Penman's method are only slightly higher than Thornthwaite's values. (From comparisons between the two methods which were made in other regions of the Mediterranean, it appears that the Penman method gives more correct values than does Thornthwaite's.)

By the end of May, the water reserves are exhausted in the dry regions of the peninsula, and reserves are available only in the Cantabrian Mountains, in the Pyrenees, the Sierra da Estrela, and in analogous rainfall zones. In contrast, in the driest regions the moisture deficit already amounts to 100 mm by this time (DANTIN and REVENGA, 1941).

**Rainfall**

The complex relief of the Iberian Peninsula gives rise to a very irregular distribution of rainfall (GONZALEZ QUIJANO, 1946). During some years, certain places receive more than 3,000 mm per annum, while others, in the southeast for example, have an average of less than 200 mm, the lowest values in Europe.

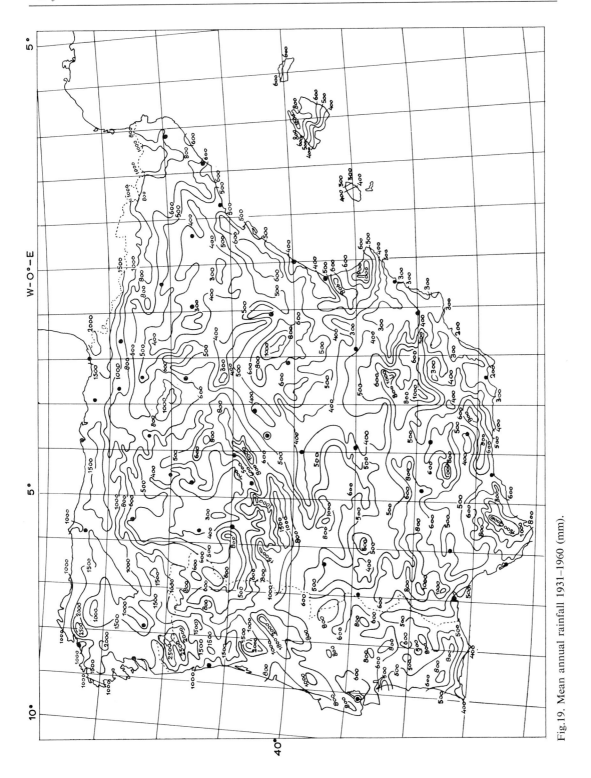

Fig.19. Mean annual rainfall 1931–1960 (mm).

Fig.19 represents a rainfall map of the peninsula for the period 1931–1960. On it may be seen some small areas where the precipitation exceeds 1,500 mm per annum. These areas are located in Galicia, Cantabria, the north of Portugal, the Sierra da Estrela, the west-central Pyrenees, the Sierra de Gredos (Central Range), and the Sierra de Grazalema (Cadiz). Except for some areas in the northwest, these high values correspond to places at altitudes of 1,500 m and over. Within these areas of high rainfall, lie very much smaller areas with values above 2,000 mm, and in the northwest there are some centres with 2,500 mm. There is very probably a small area with a maximum of 3,000 mm in the highest peaks of Sierra de Gredos, Central Range.

Surrounding the areas with values above 1,500 mm are other, much larger, areas with values above 1,000 mm. All the mountain summits lie within the 1,000 mm line, as does also Cantabria, Galicia, the northern third of Portugal, and small areas of western Andalucia. The area between the 500 and 1,000 mm lines extends around the mountain-ous regions and covers a large part of Portugal, the Guadalquivir Basin, and Cataluña. On the two plateaus, and over a large part of the basin of the Ebro and in those of the Júcar and Segura, the rainfall varies between 300 and 500 mm. Finally, values below 300 mm occur in small areas of the Ebro Valley and in the southeast, and at Cabo de Gata, near Almeria, below 200 mm. As already mentioned, this is the lowest value in Europe.

The rainfall distribution is even more complex, owing principally to the variety of the relief. Thus, for example, Navacerrada at 1,860 m, has an annual rainfall three times that of Madrid, although the distance between the two stations is only 50 km.

**Seasonal variation of rainfall**

Almost all the peninsula receives most of its rainfall in the cold half of the year (winter and autumn). Only in small mountain regions in the Iberian Range, in the watershed between the Atlantic and the Mediterranean where there is frequent thunder activity due to convection, is there more rain in the warm season than in the cold.

Fig.20–23 show the variation of rainfall during the four seasons (JUNCO REYES, 1946).

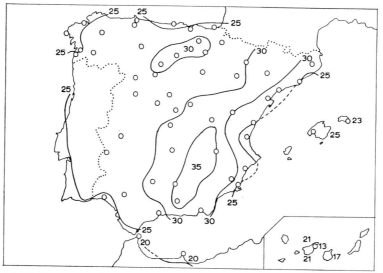

Fig.20. Percentage of total annual rain falling in spring (1901–1930).

A characteristic of the Iberian Peninsula is the occurrence of a well-marked dry season, especially in the south, on the plateau and in Aragon. It is only in isolated parts of the Iberian Range that, for reasons already mentioned, the precipitation in the three summer months amounts to 25% of the annual total. On the Mediterranean coast the most rainy season is autumn, when the sea remains warmer than the adjacent land. In autumn, the coasts of Valencia and Alicante receive 40% of their annual rainfall. On the rest of the Mediterranean coast, and in the south, 35% of the annual rainfall occurs in the autumn. During the three months of winter the rainfall in the eastern half of the peninsula decreases to about 15–25% of the annual total, although in the northwest the proportion reaches 40%. In spring, the rainfall is relatively limited on the coasts, and more abundant in the interior (LINÉS ESCARDÓ, 1959).

As already mentioned, summer is very dry in the Iberian Peninsula since it is not influenced by the westerlies and the cyclones associated with them. Only in the northeast and in the Iberian Range, is the rainfall appreciable, and it is nearly always convective in character. Summer rain is very rare in the Canaries.

In the course of the year, the Iberian Peninsula experiences two rainfall maxima, one in spring, associated with an increase in thunderstorm activity, and the other in autumn or at the beginning of winter. These rainfall maxima are associated with the period in which the jet stream, and also the westerlies appear at a latitude similar to that of the Iberian Peninsula. In Cataluña, Valencia, Castilla la Nueva, and in the north of Andalucia the autumn maximum occurs as early as October. In Castilla la Vieja and Estremadura, it is delayed until November, while in the north, in Portugal, and in the south of Andalucia the maximum is deferred until December or even January, as is also the case in parts of Galicia and in the south of Portugal. In any case, the difference between the monthly maximum of October and November is negligible.

The spring rainfall maximum occurs as early as March in the south, in Portugal, and in eastern Estremadura, but is delayed till May in Castilla la Nueva, Castilla la Vieja, Vascongadas, Cataluña, and Valencia. In the extreme south and in the southeast of the Iberian Peninsula, and in the greater part of Portugal, the two maxima occur close to

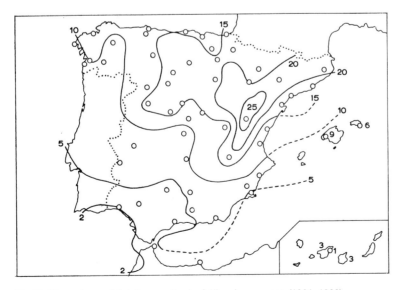

Fig.21. Percentage of total annual rain falling in summer (1901–1930).

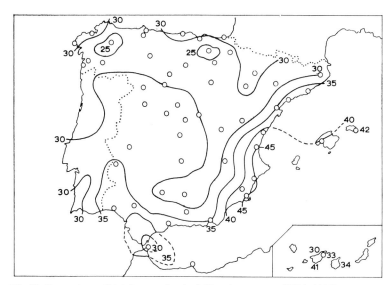

Fig.22. Percentage of total annual rain falling in autumn (1901–1930).

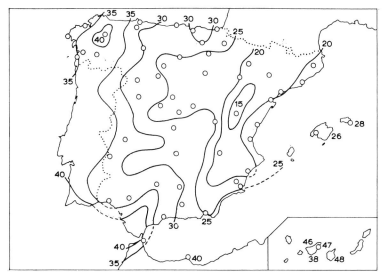

Fig.23. Percentage of total annual rain falling in winter (1901–1931).

each other and could almost be considered as one since autumn and winter between them account for almost 70% of the annual rainfall.

It seems logical that the spring rainfall maximum shows up earlier in the south since, during this season, westerlies and associated cyclones appear progressively at higher latitudes. During the second half of the spring, thunderstorms frequently appear over the interior of the peninsula, due to cold advection over warmer land areas. On account of rainfall due to instability, the maximum of rainfall over the plateau frequently occurs during May. The same could be said for the coasts of Valencia and Murcia and, beyond the peninsula, for the Balearics, Madeira and the Canaries.

**Variability of rainfall**

In general, rainfall is more uniform on the Atlantic slopes than on the Mediterranean

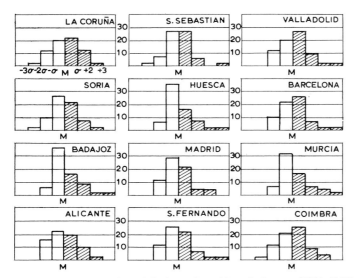

Fig.24. Distribution of rainfall about the arithmetical mean (1881–1948).

slopes, and corresponds to a larger number of days with rain in the former region than in the latter. On the Atlantic slopes, the wettest years are those in which there is considerable zonal circulation, and the jet stream lies well to the south. On the Mediterranean slopes, the wettest years are usually those in which the circulation tends to be meridional.

In the analysis of seasonal rainfall, it is more common, mainly in the autumn and winter, to find bimodal frequency distributions on the Mediterranean slopes than on those of the Atlantic.

As already mentioned, cyclones of Atlantic origin cause less rainfall over the Mediterranean areas than over the Atlantic regions. The heaviest rainfall over the Mediterranean occurs with type *B3*, *C* and *D* cyclones. These cyclones are more frequent when the general circulation is meridional than during periods of strong zonal circulation. A consequence of this can be that the Mediterranean coastal zone may have wet years when the circulation is meridional, and dry years when the circulation is zonal and the type *A* cyclones prevail. In any event, intermediate years in between wet and dry are seldom encountered. In the analysis of annual rainfall, the bimodal character is less marked, or even absent.

On the Cantabrian slopes, the rainfall figures conform moderately well to the "normal distribution". Application of Cornu's criterion, gives the following values for the ratio between the mean deviation and the standard deviation:

Gijon              0.81
Coruña             0.80
Castropol (Oviedo) 0.80
Santander          0.81
Bilbao             0.82

Fig.24 shows the frequency distribution of rainfall around the mean value for certain stations.

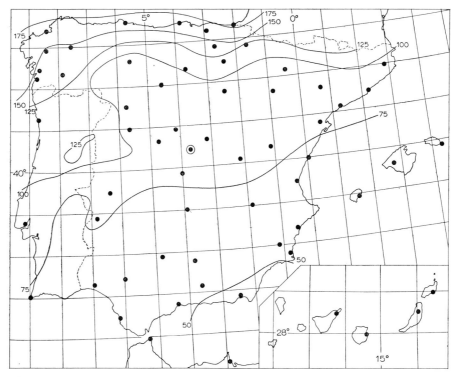

Fig.25. Mean annual number of days with rain (including snow).

## Mean annual number of days with rain

Fig.25 shows the annual number of days with rain. This may be seen to be greatest in the coastal regions of Vascongadas and Santander, with a maximum of 183 days at San Sebastian. There are more than 150 such days on the Cantabrian coasts and in the north of Galicia. In the Pyrenees, in the higher reaches of the Ebro and the Douro rivers, in Galicia, and in the northern half of Portugal, the number of days with precipitation is over 100. Almeria has the lowest number with 47.

## Extreme rainfall intensities

The heaviest downpours and the highest intensities of rainfall have all been recorded on the Mediterranean slopes, nearly always in early autumn. A case in point is the flooding of Valencia in October 1957 when Bejís (Valencia) recorded 210 mm in 90 min. and 361 mm in 24 h (Garcia Miralles and Carrasco Andreu, 1958). On the same date unofficial records gave as much as 1,000 mm in 24 h. At Sabadell (Barcelona) instantaneous intensities were recorded on 25 September 1962 of 257.4 mm/h and 95.2 mm in 44 min.

## Number of days with snow

The number of days with snow (Fig.26) is practically zero on the Mediterranean coasts, on the coasts of Portugal and Galicia and, generally, in the south of the peninsula. It

220

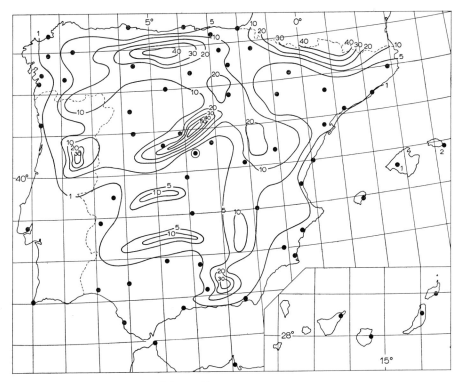

Fig.26. Mean annual number of days with snow.

varies between two and six days in Cantabria, and increases rapidly inland, especially over high ground. On the southern plateau snow falls on about three or four days per annum, while on the northern plateau the number of days with snow is very variable, ranging from three days per annum at Zamora, to seventeen at Soria and Burgos, and eighteen at Avila.

Above 2,000 m the annual number of days with snow exceeds 50. Invasions of cold air normally occur with northerly or northeasterly winds, so that hill slopes facing these directions always have a larger number of days with snow than those facing the south. In Portugal there is practically no snowfall either on the coast or in the southern half, but it is moderately frequent in the mountain regions of the northeast (32 days in Penhas Douradas).

## Duration of rainfall

Table IV gives the values of Besson's persistence coefficient for a number of Spanish stations (BROOKS and CARRUTHERS, 1953, p. 310). Except in the summer months, in areas of low rainfall where the very small rainfall values tend to give negative values for the coefficient, the values of the latter are generally high. This confirms what has already been said about the way in which cloud banks are held back as a result of the special situation and orographic peculiarities of the stations, some being situated on the coasts with pronounced mountain barriers behind them (La Coruña, Santander, San Sebastian and Málaga) and others in river valleys with steep sides which form a barrier at right

TABLE IV

COEFFICIENT OF PERSISTENCE OF RAINFALL (BESSON'S COEFFICIENT)[1]

| Location | Jan. | Feb. | Mar. | Apr. | May | June | July | Aug. | Sept. | Oct. | Nov. | Dec. | Year |
|----------|------|------|------|------|------|------|------|------|------|------|------|------|------|
| La Coruña | 1.07 | 0.47 | 1.63 | 1.33 | 0.75 | 0.93 | 0.73 | 0.79 | 0.71 | 1.77 | 1.50 | 0.94 | 1,21 |
| Santander | 0.68 | 0.78 | 0.81 | 0.78 | 0.52 | 0.53 | 0.57 | 0.45 | 0.53 | 0.71 | 0.83 | 0.99 | 0.68 |
| San Sebastian | 0.66 | 0.82 | 0.76 | 0.56 | 0.54 | 0.45 | 0.49 | 0.33 | 0.47 | 0.61 | 0.68 | 1.00 | 0.61 |
| Valladolid | 0.79 | 1.17 | 1.19 | 1.91 | 1.09 | 0.79 | 0.31 | 0.48 | 0.82 | 1.02 | 0.96 | 0.88 | 1.05 |
| Soria | 0.66 | 0.93 | 0.80 | 1.06 | 0.70 | 0.48 | 0.09 | 0.31 | 1.10 | 0.71 | 0.78 | 1.13 | 0.76 |
| Zaragoza | 0.34 | 0.60 | 0.61 | 0.75 | 0.73 | 0.71 | 0.32 | 0.41 | 0.53 | 0.63 | 0.21 | 0.42 | 0.54 |
| Madrid | 0.94 | 1.08 | 0.89 | 0.76 | 0.78 | 0.72 | 0.18 | 0.41 | 0.61 | 0.82 | 0.85 | 1.12 | 0.88 |
| Barcelona | 0.34 | 0.38 | 0.52 | 0.68 | 0.52 | 0.14 | 0.15 | 0.35 | 0.32 | 0.49 | 0.32 | 0.35 | 0.39 |
| Valencia | 0.69 | 0.59 | 0.37 | 0.82 | 0.54 | 0.46 | 0.22 | 0.30 | 0.45 | 0.68 | 0.42 | 0.66 | 0.45 |
| Sevilla | 0.88 | 0.90 | 0.84 | 0.72 | 0.49 | −0.32 | −1.00 | −1.00 | −0.55 | 0.44 | 0.66 | 1.18 | 0.85 |
| Malaga | 0.80 | 0.63 | 0.42 | 0.58 | 0.50 | 0.07 | −1.00 | −1.00 | 0.18 | 0.67 | 0.63 | 1.23 | 0.73 |
| Almeria | 0.37 | 0.47 | 0.34 | 0.44 | 0.45 | 0.15 | −0.01 | −0.03 | 0.09 | 0.27 | 0.41 | 1.12 | 0.46 |

[1] Besson's coefficient of persistence is $R_B = \dfrac{1-p}{1-p^1} - 1$, where $p$ is the probability of a day of precipitation (number of days of precipitation divided by the number of days), and $p^1$ is the probability of a day of precipitation preceded by a day of precipitation (BROOKS and CARRUTHERS, 1953).

angles to the prevailing direction of movement of the cloud banks (Madrid, Valladolid and Soria) (MATEO GONZALES, 1965; J. GARCIA SANJUAN, personal communication, 1966). In order to give an idea of the comparative values of the persistence coefficients quoted in the table, it may be stated that the value of the same coefficient for Kew (Great Britain) is 0.30, while for Paris monthly values varying between 0.30 and 0.49 have been obtained over a period of 50 years.

**Winds**

During the cold months, the peninsula remains within the zone of predominantly westerly winds. In summer, on the other hand, westerly winds are comparatively few; the winds are light, and in general converge towards the interior of the peninsula. In Cantabria, and especially on the coasts of Galicia and Portugal, moderate northerly winds predominate during the summer.

The most windy parts of the peninsula are the coasts of north Galicia and Asturias, the lower and middle Ebro Valley, and the lower Tagus Valley. The island of Minorca (Balearics) is also very windy. The wind speed is also considerable in the proximity of the Straits of Gibraltar, in the valley of the river Douro, and in the southeast. At Penhas Douradas in the Sierra da Estrela, at 1,386 m, the mean daily run of the wind is 630 km, one of the highest values recorded in the peninsula. In contrast, the wind speed is usually smaller in the valleys of the Guadiana and Guadalquivir and in the middle and upper reaches of the Tagus.

The dominant winds of the peninsula are the "levante" or easterly winds in the Strait of Gibraltar, the "cierzo" (a northwesterly wind) in Aragon, the "tramontana" (a northerly wind) in Cataluña, and the "abrego", a cool and moist south-westerly which blows on the Atlantic slopes. Finally, the "terral", a very dry, southerly wind in Cantabria should also be mentioned.

Among extreme values reached by the surface winds, may be mentioned that of 180 km/h recorded in San Sebastian, on February 15, 1941. On the same day a large part of Santander was destroyed by the wind and by a fire: the wind speed could not be recorded because the recording anemometer was destroyed by the wind. At Izaña (Canaries), at 2,367 m, a speed of 216 km/h was recorded on February 25, 1947.

## Climatic evolution

As in many other parts of the world, climatic variations have occurred in the Iberian Peninsula both in recent and in ancient times. Climatic changes prior to the eighteenth century can only be deduced indirectly from references to historical events, facts about agricultural products, cultivation of plants sensitive to changes in climate, structure of tree trunks, and so on. To throw light on prehistoric times one must turn to the evidence of the life of the period in question—to fossils and to soil analysis. Facts of special importance are of course those relating to glaciers.

The last Pleistocene glaciation ended about 20,000 years ago and was probably followed by a long period with a highly continental climate. From that time to the present day, the configuration of the coasts of the peninsula experienced far less alteration than did those of northern Europe or the British Isles.

In general, during both continental and maritime periods, the climate of the Iberian Peninsula experienced less variation than did that of central and northern Europe. During the ten centuries immediately before Christ, the climate of the Iberian Peninsula was quite favorable to human life and there must have been no lack of rain, even on the Mediterranean slopes. Some 200 years B.C., however, there was a brief dry period. Until the end of the Middle Ages a large part of the Iberian Peninsula was covered with forest which since then have rapidly decreased in extent, the decrease continuing until the beginning of the present century. Since the middle of this century there has been considerable artificial afforestation.

If one examines the data relating to floods and droughts in some of the rivers of the peninsula, one finds that the two phenomena have occurred within periods of only a few years, which makes it very difficult to differentiate between dry and wet periods within a single century. The only thing which emerges clearly is that notable climatic changes have occurred in the last ten or twelve centuries. Considering the years of flood and drought in the rivers Nervión (Vascongadas), Ebro, Douro, Guadalquivir, and Segura, prior to the present century, the following facts are forthcoming:

*Nervión*: floods in 1447, 1481, 1485, 1552, 1582, 1592, 1651, 1681, 1737, 1801 and 1831.

*Ebro*: floods in 1421, 1445, 1448, 1517, 1605, 1617, 1775, 1783, 1787, 1826, 1831, 1843, 1845, 1853, 1865, 1866, 1871 and 1884;
droughts in 1725, 1749, 1751 and 1796.

*Douro*: floods in 1143, 1168, 1201, 1256, 1264, 1286, 1297, 1310, 1402, 1435, 1476, 1488, 1500, 1511, 1543, 1556, 1586, 1603, 1615, 1626, 1635, 1645, 1692, 1725, 1736, 1747, 1768, 1777, 1793, 1801, 1821, 1831, 1842, 1858, 1879 and 1891;

droughts in 707, 846, 877, 889, 901, 982, 1172, 1213, 1257, 1302, 1333, 1355, 1473, 1503, 1513, 1546, 1550, 1595, 1605, 1616, 1629, 1640, 1650, 1680, 1703, 1715, 1726, 1737, 1749, 1764, 1775, 1789, 1803, 1858, 1868 and 1878.

*Guadalquivir*: floods in 1297, 1330, 1344, 1373, 1403, 1481, 1504, 1523, 1543, 1554, 1586, 1596, 1608, 1618, 1649, 1709, 1731, 1856 and 1876;

droughts in 1524, 1602, 1682 and 1749.

*Segura*: floods in 1485, 1615, 1651, 1753, 1783, 1834 and 1879.

**Fluctuations of rainfall over the peninsula in recent years**

Climatological variation is a problem of deep concern to both Spain and Portugal. Since these countries do not possess large reserves of mineral fuel, the major source of energy is hydraulic, so that in periods of drought the situation is serious both for agriculture and for industry.

From the existing series of climatological data, it is found that there have been frequent fluctuations in the annual rainfall during the last 150 years. In general, the latter third of the last century may be said to have been a period of above-normal rainfall for the majority of stations in the peninsula, although there was also the occasional very dry year as, for example, the agricultural year 1895–1896. After the latter year, there was a series of dry years lasting until about 1915. From 1915 to 1935 dry years predominated, with slight fluctuations. After 1935—a year of very heavy rainfall—began a period of violent fluctuations, with very marked maxima and minima, the lowest values occurring in 1944–1945. Beginning in 1958 there was a short series of exceptionally wet years, culminating in 1960. Thereafter, rainfall gradually fell off, although 1963 was also very wet (LORENTE PEREZ, 1950, 1961).

The following deductions can be made from a comparison between the total rainfall of the peninsula and that of the individual stations:

(*a*) The north and the northwest stations, such as San Sebastian and Coruña, demonstrate close agreement with the total for the peninsula as regards maxima and minima.

(*b*) The stations of the Central Plateau do not, in general, follow the fluctuations of the northern stations, except in years with particularly well-marked maxima.

(*c*) Extreme values on the Mediterranean side usually occur before those in Cantabria.

(*d*) There is some degree of parallelism between the rainfall of the south and that of the north.

**Temperature trends**

Temperature data show that between the years 1880 and 1890, there was a general lowering of temperature coincident with the increase in rainfall. The final years of the last century were hot, and coincided with the beginning of a dry period. The first fifteen years of the 20th century were rather low in temperature, with small fluctuations. In the next five years the temperature increased, and remained slightly above normal until 1936 and 1937 which were very hot. Following this there was a very short spell of cold years and, subsequently, another increase in temperature which reached its maximum in 1949, an exceptionally hot year. Later there were some very violent fluctuations, with an extremely cold year in 1956, and the years 1962 and 1963 had very cold winters (LORENTE PEREZ, 1961).

It is consequently very difficult to advance theories as to the development which may be expected in the climate of the peninsula in future decades. Writers who have studied the problem are by no means unanimous. However, the theory of Lorente has many supporters. According to this, the Iberian Peninsula is tending towards drier and colder years, that is to say towards a period with characteristics similar to those at the beginning of the present century.

## Notes accompanying the climatic tables (pp.227–239)

In all the climatic tables for Spanish stations "Mean daily temperature range" is the difference between the mean monthly maximum and minimum temperatures.

"Mean cloudiness" is given as the mean monthly number of clear, cloudy, and overcast days.

"Solar radiation on horizontal surface" is given as the mean daily total radiation.

The data for the Portuguese stations are as received from Lisbon.

## References

AMORIM FERREIRA, H. and PEIXOTO, J. P., 1962. Evaporaçao e evapotranspiraçao. *Inst. Geofis. Infante D. Luis, Publ.*, 4.

AMORIM FERREIRA, H., PEIXOTO, D. and ESPIRITU SANTO, T. R., 1965. *Balance Hídrico e Clima de Portugal Continental*. University of Lisbon, Lisbon, 50 pp.

BIEL LUCEA, A. and GARCIA PEDRAZA, L., 1962. El clima de Zaragoza y ensayo climatológico para el Valle del Ebro. *Serv. Meteorol. Nacl. (Madrid), Ser. A.*, 36: 66 pp.

BROOKS, C. E. P. and CARRUTHERS, N., 1953. *Handbook of Statistical Methods in Meteorology*. H. M. Stationery Office, London, 412 pp.

CASALS MARCÉN, J., 1956. La radiación solar en Madrid. *Calendario Meteoro-fenologico*, pp.157–159.

CASTANS CAMARGO, M., 1900. Importancia de la evaporación en las características hidrológicas de lá Península Ibérica. *Serv. Meteorol. Nacl. (Madrid), Ser. A*, 43: 22 pp.

CUSTODIO DE MORAIS, 1947. Condiçoes climaticas do trabalho ao ar livre em Portugal. Divisoes climaticas de Portugal. *Memoria Noticias*, 19: 16 pp.

ELIAS CASTILLO, F. and GIMENEZ ORTIZ, R., 1965. *Evapotranspiraciones potenciales y Balances de Agua en España. Mapa Agronómico Nacional*. Ministerio de Agricultura, Madrid, 303 pp.

DANTIN, J. and REVENGA, A., 1940. *Las Lineas y las Zonas Isóxeras de España, Segun los Índices Termopluviométricos. Avance del Estudio de la Aridez en España*. Instituto de Geografia Juan Sebastian Elcano (Consejo Superior de Investigaciones Científicas), Madrid, 6 pp.

GARCIA MIRALLES, V. and CARRASCO ANDREU, A., 1958. Lluvias de Valencia, Catellón y Alicante de los dias 13 y 14 de octubre de 1957. *Serv. Meteorol. Nacl., (Madrid) Ser. A*, 30: 65 pp.

GARMENDIA IRAUNDEGUI, J., 1964. Estudio climatológico de la provincia de Salamanca. *Publ. I.D.A.T.O.*, 230 pp.

GARMENDIA IRAUNDEGUI, J., 1967. Estudio climatológico de la provincia de Zamora. *Publ. I.D.A.T.O.*, 183 pp.

GONCALVES, C. A. and MATA, L. A., 1962. *Cinco Anos de Mediçoes de Radiaçao Solar em Superficies Verticais em Lisboa*. University of Lisbon, Lisbon.

GONZALEZ QUIJANO, P. M., 1946. *Mapa pluviométrico de España*. Instituto Juan Sebastián Elcano (Consejo Superior de Investigaciones Científicas), Madrid, 574 pp.

JANSÁ GUARDIOLA, J., 1900. Meteorologia del Mediterraneo Occidental. *Serv. Meteorol. Nacl. (Madrid), Ser. A*, 43: 30 pp.

JANSÁ GUARDIOLA, J., 1969. *Curso de Climatologia*. Servicio Meteorológico Nacional, Madrid, 455 pp.

JUNCO REYES, F., 1946. El régimen anual de lluvias en la Península Ibérica. *Calendario Meteoro-fenológico*, pp.121–127.

LINÉS ESCARDÓ, A., 1955–1965. Notas acerca de los temporales que afectan a la Península Ibérica. *Rev. Geofís.*

LINÉS ESCARDÓ, A., 1959. Singularidades en el curso anual de los fenómenos atmosféricos en España. *Rev. Geofís.*, 69: 29–34.

LINÉS ESCARDÓ, A., 1966. *Apuntes de Climatologia. (Curso de Hidrologia General y Aplicada.)* Instituto de Hidrologia, Madrid, 64 pp.

LÓPEZ GÓMEZ, J. and LÓPEZ GÓMEZ, A., 1959. *El Clima de España Segun la Clasificacion de Köppen.* Instituto Juan Sebastián Elcano (Consejo Superior de Investigaciones Científicas), Madrid, pp.167–188.

LORENTE PÉREZ, J. M., 1941. Notas acerca de la historia de la meteorologia en España. *Rev. Cienc.*, 3: 607–617.

LORENTE PÉREZ, J. M., 1946. Climas Españoles. *Rev. Geofís.*, 18: 28 pp.

LORENTE PÉREZ, J. M., 1950. ¿Está cambiando el clima? *Calendario Meteoro-fenológico*, 122–145.

LORENTE PÉREZ, J. M., 1955. La variabilidad de las precipitaciones atmosféricas sobre España peninsular. *Rev. Geofís.*, 55: 229–246.

LORENTE PÉREZ, J. M., 1956. Nuevas notas acerca de la historia de la meteorologia en España. *Rev. Cienc.*, 21 (2): 229–246.

LORENTE PÉREZ, J. M., 1961a. La variabilidad de las precipitaciones atmosféricas sobre España peninsular durante los años 1947–48 a 1960–61. *Rev. Geofís.*, 79: 229–245.

LORENTE PÉREZ, J. M., 1961b. Un siglo de observaciones de temperatura en España. *Calendario Meteoro-fenológico*, 133–139.

MATEO GONZALEZ, P., 1965. Distribución de las frecuencias de las cantidades de precipitación en el norte de España. Persistencia de los dias con precipitación y sin precipitación en Gijón. *Serv. Meteorol. Nacl. (Madrid)*, Ser. A, 39: 60 pp.

MATEO GONZALEZ, P., 1965. Distribución de las frecuencias de las cantidades de precipitación en el norte de España. Persistencia de los dias con precipitatión y sin precipitación en Gijón. *Serv. Meteorol. Nacl. (Madrid)*, Ser. A, 40: 27 pp.

MORÁN SAMANIEGO, F., 1944. Termodinamíca de la atmósfera. *Serv. Meteorol. Nacl. (Madrid)*, Ser. B, 4: 345 pp.

RODRIGUEZ FRANCO, P., 1955. Notas sobre la corriente de chorro. *Rev. Geofís.*, 56: 313–346.

ROLDÁN FLORES, A., 1956. Fechas de primeras y últimas heladas en España. *Calendario Meteoro-fenológico*, 129–141.

ROLDÁN FLORES, A., 1962. Temperaturas extremas de España. *Calendario Meteoro-fenológico*, 134–141.

SERVIÇO METEOROLOGICO NACIONAL (LISBOA) and SERVICIO METEOROLÓGICO NACIONAL (MADRID), (in preparation). *Atlas Climatológico Peninsular.*

ZIMMERSCHIED, W. and BAUR, F., 1949. Acerca de situaciones tipicas de tiempo en España. *Serv. Meteorol. Nacl. (Madrid)*, Ser. A, 20: 55 pp.

TABLE V

CLIMATIC TABLE FOR SANTANDER
Latitude 43°28′N, longitude 3°49′W, elevation 68 m

| Month | Mean sta. press. (mbar) | Mean daily temp. (°C) | Mean daily temp. range(°C) | Temp. extremes (°C) highest | lowest | Mean vapor press. (mbar) | Mean precip. (mm) | Max. precip. 24 h (mm) | Mean daily evap. (mm) |
|---|---|---|---|---|---|---|---|---|---|
| Jan. | 1011.7 | 9.3 | 5.2 | 21.0 | −2.6 | 9.2 | 118.9 | 60.5 | 2.4 |
| Feb. | 1011.7 | 9.2 | 5.4 | 22.6 | −3.8 | 9.0 | 88.5 | 61.9 | 2.4 |
| Mar. | 1008.6 | 11.5 | 6.1 | 30.0 | 0.4 | 10.1 | 74.2 | 72.5 | 2.6 |
| Apr. | 1010.3 | 12.3 | 5.8 | 33.4 | 2.0 | 11.1 | 82.4 | 64.8 | 2.3 |
| May | 1009.8 | 14.2 | 5.7 | 31.0 | 3.6 | 13.0 | 88.4 | 73.0 | 2.3 |
| June | 1012.1 | 16.9 | 4.7 | 34.0 | 7.5 | 16.1 | 65.6 | 68.8 | 1.9 |
| July | 1012.5 | 18.8 | 5.5 | 34.6 | 11.0 | 17.2 | 59.1 | 86.3 | 2.1 |
| Aug. | 1011.3 | 19.3 | 5.7 | 40.2 | 12.0 | 18.6 | 84.2 | 77.7 | 1.9 |
| Sept. | 1011.5 | 18.2 | 5.9 | 34.0 | 7.7 | 17.0 | 113.6 | 138.4 | 2.2 |
| Oct. | 1010.7 | 15.3 | 6.2 | 29.6 | 4.4 | 13.9 | 134.3 | 51.1 | 2.2 |
| Nov. | 1010.0 | 12.2 | 5.3 | 23.8 | 1.8 | 11.1 | 133.5 | 57.8 | 2.5 |
| Dec. | 1011.3 | 9.9 | 4.9 | 21.4 | −0.2 | 9.6 | 154.8 | 55.2 | 2.2 |
| Annual | 1010.9 | 13.9 | 5.5 | 40.2 | −3.8 | 12.9 | 1197.5 | 138.4 | 2.3 |

| Month | Number of days with precip. ⩾0.1 mm | thunder-storm | fog | snow | frost | Mean cloudiness (days) clear | cloudy | overcast | Mean sun-shine (h) | Most freq. wind dir. | Mean wind speed (m/sec) |
|---|---|---|---|---|---|---|---|---|---|---|---|
| Jan. | 16 | 1 | 3 | 1 | 0.8 | 3 | 13 | 15 | 85 | S | 4.5 |
| Feb. | 14 | 1 | 3 | 1 | 0.3 | 3 | 12 | 13 | 99 | S/NW | 2.5 |
| Mar. | 13 | 1 | 4 | 0 | 0.0 | 4 | 15 | 12 | 140 | S/NW | 4.2 |
| Apr. | 14 | 1 | 4 | 0 | 0.0 | 3 | 15 | 12 | 165 | NW | 3.6 |
| May | 15 | 2 | 4 | 0 | 0.0 | 3 | 15 | 13 | 187 | NW | 3.4 |
| June | 13 | 3 | 3 | 0 | 0.0 | 4 | 14 | 12 | 199 | NW | 3.4 |
| July | 12 | 2 | 4 | 0 | 0.0 | 5 | 15 | 11 | 212 | NW | 3.1 |
| Aug. | 14 | 3 | 3 | 0 | 0.0 | 4 | 16 | 11 | 197 | NW | 2.8 |
| Sept. | 14 | 2 | 4 | 0 | 0.0 | 3 | 16 | 11 | 156 | NW | 2.8 |
| Oct. | 14 | 1 | 4 | 0 | 0.0 | 3 | 16 | 12 | 134 | NW | 3.4 |
| Nov. | 16 | 1 | 4 | 0 | 0.0 | 3 | 13 | 14 | 97 | S | 4.2 |
| Dec. | 18 | 1 | 3 | 0 | 0.0 | 2 | 14 | 15 | 76 | S | 4.5 |
| Annual | 173 | 19 | 43 | 2 | 1.1 | 40 | 174 | 151 | 1747 | NW | 3.6 |

TABLE VI

CLIMATIC TABLE FOR LA CORUNA
Latitude 43°22′N, longitude 8°25′W, elevation 67 m

| Month | Mean sta. press. (mbar) | Mean daily temp. (°C) | Mean daily temp. range(°C) | Temp. extremes (°C) highest | lowest | Mean vapor press. (mbar) | Mean precip. (mm) | Max. precip. 24 h (mm) | Mean daily evap. (mm) |
|-------|------|------|------|------|------|------|------|------|------|
| Jan. | 1012.2 | 9.9 | 5.7 | 20.4 | −2.0 | 10.0 | 117.7 | 49.5 | 2.2 |
| Feb. | 1012.1 | 9.8 | 6.2 | 27.4 | −3.0 | 9.6 | 77.5 | 37.8 | 2.2 |
| Mar. | 1008.2 | 11.5 | 6.9 | 26.6 | 1.0 | 10.6 | 95.1 | 44.4 | 2.5 |
| Apr. | 1010.3 | 12.4 | 6.9 | 29.6 | 2.0 | 11.1 | 71.0 | 38.3 | 2.7 |
| May | 1009.6 | 14.0 | 6.9 | 28.5 | 3.4 | 12.8 | 55.7 | 41.0 | 2.3 |
| June | 1012.2 | 16.5 | 7.1 | 30.5 | 7.2 | 15.2 | 47.1 | 60.2 | 2.4 |
| July | 1012.5 | 18.2 | 7.3 | 33.5 | 9.9 | 16.6 | 30.0 | 32.6 | 2.6 |
| Aug. | 1011.4 | 18.9 | 7.7 | 33.6 | 9.4 | 17.3 | 44.3 | 75.4 | 2.6 |
| Sept. | 1011.4 | 17.8 | 7.6 | 30.5 | 6.7 | 16.5 | 75.5 | 63.3 | 2.3 |
| Oct. | 1011.4 | 15.3 | 6.9 | 31.0 | 4.5 | 14.0 | 89.1 | 60.3 | 2.3 |
| Nov. | 1010.3 | 12.4 | 6.1 | 25.0 | 1.0 | 11.6 | 127.5 | 72.7 | 2.2 |
| Dec. | 1011.7 | 10.2 | 5.5 | 19.8 | −1.0 | 10.6 | 138.7 | 91.4 | 2.2 |
| Annual | 1011.1 | 13.9 | 6.7 | 33.6 | −3.0 | 12.9 | 969.2 | 91.4 | 2.3 |

| Month | Number of days with precip. ≥0.1 mm | thunder-storm | fog | snow | frost | Mean cloudiness (days) clear | cloudy | overcast | Mean sun-shine (h) | Most freq. wind dir. | Mean wind speed (m/sec) |
|-------|------|------|------|------|------|------|------|------|------|------|------|
| Jan. | 18 | 1 | 3 | 0 | 0.3 | 2 | 13 | 16 | 95 | SW | 6.4 |
| Feb. | 14 | 1 | 3 | 1 | 0.3 | 3 | 12 | 13 | 117 | SW | 6.4 |
| Mar. | 16 | 1 | 3 | 0 | 0.0 | 4 | 14 | 13 | 148 | SW | 5.9 |
| Apr. | 13 | 1 | 2 | 0 | 0.0 | 4 | 15 | 11 | 192 | NE | 5.9 |
| May | 13 | 1 | 3 | 0 | 0.0 | 3 | 16 | 12 | 218 | NE | 5.9 |
| June | 10 | 1 | 5 | 0 | 0.0 | 4 | 16 | 10 | 223 | NE | 5.3 |
| July | 8 | 1 | 7 | 0 | 0.0 | 5 | 18 | 8 | 264 | NE | 4.8 |
| Aug. | 9 | 1 | 6 | 0 | 0.0 | 5 | 18 | 8 | 248 | N | 4.8 |
| Sept. | 12 | 1 | 5 | 0 | 0.0 | 4 | 17 | 9 | 191 | NE | 4.2 |
| Oct. | 14 | 1 | 4 | 0 | 0.0 | 3 | 16 | 12 | 151 | NE | 4.8 |
| Nov. | 17 | 1 | 3 | 0 | 0.0 | 3 | 13 | 14 | 111 | SW | 5.3 |
| Dec. | 19 | 1 | 3 | 0 | 0.1 | 4 | 12 | 15 | 84 | SW | 5.6 |
| Annual | 162 | 12 | 47 | 1 | 0.7 | 44 | 180 | 141 | 2042 | NE | 5.3 |

TABLE VII

CLIMATIC TABLE FOR VALLADOLID
Latitude 41°39′N, longitude 4°43′W, elevation 715 m

| Month | Mean sta. press. (mbar) | Mean daily temp. (°C) | Mean daily temp. range(°C) | Temp. extremes (°C) highest | Temp. extremes (°C) lowest | Mean vapor press. (mbar) | Mean precip. (mm) | Max. precip. 24 h (mm) | Mean daily evap. (mm) |
|---|---|---|---|---|---|---|---|---|---|
| Jan. | 936.7 | 3.3 | 7.7 | 17.9 | −10.6 | 6.7 | 30.3 | 23.7 | 1.8 |
| Feb. | 936.2 | 5.1 | 9.7 | 24.8 | −11.6 | 6.4 | 26.0 | 18.0 | 2.1 |
| Mar. | 933.5 | 8.6 | 11.2 | 26.0 | − 5.2 | 7.1 | 41.6 | 30.2 | 3.5 |
| Apr. | 935.0 | 11.0 | 12.2 | 31.0 | − 2.7 | 7.6 | 30.0 | 25.5 | 5.3 |
| May | 933.6 | 14.1 | 12.0 | 33.2 | − 1.2 | 8.9 | 35.1 | 19.1 | 6.1 |
| June | 936.0 | 18.5 | 13.7 | 35.8 | 4.3 | 10.8 | 32.7 | 32.4 | 7.5 |
| July | 936.0 | 21.3 | 15.5 | 39.0 | 6.2 | 11.0 | 12.5 | 15.1 | 9.8 |
| Aug. | 935.4 | 20.4 | 14.5 | 38.3 | 7.0 | 11.2 | 13.1 | 34.6 | 9.3 |
| Sept. | 936.4 | 17.8 | 12.7 | 36.1 | 3.6 | 11.1 | 27.5 | 13.8 | 6.6 |
| Oct. | 936.2 | 12.9 | 11.3 | 29.8 | − 2.0 | 9.7 | 33.6 | 34.9 | 4.1 |
| Nov. | 935.8 | 7.7 | 9.6 | 24.8 | − 4.6 | 7.9 | 40.0 | 44.2 | 2.6 |
| Dec. | 935.2 | 4.4 | 6.4 | 16.9 | − 7.2 | 7.2 | 39.9 | 43.8 | 1.8 |
| Annual | 935.5 | 12.1 | 11.4 | 39.0 | −11.6 | 8.8 | 362.3 | 44.2 | 5.0 |

| Month | Number of days with precip. ⩾0.1 mm | Number of days with thunder-storm | Number of days with fog | Number of days with snow | Number of days with frost | Mean cloudiness (days) clear | Mean cloudiness (days) cloudy | Mean cloudiness (days) overcast | Mean sun-shine (h) | Most freq. wind dir. | Mean wind speed (m/sec) |
|---|---|---|---|---|---|---|---|---|---|---|---|
| Jan. | 8 | 0 | 11 | 2 | 17 | 4 | 14 | 13 | 105 | N | 3.1 |
| Feb. | 6 | 0 | 5 | 3 | 15 | 4 | 15 | 9 | 142 | S | 3.4 |
| Mar. | 11 | 0 | 2 | 0 | 6 | 6 | 14 | 11 | 180 | S | 3.4 |
| Apr. | 8 | 2 | 1 | 1 | 2 | 5 | 16 | 9 | 231 | N | 3.4 |
| May | 10 | 2 | 1 | 0 | 0 | 4 | 16 | 11 | 263 | N | 3.4 |
| June | 7 | 4 | 2 | 0 | 0 | 5 | 19 | 6 | 318 | N | 3.1 |
| July | 4 | 3 | 1 | 0 | 0 | 13 | 16 | 2 | 345 | N | 3.1 |
| Aug. | 4 | 3 | 2 | 0 | 0 | 10 | 18 | 3 | 341 | N | 3.1 |
| Sept. | 6 | 2 | 1 | 0 | 0 | 6 | 18 | 6 | 243 | N | 3.4 |
| Oct. | 9 | 1 | 4 | 0 | 1 | 4 | 18 | 9 | 197 | N | 2.5 |
| Nov. | 8 | 0 | 10 | 0 | 7 | 4 | 16 | 10 | 131 | N | 2.8 |
| Dec. | 10 | 0 | 11 | 1 | 14 | 2 | 14 | 15 | 88 | N | 3.1 |
| Annual | 91 | 17 | 51 | 7 | 62 | 67 | 194 | 104 | 2584 | N | 3.1 |

TABLE VIII

CLIMATIC TABLE FOR BARCELONA
Latitude 41°24′N, longitude 2°9′E, elevation 95 m

| Month | Mean sta. press. (mbar) | Mean daily temp. (°C) | Mean daily temp. range(°C) | Temp. extremes (°C) highest | lowest | Mean vapor press. (mbar) | Mean precip. (mm) | Max. precip. 24 h (mm) | Mean daily evap. (mm) |
|---|---|---|---|---|---|---|---|---|---|
| Jan. | 1004.9 | 9.4 | 6.3 | 20.8 | −2.4 | 8.4 | 33.3 | 63.2 | 1.8 |
| Feb. | 1005.4 | 9.9 | 6.6 | 21.1 | −6.7 | 8.5 | 42.1 | 143.2 | 1.9 |
| Mar. | 1004.3 | 12.3 | 6.8 | 24.3 | 0.8 | 10.3 | 46.2 | 53.4 | 1.7 |
| Apr. | 1004.3 | 14.6 | 7.4 | 27.8 | 3.9 | 11.6 | 46.9 | 51.6 | 1.9 |
| May | 1003.7 | 17.7 | 7.5 | 32.2 | 4.8 | 14.1 | 51.8 | 76.2 | 2.3 |
| June | 1005.8 | 21.6 | 7.4 | 34.5 | 11.0 | 17.9 | 42.8 | 56.2 | 2.4 |
| July | 1005.5 | 24.4 | 7.3 | 35.4 | 14.3 | 21.0 | 29.2 | 65.0 | 2.7 |
| Aug. | 1004.7 | 24.2 | 6.9 | 36.1 | 13.2 | 21.9 | 48.4 | 82.7 | 2.4 |
| Sept. | 1005.8 | 21.7 | 6.3 | 32.6 | 10.4 | 19.6 | 77.4 | 107.2 | 2.1 |
| Oct. | 1005.1 | 17.5 | 6.0 | 27.5 | 5.0 | 15.2 | 80.2 | 106.1 | 1.7 |
| Nov. | 1005.2 | 13.5 | 5.8 | 24.5 | 2.8 | 11.5 | 48.8 | 69.6 | 1.9 |
| Dec. | 1005.2 | 10.2 | 5.5 | 20.0 | −2.5 | 9.0 | 47.1 | 91.6 | 1.8 |
| Annual | 1005.0 | 16.4 | 6.7 | 36.1 | −6.7 | 14.1 | 594.2 | 143.2 | 2.0 |

| Month | Number of days with precip. ≥0.1 mm | thunder- storm | fog | snow | frost | Mean cloudiness (days) clear | cloudy | overcast | Mean sun- shine (h) | Most freq. wind dir. | Mean wind speed (m/sec) |
|---|---|---|---|---|---|---|---|---|---|---|---|
| Jan. | 5 | 0 | 3 | 0 | 1 | 8 | 17 | 6 | 150 | W | 2.5 |
| Feb. | 5 | 1 | 3 | 1 | 1 | 8 | 15 | 5 | 164 | W | 2.5 |
| Mar. | 8 | 1 | 3 | 0 | 0 | 6 | 17 | 8 | 175 | S | 2.2 |
| Apr. | 8 | 1 | 1 | 0 | 0 | 6 | 17 | 7 | 213 | SW | 2.2 |
| May | 8 | 1 | 1 | 0 | 0 | 5 | 19 | 7 | 252 | S | 2.2 |
| June | 6 | 1 | 1 | 0 | 0 | 7 | 19 | 4 | 280 | S | 2.2 |
| July | 4 | 1 | 0 | 0 | 0 | 11 | 17 | 3 | 313 | S | 2.2 |
| Aug. | 6 | 2 | 1 | 0 | 0 | 8 | 19 | 4 | 274 | S | 2.0 |
| Sept. | 8 | 2 | 1 | 0 | 0 | 6 | 18 | 6 | 202 | S | 2.0 |
| Oct. | 9 | 1 | 1 | 0 | 0 | 6 | 17 | 8 | 175 | SW | 2.0 |
| Nov. | 5 | 0 | 2 | 0 | 0 | 7 | 17 | 6 | 150 | W/N | 2.2 |
| Dec. | 6 | 0 | 2 | 0 | 0 | 9 | 15 | 7 | 132 | NW | 2.5 |
| Annual | 79 | 11 | 19 | 1 | 2 | 87 | 207 | 71 | 2480 | S | 2.2 |

TABLE IX

CLIMATIC TABLE FOR OPORTO
Latitude 41°08′N, longitude 8°36′W, elevation 95 m

| Month | Mean sta. press. (mbar) | Mean daily temp. (°C) | Mean daily temp. range(°C) | Temp. extremes (°C) | | | | Mean precip. (mm) | Max. precip. 24 h (mm) | Total solar radiation (cal./cm²) |
|---|---|---|---|---|---|---|---|---|---|---|
| | | | | highest mean | max. | lowest mean | min. | | | |
| Jan. | 1007.5 | 9.0 | 8.5 | 13.2 | 21.7 | 4.7 | −4.1 | 158.8 | 94.6 | 4900 |
| Feb. | 1006.5 | 9.6 | 9.2 | 14.2 | 29.0 | 5.0 | −3.8 | 111.6 | 55.8 | 6763 |
| Mar. | 1002.9 | 11.9 | 8.8 | 16.3 | 28.5 | 7.5 | −1.9 | 147.2 | 43.8 | 10542 |
| Apr. | 1003.7 | 13.6 | 9.6 | 18.4 | 31.9 | 8.8 | 0.6 | 86.1 | 41.0 | 15362 |
| May | 1003.6 | 15.2 | 8.8 | 19.6 | 33.3 | 10.8 | 3.6 | 86.8 | 85.4 | 18264 |
| June | 1005.5 | 18.0 | 9.2 | 22.6 | 36.6 | 13.4 | 6.8 | 41.2 | 68.8 | 19535 |
| July | 1005.8 | 19.6 | 10.1 | 24.7 | 40.1 | 14.6 | 8.8 | 19.6 | 42.7 | 20619 |
| Aug. | 1004.9 | 19.8 | 10.4 | 25.0 | 39.4 | 14.6 | 8.8 | 26.2 | 35.5 | 18098 |
| Sept. | 1005.5 | 18.6 | 10.1 | 23.7 | 37.2 | 13.6 | 5.6 | 50.6 | 71.5 | 13173 |
| Oct. | 1005.3 | 15.8 | 10.0 | 20.8 | 34.4 | 10.8 | 1.5 | 105.2 | 60.6 | 9278 |
| Nov. | 1005.4 | 12.2 | 8.9 | 16.7 | 25.7 | 7.8 | −1.3 | 147.9 | 101.2 | 5575 |
| Dec. | 1007.0 | 9.6 | 8.3 | 13.7 | 21.9 | 5.4 | −3.7 | 168.4 | 61.5 | 4395 |
| Annual | 1005.3 | 14.4 | 9.3 | 19.1 | 40.1 | 9.8 | −4.1 | 1149.6 | 101.2 | 146504 |

| Month | Mean evap. (mm) | Number of days with | | | Mean cloudiness (days) | | | Mean sun-shine (h) | Most freq. wind dir. | Mean wind speed (km/h) |
|---|---|---|---|---|---|---|---|---|---|---|
| | | precip. ≥0.1 mm | thunder-storm | fog | clear | cloudy | overcast | | | |
| Jan. | 62.0 | 17.5 | 1.5 | 9.4 | 7 | 6 | 5 | 141.5 | E | 20.6 |
| Feb. | 68.2 | 14.9 | 1.7 | 9.8 | 6 | 6 | 5 | 163.2 | E | 20.1 |
| Mar. | 85.9 | 17.3 | 3.1 | 6.7 | 7 | 7 | 6 | 182.8 | E | 20.5 |
| Apr. | 110.3 | 12.8 | 2.6 | 5.3 | 6 | 6 | 4 | 249.4 | NW | 19.9 |
| May | 103.0 | 12.8 | 2.8 | 7.1 | 6 | 6 | 5 | 274.5 | NW | 18.9 |
| June | 110.2 | 7.4 | 2.2 | 8.2 | 6 | 5 | 5 | 297.8 | NW | 17.5 |
| July | 125.2 | 5.2 | 1.0 | 10.2 | 5 | 3 | 3 | 329.8 | NW | 16.7 |
| Aug. | 126.2 | 6.5 | 1.4 | 12.8 | 5 | 3 | 3 | 311.0 | NW | 16.5 |
| Sept. | 103.7 | 10.5 | 1.7 | 11.1 | 6 | 5 | 4 | 238.2 | NW | 15.7 |
| Oct. | 88.4 | 15.3 | 2.2 | 11.8 | 6 | 6 | 5 | 194.0 | E | 16.4 |
| Nov. | 67.5 | 17.5 | 2.1 | 11.8 | 7 | 6 | 5 | 150.8 | E | 19.2 |
| Dec. | 61.7 | 18.1 | 2.2 | 12.5 | 7 | 6 | 5 | 133.8 | E | 20.5 |
| Annual | 1112.7 | 155.8 | 24.5 | 116.7 | 6 | 5 | 5 | 2666.8 | E | 18.5 |

TABLE X

Latitude 40°25'N, longitude 3°41'W, elevation 667 m

| Month | Mean sta. press. (mbar) | Mean daily temp. (°C) | Mean daily temp. range(°C) | Temp. extremes (°C) highest | lowest | Mean vapor press. (mbar) | Mean precip. (mm) | Max. precip. 24 h (mm) | Mean daily evap. (mm) |
|---|---|---|---|---|---|---|---|---|---|
| Jan. | 941.8 | 4.9 | 7.1 | 18.0 | −10.1 | 7.2 | 38.0 | 32.8 | 1.1 |
| Feb. | 940.8 | 6.5 | 8.9 | 22.0 | − 9.1 | 6.8 | 34.0 | 40.0 | 1.5 |
| Mar. | 938.6 | 10.0 | 9.7 | 25.8 | − 3.5 | 7.9 | 44.6 | 56.0 | 2.1 |
| Apr. | 938.7 | 13.0 | 11.2 | 30.1 | − 0.6 | 8.5 | 44.1 | 35.5 | 2.8 |
| May | 938.6 | 15.7 | 11.2 | 33.4 | 0.6 | 10.3 | 43.5 | 41.0 | 3.1 |
| June | 940.5 | 20.6 | 12.4 | 38.1 | 6.4 | 12.3 | 27.4 | 48.0 | 4.2 |
| July | 940.5 | 24.2 | 13.5 | 39.1 | 8.5 | 12.9 | 11.9 | 30.0 | 5.6 |
| Aug. | 940.0 | 23.6 | 12.9 | 38.9 | 9.2 | 12.9 | 13.8 | 38.8 | 5.3 |
| Sept. | 941.2 | 19.8 | 11.2 | 35.1 | 5.0 | 12.6 | 32.2 | 53.2 | 3.5 |
| Oct. | 940.9 | 14.0 | 9.1 | 29.8 | − 0.4 | 11.0 | 53.3 | 53.5 | 1.9 |
| Nov. | 940.7 | 8.9 | 7.7 | 22.1 | − 3.0 | 8.8 | 47.1 | 65.4 | 1.3 |
| Dec. | 940.9 | 5.6 | 6.5 | 17.2 | − 6.5 | 7.3 | 47.5 | 29.7 | 1.0 |
| Annual | 940.3 | 13.9 | 10.2 | 39.1 | −10.1 | 9.9 | 435.4 | 65.4 | 2.9 |

| Month | Number of days with precip. ≥0.1 mm | thunder-storm | fog | snow | frost | Mean cloudiness (days) clear | cloudy | overcast | Mean sun-shine (h) | Most freq. wind dir. | Mean wind speed (m/sec) | Solar radiation horiz. surf. (ly/day) |
|---|---|---|---|---|---|---|---|---|---|---|---|---|
| Jan. | 7 | 0 | 9 | 1 | 11 | 8 | 13 | 10 | 153 | NE | 2.6 | 126 |
| Feb. | 6 | 0 | 6 | 1 | 7 | 7 | 14 | 7 | 173 | NE | 2.6 | 191 |
| Mar. | 10 | 0 | 3 | 0 | 2 | 6 | 15 | 10 | 187 | SW | 2.8 | 261 |
| Apr. | 9 | 1 | 1 | 0 | 0 | 7 | 14 | 9 | 235 | NE | 3.1 | 361 |
| May | 9 | 2 | 1 | 0 | 0 | 6 | 18 | 7 | 279 | SW | 2.7 | 464 |
| June | 6 | 3 | 0 | 0 | 0 | 8 | 18 | 4 | 317 | NE | 2.7 | 551 |
| July | 2 | 2 | 0 | 0 | 0 | 17 | 13 | 1 | 382 | NE | 2.9 | 559 |
| Aug. | 3 | 2 | 1 | 0 | 0 | 15 | 14 | 2 | 352 | NE | 2.9 | 438 |
| Sept. | 6 | 2 | 1 | 0 | 0 | 9 | 17 | 4 | 256 | NE | 2.5 | 311 |
| Oct. | 8 | 1 | 2 | 0 | 0 | 8 | 16 | 7 | 206 | NE | 2.4 | 176 |
| Nov. | 9 | 0 | 5 | 0 | 1 | 8 | 14 | 8 | 157 | NE | 2.3 | 104 |
| Dec. | 9 | 0 | 8 | 1 | 8 | 9 | 13 | 9 | 136 | NE | 2.5 | 91 |
| Annual | 84 | 13 | 37 | 3 | 29 | 108 | 179 | 78 | 2824 | NE | 2.7 | 303 |

TABLE XI

CLIMATIC TABLE FOR PALMA DE MALLORCA (BALEARIC ISLANDS)
Latitude 39°33′N, longitude 2°39′E, elevation 28 m

| Month | Mean sta. press. (mbar) | Mean daily temp. (°C) | Mean daily temp. range(°C) | Temp. extremes (°C) highest | lowest | Mean vapor press. (mbar) | Mean precip. (mm) | Max. precip. 24 h (mm) | Mean daily evap. (mm) |
|---|---|---|---|---|---|---|---|---|---|
| Jan. | 1014.4 | 10.1 | 8.0 | 24.0 | −3.0 | 10.1 | 38.9 | 55.0 | 1.4 |
| Feb. | 1013.9 | 10.5 | 8.5 | 22.8 | −4.0 | 10.3 | 34.0 | 31.8 | 1.6 |
| Mar. | 1012.5 | 12.2 | 8.9 | 25.0 | −1.0 | 11.5 | 36.4 | 38.5 | 2.1 |
| Apr. | 1012.3 | 14.5 | 9.0 | 26.0 | 0.5 | 12.8 | 28.2 | 25.8 | 2.4 |
| May | 1011.9 | 17.4 | 9.3 | 31.0 | 4.6 | 15.5 | 27.1 | 83.0 | 2.6 |
| June | 1013.8 | 21.4 | 9.6 | 37.8 | 8.4 | 18.8 | 19.9 | 50.3 | 3.0 |
| July | 1013.6 | 24.1 | 9.8 | 37.6 | 12.0 | 22.1 | 3.6 | 19.5 | 3.4 |
| Aug. | 1012.6 | 24.5 | 9.4 | 35.0 | 11.0 | 23.5 | 22.7 | 44.3 | 3.1 |
| Sept. | 1013.6 | 22.6 | 8.7 | 34.6 | 9.1 | 21.0 | 55.9 | 132.5 | 2.5 |
| Oct. | 1013.1 | 18.4 | 8.7 | 31.4 | 1.6 | 17.6 | 77.3 | 97.0 | 2.0 |
| Nov. | 1013.2 | 14.3 | 8.0 | 25.5 | 0.8 | 13.2 | 55.6 | 65.5 | 1.6 |
| Dec. | 1010.2 | 11.6 | 7.4 | 23.6 | −1.5 | 11.1 | 51.3 | 35.8 | 1.6 |
| Annual | 1013.1 | 16.8 | 8.8 | 37.8 | −4.0 | 15.6 | 450.9 | 132.5 | 2.3 |

| Month | Number of days with precip. ⩾0.1 mm | thunder- storm | fog | snow | frost | Mean cloudiness (days) clear | cloudy | overcast | Mean sun- shine (h) | Most freq. wind dir. | Mean wind speed (m/sec) |
|---|---|---|---|---|---|---|---|---|---|---|---|
| Jan. | 8 | 1 | 4 | 0.4 | 0.7 | 4 | 20 | 7 | 158 | N | 2.5 |
| Feb. | 6 | 0 | 3 | 0.4 | 0.7 | 5 | 18 | 5 | 171 | NW | 2.8 |
| Mar. | 8 | 1 | 3 | 0.0 | 0.0 | 6 | 19 | 6 | 200 | SW | 3.1 |
| Apr. | 5 | 1 | 2 | 0.0 | 0.0 | 6 | 19 | 5 | 229 | SW | 3.1 |
| May | 5 | 1 | 2 | 0.0 | 0.0 | 7 | 20 | 4 | 298 | SW | 2.8 |
| June | 3 | 1 | 1 | 0.0 | 0.0 | 10 | 17 | 3 | 311 | SW | 2.8 |
| July | 1 | 0 | 1 | 0.0 | 0.0 | 17 | 13 | 1 | 355 | SW | 2.5 |
| Aug. | 3 | 1 | 1 | 0.0 | 0.0 | 12 | 17 | 2 | 331 | SW | 2.5 |
| Sept. | 6 | 2 | 0 | 0.0 | 0.0 | 6 | 21 | 3 | 239 | N | 2.5 |
| Oct. | 9 | 2 | 1 | 0.0 | 0.0 | 4 | 22 | 5 | 196 | N | 2.5 |
| Nov. | 8 | 1 | 3 | 0.0 | 0.0 | 5 | 20 | 5 | 166 | NW | 2.5 |
| Dec. | 9 | 0 | 2 | 0.1 | 0.2 | 3 | 21 | 7 | 141 | NW | 2.8 |
| Annual | 71 | 11 | 23 | 0.9 | 1.6 | 85 | 227 | 53 | 2795 | SW | 2.8 |

TABLE XII

CLIMATIC TABLE FOR CAMPO MAIOR
Latitude 39°01'N, longitude 7°04'W, elevation 280 m

| Month | Mean sta. press. (mbar) | Mean daily temp. (°C) | Mean daily temp. range(°C) | Temp. extremes (°C) | | | | Mean precip. (mm) | Max. precip. 24 h (mm) | Mean evap. (mm) |
|---|---|---|---|---|---|---|---|---|---|---|
| | | | | highest | | lowest | | | | |
| | | | | mean | max. | mean | min. | | | |
| Jan. | 986.3 | 8.7 | 8.6 | 13.0 | 22.5 | 4.4 | −4.2 | 64.2 | 41.3 | 61.6 |
| Feb. | 985.5 | 10.1 | 9.8 | 15.0 | 26.0 | 5.2 | −5.0 | 55.0 | 40.9 | 72.9 |
| Mar. | 982.5 | 12.6 | 10.3 | 17.7 | 29.0 | 7.4 | −1.3 | 81.2 | 51.2 | 99.0 |
| Apr. | 982.3 | 15.1 | 11.6 | 20.9 | 34.3 | 9.3 | 1.4 | 43.5 | 64.0 | 123.6 |
| May | 982.1 | 17.6 | 12.9 | 24.0 | 38.0 | 11.1 | 3.1 | 38.3 | 36.5 | 156.5 |
| June | 983.4 | 22.1 | 15.2 | 29.7 | 41.1 | 14.5 | 6.5 | 19.5 | 40.0 | 216.1 |
| July | 983.1 | 25.1 | 17.4 | 33.8 | 45.6 | 16.4 | 8.2 | 1.5 | 6.0 | 304.0 |
| Aug. | 982.6 | 25.0 | 16.5 | 33.3 | 44.1 | 16.8 | 9.0 | 4.1 | 25.8 | 266.6 |
| Sept. | 983.6 | 22.4 | 14.3 | 29.5 | 40.1 | 15.2 | 6.3 | 28.7 | 92.0 | 208.0 |
| Oct. | 983.9 | 17.7 | 11.4 | 23.4 | 35.8 | 12.0 | 3.1 | 53.0 | 50.2 | 136.6 |
| Nov. | 984.5 | 12.6 | 9.5 | 17.4 | 26.9 | 7.9 | −1.0 | 62.0 | 93.0 | 79.4 |
| Dec. | 986.3 | 9.2 | 8.4 | 13.4 | 22.2 | 5.0 | −3.0 | 67.8 | 58.2 | 58.5 |
| Annual | 983.8 | 16.5 | 12.2 | 22.6 | 45.6 | 10.4 | −5.0 | 518.8 | 93.0 | 1782.8 |

| Month | Number of days with | | | Mean cloudiness (oktas) | | | Mean sunshine (h) | Most freq. wind dir. | Mean wind speed (km/h) |
|---|---|---|---|---|---|---|---|---|---|
| | precip. ⩾0.1 mm | thunderstorm | fog | 09h00 | 15h00 | 21h00 | | | |
| Jan. | 10.8 | 0.2 | 4.9 | 5 | 5 | 3 | 161.3 | C | 8.3 |
| Feb. | 8.6 | 0.1 | 2.9 | 5 | 5 | 3 | 183.4 | C | 8.0 |
| Mar. | 12.2 | 1.0 | 1.4 | 5 | 6 | 3 | 217.6 | W | 8.3 |
| Apr. | 8.2 | 1.3 | 0.7 | 4 | 5 | 2 | 261.3 | W | 9.0 |
| May | 7.7 | 1.2 | 0.7 | 4 | 5 | 2 | 281.0 | W | 8.7 |
| June | 3.0 | 1.1 | 0.1 | 3 | 3 | 1 | 348.4 | W | 9.0 |
| July | 0.6 | 0.6 | 0.2 | 1 | 2 | 0 | 383.9 | W | 9.7 |
| Aug. | 0.8 | 0.7 | 0.2 | 2 | 2 | 0 | 361.7 | W | 9.2 |
| Sept. | 3.3 | 1.4 | 0.5 | 3 | 4 | 2 | 276.6 | W | 7.7 |
| Oct. | 7.9 | 0.9 | 1.5 | 4 | 4 | 2 | 223.0 | W | 7.3 |
| Nov. | 9.0 | 0.2 | 3.7 | 5 | 5 | 3 | 170.8 | W | 7.4 |
| Dec. | 10.8 | 0.1 | 5.2 | 5 | 5 | 3 | 160.8 | W | 7.3 |
| Annual | 82.9 | 8.8 | 22.0 | 4 | 4 | 2 | 3029.8 | W | 8.3 |

TABLE XIII

CLIMATIC TABLE FOR LISBON
Latitude 38°43′N, longitude 9°09′W, elevation 77 m

| Month | Mean sta. press. (mbar) | Mean daily temp. (°C) | Mean daily temp. range(°C) | Temp. extremes (°C) | | | | Mean precip. (mm) | Max. precip. 24 h (mm) | Total solar radiation (cal./cm²) |
|---|---|---|---|---|---|---|---|---|---|---|
| | | | | highest | | lowest | | | | |
| | | | | mean | max. | mean | min. | | | |
| Jan. | 1009.5 | 10.8 | 6.1 | 13.9 | 20.6 | 7.8 | −0.5 | 110.6 | 59.1 | 6052 |
| Feb. | 1008.2 | 11.6 | 7.1 | 15.2 | 25.4 | 8.1 | −1.2 | 75.8 | 59.0 | 7991 |
| Mar. | 1005.1 | 13.6 | 7.3 | 17.3 | 27.4 | 10.0 | 2.8 | 108.7 | 69.8 | 11781 |
| Apr. | 1005.2 | 15.6 | 8.1 | 19.6 | 31.0 | 11.5 | 4.4 | 53.9 | 34.6 | 15814 |
| May | 1005.4 | 17.2 | 8.5 | 21.4 | 34.3 | 12.9 | 6.4 | 43.8 | 49.6 | 18728 |
| June | 1006.7 | 20.1 | 9.4 | 24.8 | 37.7 | 15.4 | 9.8 | 16.4 | 35.8 | 20205 |
| July | 1006.7 | 22.2 | 10.4 | 27.4 | 39.9 | 17.0 | 12.1 | 3.1 | 17.5 | 21870 |
| Aug. | 1005.9 | 22.5 | 10.4 | 27.7 | 40.3 | 17.3 | 13.3 | 4.3 | 24.5 | 19325 |
| Sept. | 1006.4 | 21.2 | 9.4 | 25.9 | 35.5 | 16.5 | 10.3 | 33.3 | 49.6 | 14349 |
| Oct. | 1006.7 | 18.2 | 8.1 | 22.3 | 35.3 | 14.2 | 6.7 | 62.2 | 63.6 | 10500 |
| Nov. | 1006.8 | 14.4 | 6.7 | 17.7 | 26.7 | 11.0 | 3.6 | 92.6 | 87.5 | 6752 |
| Dec. | 1008.9 | 11.5 | 6.0 | 14.5 | 20.9 | 8.5 | 0.0 | 102.8 | 78.4 | 5489 |
| Annual | 1006.8 | 16.6 | 8.1 | 20.6 | 40.3 | 12.5 | −1.2 | 707.5 | 87.5 | 158856 |

| Month | Mean evap. (mm) | Number of days with | | | Mean cloudiness (oktas) | | | Mean sun-shine (h) | Most freq. wind dir. | Mean wind speed (km/h) |
|---|---|---|---|---|---|---|---|---|---|---|
| | | precip. ⩾0.1 mm | thunder-storm | fog | 09h00 | 15h00 | 21h00 | | | |
| Jan. | 60.3 | 15.1 | 0.5 | 3.5 | 6 | 6 | 4 | 161.2 | N | 14.6 |
| Feb. | 73.4 | 11.8 | 0.7 | 2.2 | 6 | 6 | 4 | 181.7 | N | 14.5 |
| Mar. | 92.3 | 14.3 | 1.2 | 1.2 | 6 | 6 | 4 | 205.7 | N | 15.0 |
| Apr. | 131.0 | 10.2 | 1.1 | 0.4 | 6 | 5 | 4 | 264.9 | N | 15.5 |
| May | 141.6 | 9.7 | 1.2 | 0.2 | 6 | 5 | 3 | 300.7 | N | 15.1 |
| June | 167.3 | 4.5 | 0.5 | 0.2 | 5 | 4 | 3 | 329.5 | N | 16.1 |
| July | 210.2 | 1.9 | 0.3 | 0.2 | 3 | 2 | 1 | 378.4 | N | 17.1 |
| Aug. | 201.0 | 2.0 | 0.4 | 0.3 | 3 | 2 | 2 | 357.2 | N | 15.8 |
| Sept. | 157.2 | 6.1 | 1.0 | 0.5 | 4 | 4 | 2 | 279.2 | N | 14.2 |
| Oct. | 117.4 | 9.4 | 0.8 | 0.9 | 5 | 5 | 3 | 230.7 | N | 13.0 |
| Nov. | 73.0 | 12.6 | 0.8 | 2.4 | 6 | 6 | 4 | 173.9 | N | 13.4 |
| Dec. | 59.5 | 15.1 | 1.0 | 3.7 | 6 | 6 | 4 | 159.4 | N | 14.0 |
| Annual | 1484.2 | 112.7 | 9.5 | 15.7 | 5 | 5 | 3 | 3022.5 | N | 14.8 |

TABLE XIV

CLIMATIC TABLE FOR SEVILLA
Latitude 37°24′N, longitude 6°0′W, elevation 30 m

| Month | Mean sta. press. (mbar) | Mean daily temp. (°C) | Mean daily temp. range(°C) | Temp. extremes (°C) highest | Temp. extremes (°C) lowest | Mean vapor press. (mbar) | Mean precip. (mm) | Max. precip. 24 h (mm) | Mean daily evap. (mm) |
|---|---|---|---|---|---|---|---|---|---|
| Jan. | 1018.2 | 10.5 | 9.7 | 25.0 | −2.8 | 10.6 | 64.1 | 80.0 | 2.2 |
| Feb. | 1016.5 | 12.3 | 11.0 | 29.0 | −3.2 | 11.0 | 61.9 | 72.5 | 2.8 |
| Mar. | 1013.8 | 14.6 | 11.3 | 30.8 | 0.0 | 11.9 | 57.4 | 60.0 | 3.4 |
| Apr. | 1012.5 | 17.2 | 12.2 | 36.0 | 4.0 | 13.2 | 59.2 | 99.5 | 4.6 |
| May | 1012.5 | 19.9 | 13.0 | 39.5 | 2.6 | 14.6 | 38.5 | 36.7 | 5.9 |
| June | 1013.0 | 24.8 | 14.8 | 43.5 | 10.2 | 17.2 | 8.8 | 51.0 | 7.8 |
| July | 1012.3 | 27.9 | 16.2 | 45.6 | 12.5 | 19.0 | 1.0 | 7.2 | 9.2 |
| Aug. | 1011.5 | 27.8 | 16.0 | 47.0 | 12.1 | 19.1 | 3.8 | 36.5 | 8.9 |
| Sept. | 1013.1 | 24.8 | 14.1 | 40.8 | 9.0 | 17.6 | 20.4 | 48.5 | 6.6 |
| Oct. | 1014.2 | 19.8 | 11.7 | 39.0 | 5.2 | 15.9 | 66.3 | 68.5 | 4.4 |
| Nov. | 1015.4 | 15.0 | 10.1 | 32.9 | 0.2 | 13.6 | 69.5 | 77.5 | 2.8 |
| Dec. | 1017.1 | 11.4 | 9.1 | 24.0 | −2.8 | 11.2 | 84.0 | 68.8 | 2.1 |
| Annual | 1014.2 | 18.8 | 12.4 | 47.0 | −3.2 | 14.6 | 534.9 | 99.5 | 5.1 |

| Month | Number of days with precip. ⩾0.1 mm | Number of days with thunder-storm | fog | snow | frost | Mean cloudiness (days) clear | Mean cloudiness (days) cloudy | Mean cloudiness (days) overcast | Mean sun-shine (h) | Most. freq. wind dir. | Mean wind speed (m/sec) | Solar radiation horiz. surf. (ly/day) |
|---|---|---|---|---|---|---|---|---|---|---|---|---|
| Jan. | 8 | 0 | 9 | 0 | 2 | 8 | 14 | 9 | 182 | NE | 1.4 | 174 |
| Feb. | 6 | 0 | 6 | 0 | 1 | 9 | 12 | 7 | 190 | SW | 1.7 | 273 |
| Mar. | 9 | 1 | 6 | 0 | 0 | 7 | 15 | 9 | 189 | SW | 2.0 | 360 |
| Apr. | 7 | 1 | 2 | 0 | 0 | 9 | 14 | 7 | 235 | SW | 1.7 | 476 |
| May | 5 | 1 | 3 | 0 | 0 | 8 | 16 | 7 | 292 | SW | 1.7 | 532 |
| June | 1 | 1 | 2 | 0 | 0 | 13 | 15 | 2 | 332 | SW | 1.7 | 567 |
| July | 0 | 0 | 2 | 0 | 0 | 22 | 8 | 1 | 360 | SW | 1.7 | 558 |
| Aug. | 0 | 0 | 2 | 0 | 0 | 19 | 11 | 1 | 328 | SW | 1.4 | 505 |
| Sept. | 2 | 1 | 5 | 0 | 0 | 12 | 15 | 3 | 242 | SW | 1.4 | 413 |
| Oct. | 5 | 0 | 6 | 0 | 0 | 9 | 16 | 6 | 207 | SW | 1.4 | 294 |
| Nov. | 6 | 0 | 7 | 0 | 0 | 7 | 16 | 7 | 166 | NE | 1.4 | 196 |
| Dec. | 8 | 0 | 10 | 0 | 2 | 9 | 13 | 9 | 155 | SW | 1.4 | 154 |
| Annual | 57 | 5 | 60 | 0 | 5 | 132 | 165 | 68 | 2878 | SW | 1.7 | 375 |

TABLE XV

CLIMATIC TABLE FOR PRAIA DA ROCHA
Latitude 37°07′N, longitude 8°32′W, elevation 19 m

| Month | Mean sta. press. (mbar) | Mean daily temp. (°C) | Mean daily temp. range(°C) | Temp. extremes (°C) | | | | Mean precip. (mm) | Max. precip. 24 h (mm) | Mean evap. (mm) |
|-------|------|------|------|------|------|------|------|------|------|------|
| | | | | highest mean | max. | lowest mean | min. | | | |
| Jan. | 1017.7 | 11.6 | 7.3 | 15.3 | 20.8 | 8.0 | 0.0 | 58.6 | 55.8 | 74.7 |
| Feb. | 1016.4 | 12.1 | 7.6 | 15.9 | 22.6 | 8.3 | −1.9 | 37.9 | 26.8 | 57.5 |
| Mar. | 1013.8 | 13.6 | 6.9 | 17.1 | 25.5 | 10.2 | 3.5 | 69.4 | 48.4 | 62.4 |
| Apr. | 1013.2 | 15.4 | 7.7 | 19.3 | 30.3 | 11.6 | 5.0 | 31.3 | 52.5 | 77.8 |
| May | 1013.2 | 16.8 | 8.9 | 21.3 | 33.9 | 12.4 | 6.1 | 24.2 | 42.5 | 94.1 |
| June | 1014.1 | 20.4 | 8.5 | 24.7 | 35.1 | 16.2 | 9.7 | 6.7 | 30.0 | 121.4 |
| July | 1013.6 | 22.8 | 9.6 | 27.6 | 40.6 | 18.0 | 12.2 | 2.1 | 34.0 | 142.3 |
| Aug. | 1013.0 | 23.0 | 9.6 | 27.8 | 37.7 | 18.2 | 13.6 | 0.7 | 8.8 | 147.7 |
| Sept. | 1014.0 | 21.3 | 8.2 | 25.4 | 34.5 | 17.2 | 10.5 | 17.8 | 56.6 | 102.8 |
| Oct. | 1014.3 | 18.3 | 7.4 | 22.0 | 32.2 | 14.6 | 7.2 | 43.9 | 37.8 | 80.6 |
| Nov. | 1015.1 | 15.1 | 7.0 | 18.6 | 27.5 | 11.6 | 4.2 | 58.0 | 84.0 | 56.6 |
| Dec. | 1017.1 | 12.5 | 7.6 | 16.3 | 21.4 | 8.7 | 0.0 | 66.0 | 58.4 | 54.3 |
| Annual | 1014.6 | 16.9 | 8.0 | 20.9 | 40.6 | 12.9 | −1.9 | 416.6 | 84.0 | 1072.2 |

| Month | Number of days with | | | Mean cloudiness (oktas) | | | Mean sun-shine (h) | Most freq. wind dir. | Mean wind speed (km/h) |
|-------|------|------|------|------|------|------|------|------|------|
| | precip. ⩾0.1 mm | thunder-storm | fog | 09h00 | 15h00 | 21h00 | | | |
| Jan. | 10.4 | 0.6 | 1.3 | 5 | 5 | 4 | 174.5 | NE | 16.3 |
| Feb. | 8.2 | 0.5 | 0.9 | 4 | 4 | 3 | 190.6 | NE | 16.2 |
| Mar. | 10.8 | 1.3 | 1.1 | 5 | 5 | 3 | 218.9 | NW | 16.5 |
| Apr. | 7.0 | 1.2 | 0.4 | 4 | 4 | 3 | 275.5 | NW | 15.9 |
| May | 5.2 | 0.9 | 0.4 | 4 | 3 | 2 | 321.8 | NW | 14.6 |
| June | 2.1 | 0.5 | 0.5 | 3 | 2 | 2 | 351.5 | NW | 14.8 |
| July | 0.5 | 0.2 | 0.5 | 2 | 1 | 1 | 386.8 | NW | 14.1 |
| Aug. | 0.3 | 0.1 | 0.7 | 2 | 1 | 1 | 363.4 | NW | 13.6 |
| Sept. | 3.1 | 0.9 | 1.1 | 3 | 3 | 2 | 281.3 | NW | 13.4 |
| Oct. | 7.0 | 0.9 | 0.9 | 4 | 4 | 2 | 235.0 | NW | 12.9 |
| Nov. | 9.0 | 1.1 | 1.0 | 4 | 4 | 2 | 181.3 | N | 14.9 |
| Dec. | 10.6 | 1.0 | 1.6 | 4 | 4 | 2 | 179.6 | N | 15.3 |
| Annual | 74.2 | 9.2 | 10.4 | 4 | 3 | 2 | 3160.2 | NW | 14.9 |

TABLE XVI

CLIMATIC TABLE FOR ALMERIA
Latitude 36°50′N, longitude 2°28′W, elevation 7 m

| Month | Mean sta. press. (mbar) | Mean daily temp. (°C) | Mean daily temp. range(°C) | Temp. extremes (°C) highest | Temp. extremes (°C) lowest | Mean vapor press. (mbar) | Mean precip. (mm) | Max. precip. 24 h (mm) | Mean daily evap. (mm) |
|---|---|---|---|---|---|---|---|---|---|
| Jan. | 1018.2 | 11.7 | 7.5 | 22.6 | 1.9 | 10.6 | 31.0 | 51.6 | 1.9 |
| Feb. | 1017.8 | 11.8 | 7.6 | 25.7 | 0.2 | 10.8 | 20.7 | 23.7 | 2.1 |
| Mar. | 1015.7 | 14.1 | 7.3 | 26.6 | 2.6 | 11.9 | 20.1 | 60.0 | 2.3 |
| Apr. | 1014.6 | 16.1 | 7.3 | 29.7 | 5.3 | 13.3 | 28.4 | 54.4 | 2.5 |
| May | 1014.0 | 18.4 | 7.2 | 34.8 | 8.2 | 15.6 | 17.1 | 33.9 | 2.7 |
| June | 1015.3 | 22.0 | 7.4 | 35.9 | 12.7 | 19.2 | 3.5 | 24.7 | 3.0 |
| July | 1014.6 | 24.7 | 7.6 | 37.7 | 14.6 | 22.7 | 0.2 | 2.2 | 3.0 |
| Aug. | 1013.8 | 25.3 | 7.4 | 37.4 | 15.5 | 23.9 | 5.3 | 41.5 | 2.8 |
| Sept. | 1015.4 | 23.4 | 7.2 | 36.0 | 10.1 | 21.4 | 15.5 | 98.0 | 2.5 |
| Oct. | 1015.9 | 19.4 | 7.3 | 31.5 | 7.6 | 17.2 | 25.8 | 58.0 | 2.1 |
| Nov. | 1016.7 | 15.6 | 7.1 | 26.7 | 4.5 | 13.9 | 27.3 | 73.9 | 1.8 |
| Dec. | 1017.5 | 12.8 | 7.5 | 25.3 | 2.5 | 11.3 | 35.7 | 58.7 | 1.8 |
| Annual | 1015.8 | 18.0 | 7.4 | 37.7 | 0.2 | 16.0 | 230.6 | 98.0 | 2.4 |

| Month | Number of days with precip. ≥0.1 mm | Number of days with thunder-storm | fog | snow | frost | Mean cloudiness (days) clear | Mean cloudiness (days) cloudy | Mean cloudiness (days) overcast | Mean sun-shine (h) | Most freq. wind dir. | Mean wind speed (m/sec) | Solar radiation horiz. surf. (ly/day) |
|---|---|---|---|---|---|---|---|---|---|---|---|---|
| Jan. | 7 | 0 | 1 | 0 | 0 | 8 | 19 | 4 | 189 | W | 3.1 | 215 |
| Feb. | 4 | 0 | 2 | 0 | 0 | 7 | 17 | 4 | 190 | W | 3.4 | 298 |
| Mar. | 5 | 1 | 1 | 0 | 0 | 7 | 20 | 4 | 227 | W | 3.6 | 402 |
| Apr. | 5 | 0 | 1 | 0 | 0 | 8 | 18 | 4 | 260 | W | 3.6 | 502 |
| May | 4 | 1 | 1 | 0 | 0 | 9 | 19 | 3 | 309 | W | 3.6 | 559 |
| June | 2 | 1 | 1 | 0 | 0 | 13 | 16 | 1 | 331 | W | 3.4 | 588 |
| July | 1 | 1 | 1 | 0 | 0 | 21 | 10 | 0 | 362 | E/W | 2.8 | 596 |
| Aug. | 1 | 1 | 1 | 0 | 0 | 18 | 12 | 1 | 336 | E | 3.1 | 539 |
| Sept. | 3 | 1 | 1 | 0 | 0 | 11 | 16 | 3 | 264 | S | 2.8 | 439 |
| Oct. | 5 | 1 | 1 | 0 | 0 | 7 | 20 | 4 | 227 | WSW | 2.8 | 340 |
| Nov. | 5 | 1 | 1 | 0 | 0 | 6 | 19 | 5 | 185 | WSW/E | 2.8 | 242 |
| Dec. | 6 | 1 | 1 | 0 | 0 | 8 | 19 | 4 | 173 | N | 3.1 | 185 |
| Annual | 48 | 9 | 13 | 0 | 0 | 123 | 205 | 37 | 3053 | W | 3.1 | 409 |

TABLE XVII

CLIMATIC TABLE FOR MÁLAGA
Latitude 36°43′N, longitude 4°25′W, elevation 34 m

| Month | Mean sta. press. (mbar) | Mean daily temp. (°C) | Mean daily temp. range(°C) | Temp. extremes (°C) highest | lowest | Mean vapor press. (mbar) | Mean precip. (mm) | Max. precip. 24 h (mm) | Mean daily evap. (mm) |
|---|---|---|---|---|---|---|---|---|---|
| Jan. | 1017.8 | 12.5 | 8.0 | 29.0 | 0.0 | 10.4 | 58.7 | 71.1 | 2.8 |
| Feb. | 1016.1 | 12.9 | 7.8 | 27.0 | 0.0 | 11.1 | 49.0 | 58.0 | 2.9 |
| Mar. | 1013.6 | 15.0 | 7.7 | 27.8 | 3.0 | 12.0 | 62.4 | 60.5 | 3.2 |
| Apr. | 1012.6 | 16.3 | 7.7 | 31.8 | 5.2 | 12.9 | 45.5 | 49.7 | 3.5 |
| May | 1012.5 | 19.3 | 8.1 | 33.6 | 7.6 | 14.6 | 24.7 | 36.5 | 4.0 |
| June | 1013.2 | 22.8 | 7.5 | 39.0 | 12.0 | 17.7 | 5.5 | 33.8 | 4.7 |
| July | 1013.0 | 25.2 | 7.9 | 40.6 | 12.0 | 20.6 | 1.1 | 23.5 | 5.0 |
| Aug. | 1011.9 | 25.6 | 8.2 | 40.4 | 12.0 | 21.0 | 3.2 | 28.0 | 4.8 |
| Sept. | 1013.4 | 23.5 | 7.9 | 39.6 | 11.0 | 19.9 | 28.3 | 60.0 | 4.3 |
| Oct. | 1013.8 | 19.7 | 7.5 | 34.6 | 4.2 | 16.8 | 62.0 | 88.0 | 3.1 |
| Nov. | 1015.0 | 15.8 | 7.8 | 29.4 | 4.0 | 13.4 | 63.4 | 124.9 | 2.8 |
| Dec. | 1016.2 | 13.3 | 7.7 | 29.2 | 2.0 | 11.2 | 65.6 | 84.5 | 3.0 |
| Annual | 1015.1 | 18.5 | 7.8 | 40.6 | 0.0 | 15.2 | 469.4 | 124.9 | 3.7 |

| Month | Number of days with precip. ≥0.1 mm | thunder- storm | fog | snow | frost | Mean cloudiness (days) clear | cloudy | overcast | Mean sun- shine (h) | Most. freq. wind dir. | Mean wind speed (m/sec) |
|---|---|---|---|---|---|---|---|---|---|---|---|
| Jan. | 6 | 1 | 1 | 0 | 0.2 | 9 | 16 | 6 | 185 | W | 2.1 |
| Feb. | 6 | 0 | 1 | 0 | 0.0 | 7 | 15 | 6 | 179 | NW | 2.2 |
| Mar. | 7 | 1 | 1 | 0 | 0.0 | 6 | 17 | 8 | 192 | W | 2.1 |
| Apr. | 6 | 1 | 1 | 0 | 0.0 | 7 | 17 | 6 | 236 | SE | 2.2 |
| May | 4 | 1 | 1 | 0 | 0.0 | 8 | 18 | 5 | 299 | SE | 2.1 |
| June | 1 | 1 | 1 | 0 | 0.0 | 12 | 16 | 2 | 344 | SE | 1.9 |
| July | 0 | 0 | 2 | 0 | 0.0 | 19 | 11 | 1 | 354 | SE | 1.9 |
| Aug. | 1 | 1 | 1 | 0 | 0.0 | 17 | 13 | 1 | 326 | SE | 1.9 |
| Sept. | 2 | 1 | 1 | 0 | 0.0 | 10 | 17 | 3 | 243 | SE | 2.0 |
| Oct. | 5 | 1 | 1 | 0 | 0.0 | 7 | 17 | 7 | 209 | SE | 1.8 |
| Nov. | 7 | 1 | 1 | 0 | 0.0 | 5 | 18 | 7 | 178 | W | 1.9 |
| Dec. | 7 | 0 | 0 | 0 | 0.0 | 7 | 17 | 7 | 167 | NW | 2.3 |
| Annual | 52 | 9 | 12 | 0 | 0.2 | 114 | 192 | 59 | 2912 | SE | 2.1 |

# References Index

*241*

# Geographical Index

# Subject Index